Nucleic Acids and Molecular Biology

16

Series Editor

H. J. Gross

Marlene Belfort, Barry L. Stoddard,
David W. Wood, Victoria Derbyshire (Eds.)

Homing Endonucleases and Inteins

With 80 Figures, 25 of Them in Color, and 7 Tables

Dr. Marlene Belfort
New York State Department
of Health
Wadsworth Center
P.O. Box 22002
Albany NY 12201-2002
USA

David W. Wood
Dept. of Chemical Engineering
University of Princeton
A417 Engineering Quadrangle
Princeton, NJ 08544
USA

Barry L. Stoddard
Fred Hutchison Cancer
Research Center,
1100, Fairview Ave.N.
Mailstop A3-023
Seattle, WA, USA

Dr. Victoria Derbyshire
New York State Department
of Health
Wadsworth Center
P.O. Box 22002
Albany NY 12201-2002
USA

Hard cover version ISBN 978-3-540-25106-4

ISBN 978-3-540-85235-3 e-ISBN 978-3-540-29474-0

Library of Congress Control Number: 2008932933

Cover design: WMX Design GmbH, Heidelberg

Printed on acid-free paper

9 8 7 6 5 4 3 2 1

springer.com

Contents

Contributors

GEORGES BELFORT (e-mail: belfog@rpi.edu)

Department of Chemical and Biological Engineering,
Rensselaer Polytechnic Institute, Troy, New York 12180, USA

MARLENE BELFORT (e-mail: belfort@wadsworth.org)

Wadsworth Center, New York State Department of Health,
Center for Medical Science, 150 New Scotland Avenue, Albany,
New York 12208, USA

STEPHEN J. BENKOVIC (e-mail: sjb1@chem.psu.edu)

Department of Chemistry, Pennsylvania State University, University Park,
Pennsylvania 16802, USA

MARK G. CAPRARA (e-mail: mgc3@po.cwru.edu)

Center for RNA Molecular Biology, Case Western Reserve University School
of Medicine, Cleveland, Ohio 44106, USA

BRETT CHEVALIER (e-mail: chevalier@wi.mit.edu)

Division of Basic Sciences, Fred Hutchinson Cancer Research Center,
1100 Fairview Ave. N. A3–025, Seattle, Washington 98109, USA

SHAORONG CHONG (e-mail: chong@neb.com)

32 Tozer Road, New England Biolabs, Inc., Beverly, Massachusetts 01915,
USA

BAREKET DASSA (e-mail: bareket@wicc.weizmann.ac.il)

Department of Molecular Genetics, Weizmann Institute of Science, Rehovot,
Israel, 76100

Victoria Derbyshire (e-mail: vicky.derbyshire@wadsworth.org)

Wadsworth Center, New York State Department of Health,
Center for Medical Science, 150 New Scotland Avenue, Albany,
New York 12208, USA

Bernard Dujon (e-mail: bdujon@pasteur.fr)

Unité de Génétique moléculaire des levures (URA2171 CNRS and UFR927
Univ. P. M. Curie), Institut Pasteur, 25 rue du Dr Roux,
75724 Paris Cedex 15, France

David R. Edgell (e-mail: dedgell@uwo.ca)

Department of Biochemistry, University of Western Ontario, London,
ON N6A 5C1, Canada

Eric A. Galburt (e-mail: eagalburt@lbl.gov)

Physical Biosciences, Lawrence Berkeley National Laboratory,
1 Cyclotron Road Mail Stop 20A-355, Berkeley, California 94720, USA

Frederick S. Gimble (e-mail: fgimble@ibt.tamhsc.edu)

Center for Genome Research, Institute of Biosciences and Technology,
Texas A&M University System Health Science Center,
2121 W. Holcombe Blvd., Houston, Texas 77096, USA

James E. Haber (e-mail: haber@brandeis.edu)

Department of Biology and Rosenstiel Center, Brandeis University, Waltham,
Massachusetts 02454–9110, USA

Sarah W. Harcum (e-mail: harcum@clemson.edu)

Department of Bioengineering, Clemson University, Clemson,
South Carolina 29634, USA

Melissa S. Jurica (e-mail: jurica@biology.ucsc.edu)

Molecular, Cell and Developmental Biology, Center for Molecular
Biology of RNA, UC Santa Cruz, 1156 High Street, Santa Cruz,
California 95064, USA

Anthony H. Keeble (e-mail: ak28@york.ac.uk)

Department of Biology, Area 10, University of York, Heslington,
York YO10 5YW, UK

Colin Kleanthous (e-mail: ck11@york.ac.uk)

Department of Biology, Area 10, University of York, Heslington, York YO10 5YW, UK

Alan M. Lambowitz (e-mail: lambowitz@mail.utexas.edu)

Institute for Cellular and Molecular Biology, Department of Chemistry and Biochemistry, and Section of Molecular Genetics and Microbiology, School of Biological Sciences, University of Texas at Austin, Austin, Texas 78712, USA

María J. Maté (e-mail: mjmp1@york.ac.uk)

Department of Biology, Area 10, University of York, Heslington, York YO10 5YW, UK

Kenneth V. Mills (e-mail: kmills@holycross.edu)

College of the Holy Cross, Department of Chemistry, 1 College Street, Worcester, Massachusetts 01610, USA

Georg Mohr (e-mail: georgius@mail.utexas.edu)

Institute for Cellular and Molecular Biology, Department of Chemistry and Biochemistry, and Section of Molecular Genetics and Microbiology, School of Biological Sciences, University of Texas at Austin, Austin, Texas 78712, USA

Raymond J. Monnat, Jr. (e-mail: monnat@u.washington.edu)

Departments of Pathology and Genome Sciences, Box 357470, University of Washington, Seattle, Washington 98195, USA

Carmen M. Moure (e-mail: mmoure@bcm.tmc.edu)

Department of Biochemistry and Molecular Biology, Baylor College of Medicine, Houston, Texas 77030, USA

Todd A. Naumann

Department of Chemistry, Pennsylvania State University, University Park, Pennsylvania 16802, USA

Takeaki Ozawa

Department of Chemistry, School of Science, The University of Tokyo, Hongo, Bunkyo-ku, Tokyo 113–0033, Japan, and Japan Science and Technology Agency, Tokyo, Japan

Henry Paulus (e-mail: paulus@bbri.org)

Boston Biomedical Research Institute, 64 Grove Street, Watertown, Massachusetts 02472, USA

Francine B. Perler (e-mail: perler@neb.com)

New England Biolabs, Inc., 32 Tozer Road, Beverly, Massachusetts 01915, USA

Shmuel Pietrokovski (e-mail: shmuel.pietrokovski@weizmann.ac.il)

Department of Molecular Genetics, Weizmann Institute of Science, Rehovot, Israel, 76100

Florante A. Quiocho (e-mail: faq@bcm.tmc.edu)

Howard Hughes Medical Institute, Baylor College of Medicine, Houston, Texas 77030, USA

Georgios Skretas (e-mail: gskretas@princeton.edu)

Department of Chemical Engineering, Princeton University, Princeton, New Jersey 08544, USA

Barry L. Stoddard (e-mail: bstoddar@fhcrc.org)

Division of Basic Sciences, Fred Hutchinson Cancer Research Center, 1100 Fairview Ave. N. A3–025, Seattle, Washington 98109, USA

Ali Tavassoli

Department of Chemistry, Pennsylvania State University, University Park, Pennsylvania 16802, USA

Yoshio Umezawa (e-mail: umezawa@chem.s.u-tokyo.ac.jp)

Department of Chemistry, School of Science, The University of Tokyo, Hongo, Bunkyo-ku, Tokyo 113–0033, Japan, and Japan Science and Technology Agency, Tokyo, Japan

Patrick Van Roey (e-mail: vanroey@wadsworth.org)

Wadsworth Center, New York State Department of Health,
Center for Medical Sciences, 150 New Scotland Avenue, Albany,
New York 12208, USA

Richard B. Waring (e-mail: waring@temple.edu)

Department of Biology, Temple University, Philadelphia,
Pennsylvania 19122, USA

Kenneth H. Wolfe (e-mail: khwolfe@tcd.ie)

Department of Genetics, Smurfit Institute, University of Dublin,
Trinity College, Dublin 2, Ireland

David W. Wood (e-mail: dwood@princeton.edu)

Department of Chemical Engineering, Princeton University,
Princeton, New Jersey 08544, USA

Ming-Qun Xu (e-mail: xum@neb.com)

32 Tozer Road, New England Biolabs, Inc., Beverly, Massachusetts 01915,
USA

Steven Zimmerly

University of Calgary, Department of Biological Sciences,
2500 University Drive NW, Calgary, AB, T2N 1N4, Canada

Back to Basics: Structure, Function, Evolution and Application of Homing Endonucleases and Inteins

MARLENE BELFORT

It is a profound and necessary truth that the deep things in science are not found because they are useful; they are found because it was possible to find them.

Robert Oppenheimer 1904–1967

1 Introduction: Back to Basics

Oppenheimer's words resonate with this book's theme, which is how both applied and theoretical science can emanate from answers to basic questions – whether they are being asked in model organisms or in test tubes – about molecular structure and mechanism. Thus, fundamental work on DNA, RNA and proteins, which encode or constitute homing endonucleases and inteins, is leading to refined theories of evolution in prokaryotes and eukaryotes on the one hand, and the development of laboratory tools and health-care reagents on the other.

Homing endonucleases and inteins, sometimes referred to as "protein introns", are linked at many levels. First, homing endonucleases are frequently encoded by introns that self-splice at the RNA level, in analogy to inteins that self-splice at the protein level. Second, homing endonucleases similar to those encoded by introns are often found embedded within and co-translated with inteins. Third, both types of intervening sequence are mobile elements, capable of movement from genome to genome. Fourth, the endonuclease component of both introns and inteins imparts their mobility. Fifth, each of these mobile intervening sequences is thought to have originated from invasion of the gene encoding the self-splicing element, the intron or intein, by

M. Belfort (e-mail: belfort@wadsworth.org)
Wadsworth Center, New York State Department of Health, Center for Medical Sciences, 150 New Scotland Avenue, Albany, New York 12208, USA

Nucleic Acids and Molecular Biology, Vol. 16
Marlene Belfort et al. (Eds.)
Homing Endonucleases and Inteins

an endonuclease gene, the primordial mobile element (Fig. 1). The final unifying theme is the exploitation of introns, inteins and homing endonucleases by chemists, geneticists, structural biologists and engineers, to generate reagents and tools that are useful in basic research, biotechnology and medicine. This chapter serves as an introduction to the volume entitled *Homing Endonucleases and Inteins*, which provides a wonderful illustration of the point that fundamental studies of structure and mechanism fuel evolutionary theory and technology development alike.

2 What Is a Homing Endonuclease?

Homing endonucleases are rare-cutting enzymes that are most often encoded by introns or inteins, but they can also be free-standing, occurring between genes. The genesis of the homing endonuclease field dates back to 1970, with the observation, in genetic crosses between yeast mitochondria, of a significant polarity of recombination for markers of an rRNA gene (Dujon, this Vol.). In 1985, this phenomenon became attributable to an intron-encoded homing endonuclease that initiated recombination within the rRNA gene. Minimally, homing endonucleases are protein enzymes that make a site-specific double-strand break (DSB) at the "homing" site in intron-less or intein-less alleles, thereby initiating a gene conversion event through which the intron or intein is copied into the break site (Fig. 2A, B; reviewed by Chevalier and Stoddard 2001; Belfort et al. 2002; Dujon, this Vol.). For the group I and archaeal intron endonucleases and inteins, the recombinogenic ends created at the DSB engage in a strictly DNA-dependent recombination process that duplicates

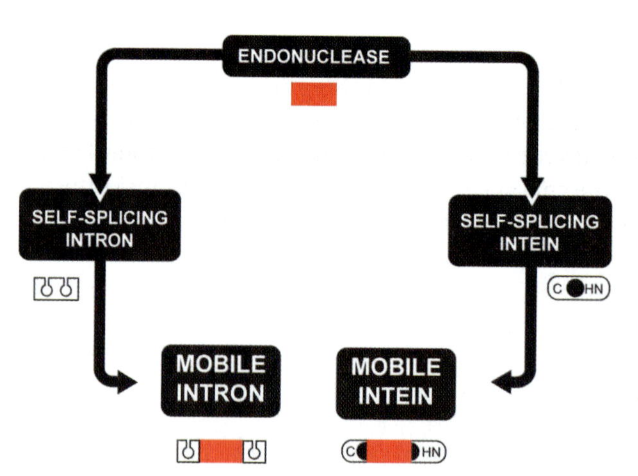

Fig. 1. Evolution of mobile introns and inteins. Endonuclease genes (*red*) are proposed to have invaded DNA encoding self-splicing introns or inteins, to generate mobile genetic elements. (Dassa and Pietrokovski, this Vol.)

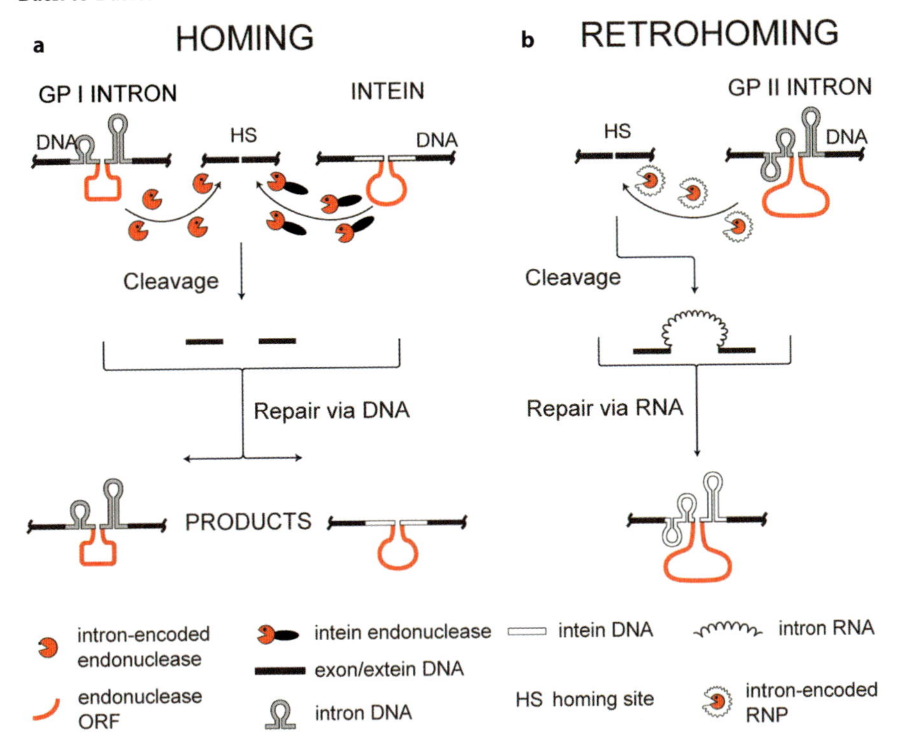

Fig. 2. Mobility of introns and inteins. **a** DNA-based homing of group I introns and inteins. The intron or intein endonuclease cleaves the homing site of a cognate intron- or intein-less allele. Gene conversion repairs the break to generate intron- or intein-containing products (Dujon, this Vol.). **b** Retrohoming of a group II intron. In this case, the homing site DNA is invaded by intron RNA, and the opposite strand is cleaved by an intron-encoded protein, which is part of an RNP complex. The intron is copied into cDNA to generate the intron-containing product (Lambowitz et al., this Vol.).

the intron or intein (Fig. 2A). The group II intron-encoded proteins are more complex, forming a ribonucleoprotein (RNP) particle with the intron RNA (Fig. 2B). The intron invades the DNA sense strand (mRNA-like strand) by reverse-splicing, whereas the endonuclease domain of the protein nicks the antisense strand. The intron acquisition event is completed with a cDNA copy of the intron, in a process termed retrohoming (Lambowitz et al., this Vol.).

The homing endonucleases fall within four families, characterized by the sequence motifs LAGLIDADG (Caprara and Waring, this Vol.; Chevalier et al., this Vol.; Dujon, this Vol.; Haber and Wolfe, this Vol), GIY-YIG (Edgell, this Vol.; Van Roey and Derbyshire, this Vol.), His-Cys box (Galburt and Jurica, this Vol.; Keeble et al., this Vol.) and HNH (Keeble et al., this Vol.). However, recent structural data support the hypothesis that the His-Cys box and HNH enzymes share features at their active sites, and should be considered a single family, called ββα-Me (Keeble et al., this Vol.). All of the homing endonucleases recognize lengthy asymmetric or pseudosymmetric DNA sequences, ranging from a 14-bp homing site for I-DmoI, a member of the LAGLIDADG family, to a 40-bp site for I-TevI, a member of the GIY-YIG family (described in Chevalier et al., this Vol.; Van Roey and Derbyshire, this Vol., respectively). In addition, the enzymes exhibit varying degrees of sequence tolerance, with I-TevI again being exceptional, in this case in its promiscuity (Van Roey and Derbyshire, this Vol.). The conserved sequences and substrate-recognition characteristics stand in contrast to the properties of the restriction endonucleases, which usually recognize short palindromic DNA sequences with absolute sequence specificity. Thus, while both types of endonuclease cleave DNA, they have evolved independently. A growing understanding of the structures and mechanisms of some of these enzymes is facilitating their engineering for genomic applications (Dujon, this Vol.; Gimble, this Vol.).

3 What Is an Intein?

The discovery of inteins and protein splicing represented a breakthrough in our concept of the catalytic repertoire of proteins and of post-translational modification (Perler, this Vol.). Conserved residues at the intein–extein junctions facilitate splicing (Mills and Paulus, this Vol.; Perler, this Vol.). Several inteins are bifunctional proteins that not only catalyze protein splicing, but also function as endonucleases, to initiate homing of the intein gene (Figs. 1 and 2A). Additionally, several inteins have motifs suggesting an evolutionary relationship to intron-encoded homing endonucleases. The endonuclease and the protein-splicing component are genetically, structurally, and functionally separable (Dassa and Pietrokovski, this Vol.; Moure and Quiocho, this Vol.; Perler, this Vol.), supporting the hypothesis that the endonuclease genes invaded the genes of these self-splicing elements, which provided safe havens, while themselves acquiring mobile properties (Fig. 1).

4 Inteins and Homing Endonucleases as Molecular Mosaics

The invasion of self-splicing introns and inteins by endonuclease genes appears to have occurred multiple times, given that these elements encode endonucleases of different families and that some endonuclease genes of the same family emanate from different positions of group I introns. It appears that some of these endonucleases then adapted to function in other processes, e.g., repression of transcription (Van Roey and Derbyshire, this Vol.), and promotion of splicing through acquisition of maturase activity (Caprara and Waring, this Vol.).

Inteins share an ancestry with metazoan hedgehog proteins, which undergo a self-cleavage reaction. The common Hint (*hedgehog/intein*) domain is a structural unit with a mechanistic identity (Dassa and Pietrokovski, this Vol.). It is apparent that composite elements like mobile introns, inteins and hedgehog proteins have interchanged functional domains in the course of evolution.

Endonucleases themselves have evolved specificity by fusion of a catalytic domain, containing the conserved motif, with variant DNA-binding domains, e.g., the GIY-YIG and HNH endonucleases (Van Roey and Derbyshire, this Vol.). The modular nature of these enzymes is further illustrated for I-TevI, in which the DNA-binding domain is itself an assembly of small DNA-binding units, some of which are present in other homing endonucleases. These enzymes have evolved a broad range of binding specificities, through the shuffling of catalytic cartridges with DNA-binding cassettes. We can only speculate as to how such molecular mosaics are formed. The most popular view is that proposed for hybrid bacteriophage genomes, in which "illegitimate recombination takes place quasi-randomly along the recombining genomes, generating an unholy mélange of recombinant types" (Pedulla et al. 2003). This sloppy, non-homologous recombination would generate a mound of genetic junk, with only a miniscule number of recombinants being selected, on the basis of their function and/or viability.

A lingering question is whether homing endonucleases and their genes are maintained specifically to promote their own selfish lifestyles and that of their host elements (introns and inteins), or whether they additionally serve some useful function for the organism. While their invasiveness and success as selfish intruders are undisputed, a potential advantage to the host organism has been observed, in experiments with phage T4 and its relative T2 (Edgell, this Vol.). Here, GIY-YIG homing endonucleases act to promote the spread of genes from their host organism to its relatives. This is a satisfying observation, considering that 8% of the phage T4 genome comprises endonuclease genes. Another "useful" homing enzyme is HO endonuclease, the first member of the LAGLIDADG family to be discovered, and the first shown to make a DSB. This intriguing enzyme catalyzes mating-type switching in yeast (Haber and Wolfe, this Vol.).

5 Applications: "Turning Junk into Gold"

The title of this subsection has been borrowed from that of an essay on the practical application of introns and inteins (Wickelgren 2003). Basically, the topic breaks down into the utility of homing endonucleases, group II intron RNPs, and inteins.

5.1 Site-Specific Group I Intron and Intein Endonucleases

The engineering of DNA often demands cleavage at rare sites and, to some extent, the highly site-specific endonucleases of group I introns and inteins fulfill the requirement. One need simply open the catalog of a molecular cloning company to identify those enzymes that cut the bacterial genome of ~5000 kb seven times, or a yeast genome of ~13,000 kb only once. They sound like restriction enzymes, bearing odd names like I-CreI and PI-PspI, with the prefixes I and PI, respectively, designating intron-encoded and protein intron-encoded (Perler, this Vol.). The engineering of single sites of the historic ω intron-endonuclease, I-SceI, into everything from yeast to mammals has facilitated not only genome sequencing projects, but also studies of DSB repair and non-homologous end joining (Dujon, this Vol.). However, the dream is not only for rare cleavage, but also for customized recognition sequence specificity. LAGLIDADG endonucleases are indeed starting to be engineered to alter their recognition-site specificity, through domain shuffling, and selection for new amino acid-DNA contacts, as described by Gimble later in this volume (Fig. 3A).

5.2 Gene Targeting by a Group II Intron RNP Complex

The novel DNA insertion mechanism of group II introns, via intron RNA-target DNA base pairing, has enabled the development of group II introns as site-specific gene targeting agents (Lambowitz et al., this Vol.). Through genetic manipulation, the intron RNA sequences that base pair with DNA, the so-called exon binding sequences, are changed (Fig. 3B). Thereby the group II Ll.LtrB intron has been re-targeted to specific genes in several bacteria. Additionally, this intron has been directed to HIV provirus and a plasmid-borne HIV coreceptor, and remained functional in transfection assays. This group II intron system is poised for different kinds of targeted gene manipulation (Fig. 3B), including gene disruption by integration into the antisense of an expressed gene, and conditional disruption via insertion into the sense strand and regulation of splicing. Such approaches will greatly facilitate functional

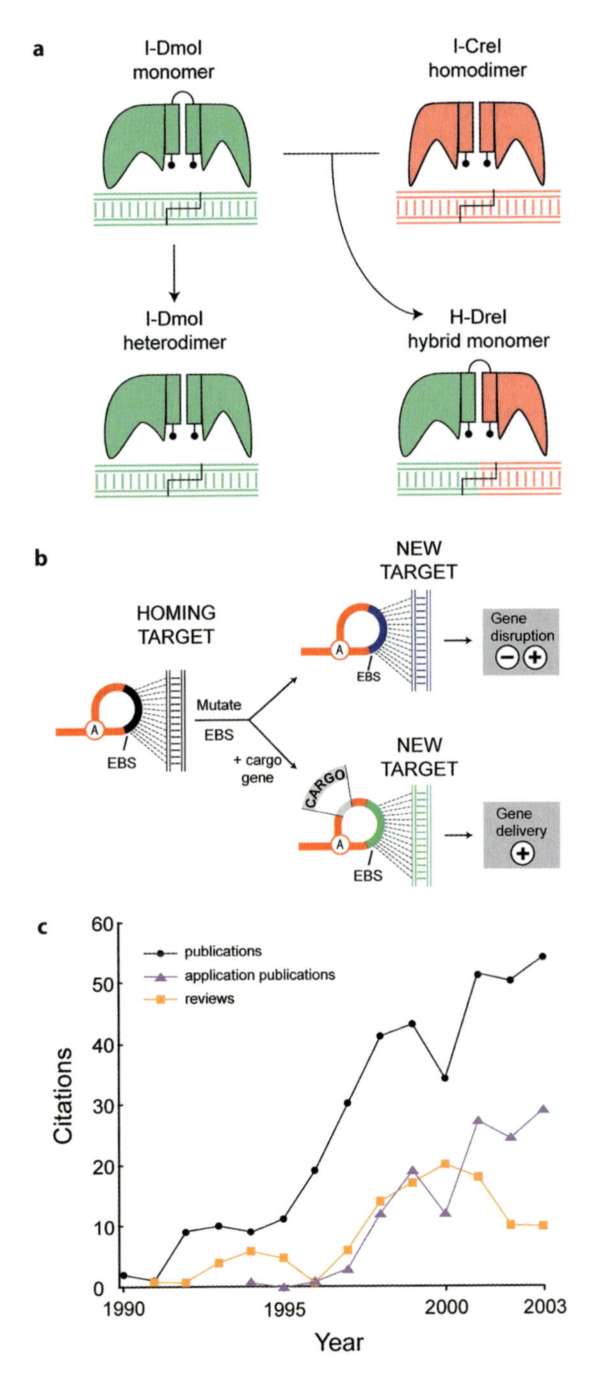

Fig. 3. Practical application of introns and inteins. **a** Domain manipulation of group I intron endonucleases. The splitting of monomers into heterodimers and the generation of hybrid nonomers pave the way for increasing the substrate repetoire of LAGLIDADG homing endonucleases (Gimble, this Vol.). **b** Group II introns as gene-targeting agents. The wild-type intron lariat targeting its natural homing site through its exon binding sequence (EBS; both in *black*) is shown on the *left*. Mutation of the EBS to target new sites is reflected on the *right* by a color change of the EBS that matches that of the new site (*blue* or *green*). Gene disruption, which can be either absolute (–) or conditional (+), is shown at the *top*, and delivery of a foreign cargo gene, with maintenance of gene function (+), is shown at the *bottom* (Lambowitz et al., this Vol.). **c** Burgeoning intein technology since the discovery of inteins in 1990 (Perler, this Vol.).

genomics. Gene therapy, involving delivery of a foreign sequence, is a major goal, pending delivery to mammalian cells, transport through the nuclear envelope, and integration into the desired chromosomal sites.

5.3 "Inteins ... Nature's Gift to the Protein Chemist"

The metaphor in this subheading derives from a recent comment made by intein researcher Tom Muir (2004, pers. comm.). In the short time span since their discovery barely 15 years ago, intein activity has been harnessed and controlled to build an entirely new area of biotechnology (Fig. 3C), which is predicted to explode in the coming years. Intein technology has already facilitated protein purification, as well as providing tools for the study of proteins both in vitro and in vivo.

Switches that control splicing range from temperature and pH shifts (Wood et al., this Vol.) to the addition of small molecules, such as the reducing agent dithiothreitol (Chong and Xu, this Vol.), rapamycin and 4-hydroxytamoxifen (reviewed by Ozawa and Umezawa, this Vol.). In all cases, the intein was modified in order to impart controllability. For facilitation of such intein engineering, and for the monitoring of intein activity in vitro and in vivo, a battery of different reporter systems have been developed (Chong and Xu, this Vol.; Wood and Skretas, this Vol.).

By using the basic chemistry of the intein, mutant intein–protein fusions can be chemically ligated to a protein or peptide with an N-terminal cysteine residue. This process has been variously termed expressed protein ligation (EPL) or intein-mediated protein ligation (IPL) (reviewed by Perler later in this volume). This EPL/IPL technology has already had enormous application, as for example in protein stabilization and potentiation by cyclization (Tavassoli et al., this Vol.), segmental labeling for NMR, or tagging with different reporters (Muir 2003). Equally impressive are in vivo intein-based technologies, many of them based on split inteins, and many using fluorescence or light as the reporter (Ozawa and Umezawa, this Vol.). These include monitoring of intracellular protein–protein interactions; identification of proteins specifically imported into mitochondria, endoplasmic reticulum, or nucleus; and even generation of safer transgenic plants.

This staggering array of technology that has developed recently is underscored by the increasing numbers of papers dedicated to the subject (Fig. 3C). Looking into the crystal ball, we can envision the proliferation of intein-based biosensors (Chong and Xu, this Vol.), proteomics utilizing peptide arrays (Tavassoli et al., this Vol.; Wood et al., this Vol.), and even functional proteomics in whole animals (Ozawa and Umezawa, this Vol.). Nature's gift, indeed!

6 The Message

What, then, are the lessons to be learned from these mobile elements that roam diverse genomes? To ponder their origins, we must first know their substance; to harness their ingenuity, we must first understand their action. The investment in discerning their substance and their action is yielding high payoffs. We are already gaining insight into how inteins, introns and homing endonucleases may have evolved, and how they may influence the evolution of genomes. Also, we have exploited the exquisite specificity of the endonucleases to manipulate DNA in vitro and in vivo. Finally, we have manipulated the activity of inteins to purify proteins, study them in test tubes, cells and multicellular organisms, and to build safer transgenic plants, as inteins provide the prospect of performing functional proteomics in whole animals.

Among the ultimate satisfactions for a scientist is to achieve mechanistic and evolutionary insights. To then apply these insights to practical problem-solving is frosting on the cake. Where, then, do we start? What is the message to our students, who, in their youthful enthusiasm, are likely to want both the thrill of discovery and the satisfaction of application? Don't start with the frosting! Study the fundamentals, and be vigilant! In that way, the potential for doing it all well will be maximized, and further compelling examples of the unpredicatable value of basic research will continue to emerge. We must keep bearing Oppenheimer's words in mind, that "the deep things in science are not found because they are useful". A terrific case in point is the study of a bizarre genetic phenomenon in 1970, which led to the discovery of the first homing endonuclease in 1985 that in turn resulted in a reagent, I-SceI, that is widely used in megasequencing projects of today (Dujon, this Vol.).

Acknowledgments. I dedicate this chapter to Bernard Dujon, who gave birth to this field, and who beautifully illustrated the message, that is, to study the fundamentals, to be vigilant, and then to seize the opportunities for application. I am grateful to all of the authors who contributed to this book and particularly to my coeditors Vicky Derbyshire, Barry Stoddard and Dave Wood, who have, through their hard work and shared insights, helped make this volume come to life. I thank Maryellen Carl for preparing this manuscript and handling the 19 others, and John Dansereau for providing the figures. Work in our laboratory is supported by NIH grants GM39422 and GM44844 and NSF grant NIRT0210419.

References

Belfort M, Derbyshire V, Cousineau B, Lambowitz A (2002) Mobile introns: pathways and proteins. In: Craig N, Craigie R, Gellert M, Lambowitz A (eds) Mobile DNA II. ASM Press, Washington, DC, pp 761–783

Chevalier B, Stoddard BL (2001) Homing endonucleases: structural and functional insight into the catalysis of intron/intein mobility. Nucleic Acids Res 29:3757–3774

Muir TW (2003) Semisynthesis of proteins by expressed protein ligation. Annu Rev Biochem 72:249–289

Pedulla ML, Ford ME, Houtz JM, Karthikeyan T, Wadsworth C, Lewis JA, Jacobs-Sera D, Falbo J, Gross J, Pannunzio NR, Brucker W, Kumar V, Kandasamy J, Keenan L, Bardarov S, Kriakov J, Lawrence JG, Jacobs WR Jr, Hendrix RW, Hatfull GF (2003) Origins of highly mosaic mycobacteriophage genomes. Cell 113:171–182

Wickelgren I (2003) Spinning junk into gold. Science 300:1646–1649

Homing Endonucleases and the Yeast Mitochondrial ω Locus – A Historical Perspective

BERNARD DUJON

It was May 1985. Outside the laboratory, near Paris, nature was exulting in its colorful mid-spring glory. The last technical details and experimental pitfalls had been fixed in the preceding weeks. Now, the site-specific endonucleolytic activity of the intron-encoded protein that I had carefully engineered to express in *Escherichia coli* was detectable. According to my autoradiogram, it was cleaving the intron-less DNA exactly where I expected. This experiment opened the way to a series of yet unexpected developments, but for me it concluded a long and rather solitary quest. Long, because the route that led to the first intron-encoded homing endonuclease, I-SceI according to the present nomenclature, had started no less than 15 years before, from a peculiarity of mitochondrial inheritance in yeast. Solitary, because over this long period, the phenomenon that led to this discovery had remained a unique oddity of nature, limiting its interest for many. Indeed, after the discovery of I-SceI, it took 3 additional years before the next examples of homing endonucleases could be identified, suggesting the generality of the phenomenon. By this time, the enzymatic properties of I-SceI and its unusual specificity were already characterized, and it was clear that we were in the presence of a novel class of enzymes.

1 The First Years of Yeast Mitochondrial Genetics

The story of intron-encoded homing endonucleases started in 1969 at the Centre de Génétique Moléculaire of CNRS in Gif-sur-Yvette. It is an example of a story that could hardly take place today, given the shortsighted, program-oriented research policies that unfortunately now prevail throughout Europe.

B. Dujon (e-mail: bdujon@pasteur.fr)
Unité de Génétique moléculaire des levures (URA2171 CNRS and UFR927 Univ. P.M. Curie), Institut Pasteur, 25 rue du Dr Roux, 75724 Paris Cedex 15, France

Nucleic Acids and Molecular Biology, Vol. 16
Marlene Belfort et al. (Eds.)
Homing Endonucleases and Inteins

Research at this time was motivated by intellectual curiosity and made possible by the freedom that CNRS and French Universities offered. I was a young student entering my fourth and last year of scholarship at Ecole Normale Supérieure in Paris. The laboratory of Piotr Slonimski had just performed the first crosses of *Saccharomyces cerevisiae* demonstrating the existence of recombination between mitochondrial genomes (Coen et al. 1970). Nothing was known about the genetic content of mitochondrial DNA. However, on the woody slopes of the Vallée de Chevreuse, serendipity was such that the two novel genetic markers used to study mitochondrial recombination were mutations flanking a mobile group I intron inserted within the gene for the large ribosomal RNA, as we learned much later. Those mutations conferred to yeast the ability to grow in the presence of either erythromycin (for the mutations designated E^R) or chloramphenicol (for the C^R mutations), two inhibitors of mitochondrial protein synthesis whose site of action is the peptidyl transferase center of the ribosomes, as we also learned later.

The oddity came from the fact that, when following the inheritance of these markers in the progeny of crosses, very different results were obtained depending on which yeast strains were used for the genetic crosses. It was as if the mitochondria of each strain were marked by a "factor" that, for some time, we used to call "sex" (Bolotin et al. 1971). In some crosses (that we thus called "homosexual" in the original literature), the two reciprocal recombinants between the two genetic loci (marked by the allelic pairs E^R/E^S and C^R/C^S, respectively) were found in approximately equal proportions in the progenies. In other crosses (hence called "heterosexual"), the same two reciprocal recombinants appeared in drastically different proportions in the progeny, a phenomenon that we called "polarity of recombination". At the same time, at each locus, one allele was very poorly represented while the other was highly overrepresented (Netter et al. 1974). By contrast, the markers of other regions of the map were not affected by this mysterious "polarity" gradient (Avner et al. 1973; Wolf et al. 1973; Dujon et al. 1975; Dujon and Slonimski 1976). The origin of the "polarity" gradient was a novel locus that one day, after a lengthy discussion, we decided to designate "ω" just a few minutes before the closing time of the campus cafeteria. This was 1971 and all of us, except Piotr Slonimski, were in our twenties and hungry. After all, if the yeast mating type was already designated by the Greek letter α, why not use ω for the mitochondrial "sex", and go for lunch? Consequently, some of our strains were classified as ω^+ and the others as ω^-, and this remained in the literature for years. The nature of the allele at the ω locus did not confer any measurable phenotype to the strains but it could be traced in the progeny of crosses. In crosses between ω^+ by ω^- mitochondria, recombinants were always ω^+, and it was always the alleles originating from the ω^+ parent that were over-represented in the progeny. The molecular nature of the ω locus could not be clarified until restriction enzymes could be used.

2 Genetic and Molecular Characterization of the ω Locus

In the mid-1970s, ideas about the mechanisms of genetic recombination were progressing. The "polarity" gradient affecting the region of the C^R/C^S and E^R/E^S allelic pairs of mitochondrial DNA was interpreted as a site-specific conversion event at the ω^- locus in the crosses between ω^+ and ω^- strains (Dujon et al. 1974). It was postulated that a DNA break at the ω^- locus was responsible for the initiation of the gene conversion tract (see Fig. 1A). The direct demonstration of this break took 10 more years, however (see Sect. 5, below). For one thing, the molecular techniques were not available. For another, double-strand breaks were not popular at that time. The initiation of genetic recombination had just been explained by a single-strand repair mechanism originating at a single-strand nick, not a double-strand break (Meselson and Radding 1975).

During these years most of the progress on the ω locus came from the genetic analysis of mutants. Back in 1970, Piotr Slonimski's lab had already noted the existence of C^R mitochondrial mutants, isolated from ω^- strains, that showed no or low "polarity" in crosses (Coen et al. 1970). Additional mutants which, after extensive genetic analysis, appeared to be second site suppressors of another and specific chloramphenicol mutation, played a critical role in our understanding of the phenomenon (Dujon et al. 1976) and, subsequently, in the discovery of the first homing endonuclease (see Sect. 6, below). They were also arising only from ω^- parental strains, never from the ω^+ ones. It was as if the ω^- allele could be inactivated by mutation but not the ω^+ allele. Mutants that had lost the ω locus were called ω^n for neutral. Despite the extensive genetic analysis of the ω^n mutants, the nature of these mutations could never have been solved without the help of molecular methods (see Fig. 2).

Introns were discovered in 1977; unfortunately, not in yeast mitochondria where, retrospectively, they are so obvious! It took less than a year, using the then novel molecular mapping methods, to understand that the difference between ω^+ and ω^- mitochondrial DNA resulted from the presence of a ca. 1.1-kb insert in the former, absent from the latter (Bos et al. 1978; Faye et al. 1979). This insert, located within the large rRNA coding gene, had the characteristics of an intron as judged from electron micrographs of RNA/DNA hybrids.

I sequenced the intron of the yeast mitochondrial rRNA gene in Walter Gilbert's lab at Harvard University where I had just arrived in the fall of 1978 as a young visiting scientist, eager to learn molecular methods. Alan Maxam and Walter Gilbert had just developed a new technique to sequence DNA, and this

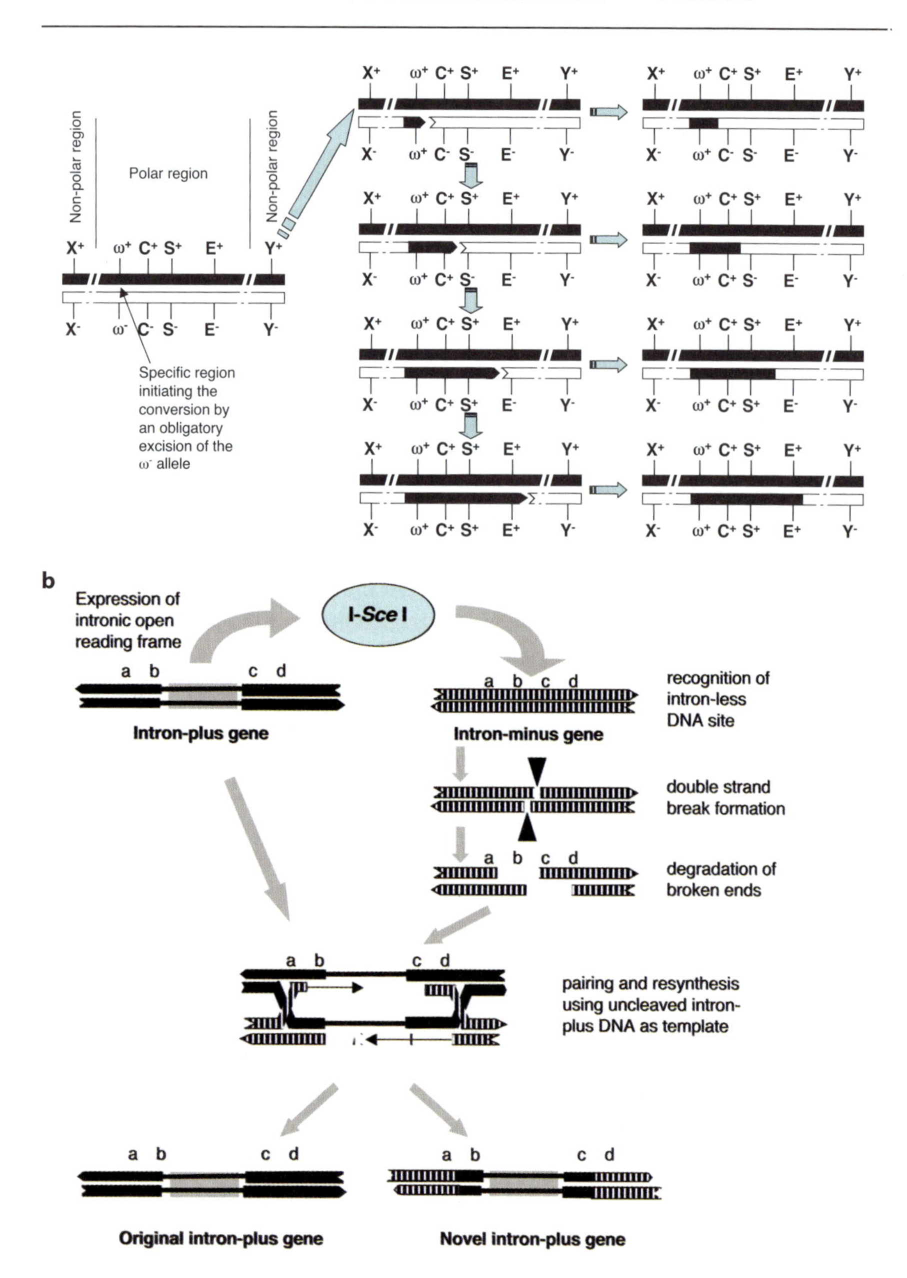
a
HETEROSEXUAL PAIRING EVENT
SEQUENTIAL DISSYMETRICAL GENE CONVERSION PROCESS
MOLECULAR PRODUCTS
Non-polar region
Polar region
Non-polar region
Specific region initiating the conversion by an obligatory excision of the ω⁻ allele
b
Expression of intronic open reading frame
I-SceI
Intron-plus gene
Intron-minus gene
recognition of intron-less DNA site
double strand break formation
degradation of broken ends
pairing and resynthesis using uncleaved intron-plus DNA as template
Original intron-plus gene
Novel intron-plus gene

laboratory was very interested in introns. The ω intron was appealing in that it did not follow the emerging theories of the time. It was associated with this strange phenomenon of "polarity", which it seemed responsible for, and during which it was propagated. The intron is 1143 bp long and, together with its flanking exons and the corresponding regions of the intron-less gene in an ω^- strain and an ω^n mutant, I sequenced less than 3000 bp in total. This, however, took me nearly a year of work, given the cloning and sequencing methods of the time. Nucleotides were spelled out one by one from gel electrophoresis lanes, sequenced fragments were assembled and verified manually. However, my efforts were extremely rewarding (Dujon 1980). First, the nature of the ω^n and C^R mutations was solved (Fig. 2). Each corresponds to a single base-pair substitution in a highly significant location of the rRNA molecule. The proximity of the ω^n mutation to the intron insertion site was remarkable (only 3 nt away), explaining why such mutations never appeared from the ω^+ mitochondria due to their possible interference with RNA splicing. Second, the intron was inserted in a region of the rRNA gene corresponding to one of the most highly conserved regions of the peptidyl transferase center of the ribosome, according to the secondary structure of the *E. coli* large rRNA molecule (for review, see Dorner et al. 2002). But my best reward was elsewhere. Within the ω intron was a 708-bp open reading frame, able to encode a 235 amino-acid protein. This was unprecedented. If this reading frame were translated, it meant that an intron of an RNA gene would have a protein product. At this time, the sequence had no similarity in the meager databases of the time. What was this strange protein?

◄───

Fig. 1. From the polarity of recombination to the mechanism of intron homing. **a** The first model proposed to explain the "polarity" gradient affecting the mitochondrial markers closely linked to the ω locus. Figure redrawn from Dujon et al. (1974). Each *bar* represents a double-stranded DNA molecule (*solid bar* ω^+, *open bar* ω^-). During a cross, a hypothetical break was postulated to occur at the ω^- site, initiating a gene conversion process that extends over the flanking genetic markers. All genetic markers were believed to lie on the same side of the ω locus (compare to the reality shown in Fig. 2). **b** Schematic representation of the mechanism of intron homing. Each *bar* represents a DNA strand (*solid bar* intron-plus gene, *striped bar* intron-minus gene). The intronic sequence is symbolized by a *thinner bar* and the reading frame encoding the homing-endonuclease is *shaded*. The intron-encoded endonuclease (here, I-SceI) recognizes and cleaves the intron-less gene (*arrowheads*), hence initiating double-strand break repair that uses the uncleaved intron-plus gene as the donor of genetic information. The conversion of flanking exon sequences (and the genetic markers lying in them, *letters*) is explained by the degradation of the free DNA ends prior to or during the repair. The probability of exon conversion diminishes with distance from the break

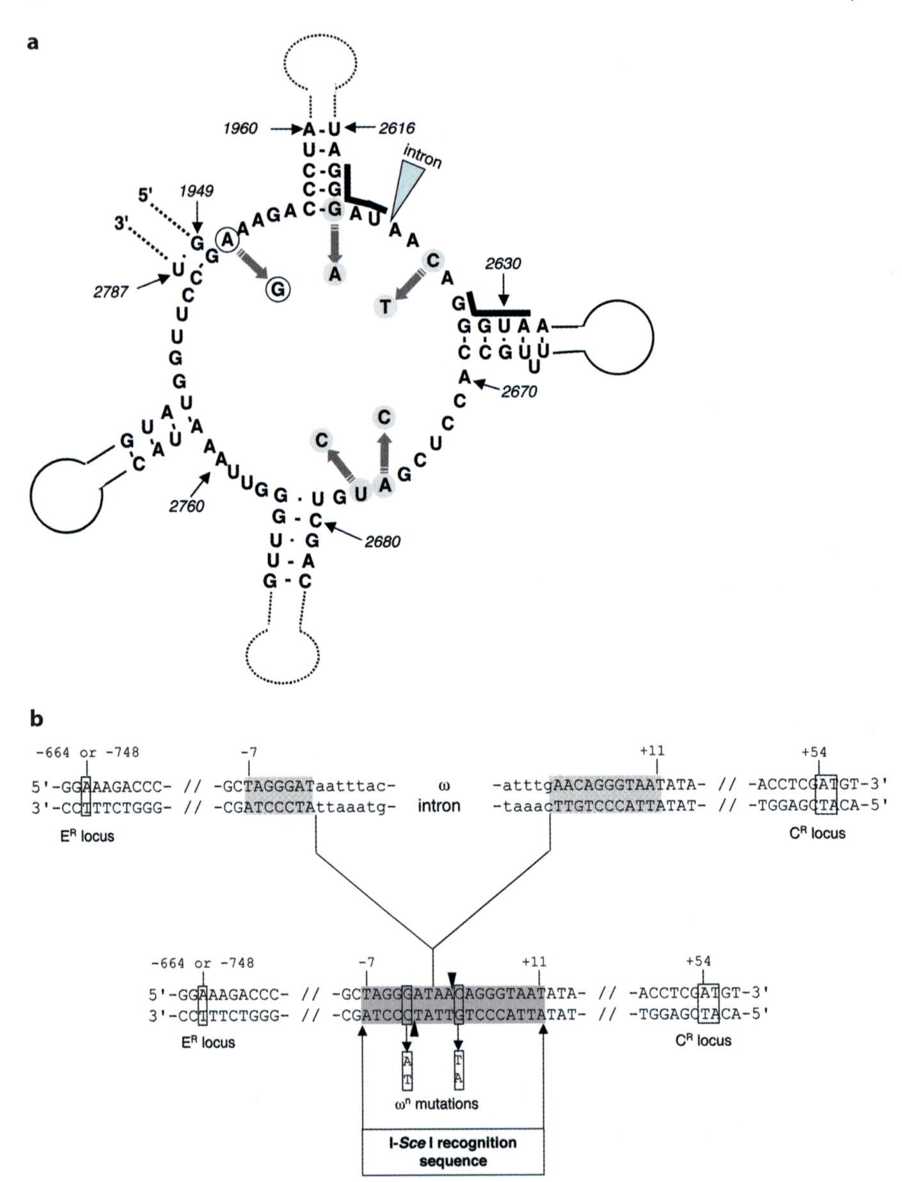

Fig. 2. Discovery of the ω locus and recognition site of I-SceI. **a** Pertinent mitochondrial mutations. Part of the conserved sequence of the yeast mitochondrial 21S rRNA molecule in the region of the peptidyl transferase center corresponds to the intron-less ω^- RNA. The position of the 1143 nt group I intron in the ω^+ gene is indicated by an *arrowhead*. The "guide" RNA necessary for intron splicing (P1 stem) is shown by a *bar*. Coordinates (*italics*) refer to positions in the mature rRNA molecule. Mutations conferring resistance to chloramphenicol (C^R) are shown in *gray circles*. The mutation conferring resistance to erythromycin (E^R) is *circled*. Data taken from Dujon (1980), Michel et al. (1982), Sor and Fukuhara (1982), and Dujon et al. (1985). **b** I-SceI recognition site. →

The mitochondrial gene in the ω^+ strain (intron borders are shown in *lowercase*) is shown. The extension of the recognition site of the intron-encoded I-SceI endonuclease is *shaded*. Cleavage of each DNA strand is indicated by *arrowheads*. Coordinates are relative to the intron insertion site. Data taken from Dujon (1980), Dujon et al. (1985) and Colleaux et al. (1988)

3 The Problems of Mitochondrial Intronic Reading Frames and Their Products

Two important points need to be mentioned here to fully appreciate the situation in 1980–1981.

First, the ω intronic reading frame was not written using the universal genetic code. Instead, it contains five UGA "stop" codons, shown to encode tryptophan in mitochondria (Barell et al. 1979; Macino et al. 1979), and eight other codons subsequently shown to be non-universal (Bonitz et al. 1980). With these differences from the normal code, I had no hope of expressing an active protein in *E. coli* or the in vitro systems of the time.

Second, genetic experiments using novel respiratory deficient mitochondrial mutations had led to the idea that some of these mutations affected the expression of their gene in an unusual manner (Slonimski and Tzagoloff 1976). With the discovery of introns, it became rapidly clear that the two genes concerned (encoding apocytochrome b and subunit I of cytochrome oxidase, respectively) were mosaic and that the unusual mutations were in their introns (Slonimski et al. 1978; reviewed in Dujon 1979, 1981) and prevented splicing (Church et al. 1979). Interestingly, some of them were complemented in *trans* by a wild-type intron, as if an intron product existed. In particular, the presence of a long open reading frame in the sequence of the second intron of the cytochrome b gene strongly suggested that the *trans*-active component necessary for proper gene expression, and altered in the mutants, was the intron translation product, inferring that the intron product would help the splicing of its own intron RNA. The putative intron-encoded protein, called a "maturase" (Jacq et al. 1980; Lazowska et al. 1980), was linked in-frame with its upstream exon, a situation now known to be common to many intron-encoded proteins but different from the ω intron. In addition, the deduced sequence of the cytochrome b maturase showed very little resemblance with the potential product of the ω intron, except for a few short C-terminal motifs.

Indeed, it took a few more years, and the sequencing of several additional introns, to recognize that many intron-encoded proteins share two typical, but short, motifs that were originally designated P1 and P2 (Michel et al. 1982) or LAGLI-DADG (Hensgens et al. 1983) from their amino-acid consensus sequence. This motif, in reality, is part of the dodecapeptide strings that form helices located at the interface between two domains and contain the two conserved catalytic acidic residues characteristic of this large family of proteins, now commonly referred to as LAGLIDADG homing endonucleases (see Chevalier et al., this Vol.).

Despite the numerous publications that appeared in the years 1981–1984 indicating the generality of the maturase model in yeast and other fungi (for review, see Lambowitz and Perlman 1990), the model did not apply to the ω intron. The first reason was that RNA splicing of this intron did not require any mitochondrially translated protein. Indeed, the ω intron was the second intron shown to be self-splicing in vitro (Tabak et al. 1984), soon after the discovery of the catalytic activity of the *Tetrahymena* intron (see Sect. 4, below). The second reason was that the maturase model concerned RNA splicing but did not address the polarity of recombination in crosses, which was the hallmark of the ω intron. The discovery of I-SceI had to await yet another strange turn.

4 Group I and Group II Introns, and RNA Splicing

In the years 1981–1984, most of the research activity on introns was concentrated on RNA splicing. Two major discoveries were made: the distinction between group I and group II introns based on their proposed secondary structures, and the existence of RNA catalysis. My lab was highly involved in the first. Back in Gif-sur-Yvette in 1981, I had started to examine the evolutionary conservation of the ω intron in other yeast species (see Sect. 5, below). Having joined my team, François Michel first noted that the sequences of some yeast mitochondrial introns (a handful) allowed the prediction of possible common RNA structures. The situation was complicated by the fact that the intronic reading frames are not always at the same place in the different introns. Among all possible hairpin structures, François Michel selected those that would be common to all introns. The logic was that the primary sequences need not be conserved between different introns so long as they allow the formation of equivalent double-stranded stems in a secondary structure model. The result was astonishing (Michel et al. 1982). Seven of the available yeast intron sequences shared a common secondary structure. The three others shared another structure. We decided to designate the first introns "group I" (the ω intron was among them) and the others "group II". This nomenclature is now universally used. Interestingly, the *Tetrahymena* intron, which had just been proven to be a ribozyme (Kruger et al. 1982), could also be folded in a secondary structure typical of group I introns (Michel and Dujon 1983). Using novel sequences from the fungus *Emericella nidulans*, a similar structural model was proposed for the group I introns by Davies et al. (1982), who proposed exon–intron pairings as a guide for RNA splicing (the so-called P1 stem, critical for splice site recognition).

The ω intron was a founding member of the group I introns, but was still the only one known to propagate in crosses. What was the origin of its "mobility"?

5 First Clues to the Function of the Translation Product of the ω Intronic Reading Frame

The first, indirect clue to the functional role of the ω intron-encoded protein came from its evolutionary conservation in other yeast species. In 1981, I had started to screen a large collection of yeast isolates (65) representing numerous (48) distinct species (their phylogeny was unclear at this time) for the possible presence of homologues to the ω intron of *S. cerevisiae*. Surprisingly, all the *Kluyveromyces* species responded positively whereas only few *Saccharomyces* did, and none of the other yeasts. Alain Jacquier, who started his Ph.D. work in my lab at this time, sequenced the corresponding piece of mitochondrial DNA in *Kluyveromyces thermoloterans* and discovered that, not only was an intron inserted within the mitochondrial rRNA gene at the exact same site as the ω intron in *S. cerevisiae*, but there was also a long open reading frame at the same position (Jacquier and Dujon 1983). We now know that *K. thermotolerans* and *S. cerevisiae* are farther apart in terms of molecular evolution than man and fishes (Dujon et al. 2004, and unpubl. data). However, even without this notion, the evolutionary conservation of the ω intronic reading frame in *K. thermotolerans* played a considerable role in the discovery of the first homing endonuclease, helping me to design the artificial gene (see Sect. 6, below).

The second clue for the direct action of the ω intron-encoded protein in the propagation of its own intron in crosses came from an observation made several years earlier by Julius Subik in Bratislava (Subik et al. 1977). I was very intrigued by his results because he had isolated a ω^n mutant from a strain that I showed clearly possessed the ω intron, whereas all previous ω^n mutants were missing it. The sequence once again was immediately informative (Jacquier and Dujon 1985). The intron and flanking exons were normal, but a single-base deletion was obvious in the middle of the intron reading frame, generating a −1 frameshift mutation leading to a stop codon immediately downstream. This was better than I could have ever expected. I called this mutation ω^d (for deficient) to differentiate it from the ω^n mutations which were single nucleotide substitutions in the intron-less gene next to the site of insertion of the intron (see Sect. 2, above). The ω^d mutation was the proof that the translation product of the ω intron was necessary for the "mobility" of this intron.

Together with polymorphic variations in exons used as molecular markers, this mutant also allowed us to show that the insertion of the intron in the intron-less gene was determining the co-conversion of the flanking exon sequences with an efficiency decreasing with distances. The ancient "polarity" of recombination was explained by a bidirectional repair mechanism initiated at the site of the intron insertion. After the discovery of I-SceI and its exceptional cleavage specificity (see Sect. 7, below), I used the word *homing* to designate the "transposition" of mobile introns to cognate sites, hence the term homing endonucleases now widely used.

The last clue about the mechanism responsible for the propagation of the ω intron in crosses was the demonstration that a double-strand break is transiently formed in mitochondrial DNA during crosses between ω^+ and ω^- yeast strains. The idea of a break was not novel (see Sect. 2, above). However, with the analogy to the double-strand break generated by the HO endonuclease to initiate mating-type switching in yeast (Strathern et al. 1982), we again insisted on the similarity to the ω intron propagation (Dujon and Jacquier 1983).

In these years, the competition between my laboratory and that of Ronald Butow was fierce. Using properly synchronized mating populations with very sensitive molecular methods, they were able to demonstrate the transient appearance of a double-strand break at the intron insertion site of the intron-less gene in crosses with a wild-type ω^+ strain (Zinn and Butow 1985). Meanwhile, my laboratory had also observed the break using less sophisticated techniques (Dujon et al. 1985). The break was specific to the wild-type ω^- DNA after crosses with a wild-type ω^+ strain but was absent from ω^n DNA. Using an imaginative combination of mutagenesis of the ω intron during its transfer to the intron-less gene, Butow's group was able to mutate the intronic reading frame, creating mutant strains that had lost the "polarity" of recombination (Macreadie et al. 1985). This was not different from the ω^d mutation that nature had spontaneously created in Subik's strain (Jacquier and Dujon 1985). However, with their intron mutants, they showed that the double-strand break was not formed at the ω^- site.

How to demonstrate the exact role played by the ω intron-encoded protein in the formation of the DNA break? By the end of 1983, before any of these results were published, I had speculated that, for a number of logical reasons, the intronic protein itself, and not cellular cofactors, had to be the active endonuclease. Consequently, I had decided to engineer the intronic reading frame to express it in *E. coli*.

6 Expressing the ω Intron-Encoded Protein in a Heterologous System

In 1984, in vitro mutagenesis was a technical challenge, and, therefore, creating a universal code equivalent of the ω protein was not trivial. Novel vectors based on the M13 phage had just been developed and I used them to insert the natural ω intron reading frame in order to facilitate the replacement of its non-universal codons, 13 in total, by their equivalents in the universal code. Synthetic oligonucleotides were the next problem. Young scientists, used to DNA arrays and purchase orders on the internet, may be surprised that only 20 years ago synthetic oligonucleotides were very valuable materials that were not commercially available. Francis Galibert in the Hôpital Saint Louis in Paris had clearly understood the power of synthetic oligonucleotides for molecular genetics and just equipped his laboratory with one of the DNA synthesis machines that were appearing on the market. Using this equipment, Luc d'Auriol, a young CNRS scientist, was experimenting with the then novel phosphoramidite chemistry. Synthesis was labor intensive, with uncertain yield and high failure rate. We carefully selected the minimum set of necessary oligonucleotides. The mutagenesis was done step by step, and, after each step, the M13 inserts were resequenced entirely to verify that the desired mutation was properly introduced and that no other mutation had occurred in other parts of the insert. Laurence Colleaux and Mireille Bétermier, two undergraduate students, helped me in some of these rounds. In total, the mutagenesis took a year. By the winter of 1985, the universal code equivalents of the wild-type ω intron reading frame and its ω^d mutant (used as control) were available in an IPTG-inducible expression plasmid, and assayed in *E. coli* maxicells. Proteins of the expected sizes were visible as radioactive bands (Dujon et al. 1986).

Instead of trying to purify them, I decided to go directly for the expected endonucleolytic activity in vivo in *E. coli*. To set up the assay, I inserted the piece of DNA corresponding to the intron insertion site from a ω^- mitochondria into the *E. coli* plasmid vector expressing the artificial protein. For controls, I inserted the same piece from each of the two different ω^n mutants (see Fig. 1), and did the same with the ω^d construct. In total, there were four plasmid constructs, one of which was supposed to self-linearize at the ω^- site upon IPTG induction. This is how the endonucleolytic activity of the ω intron-encoded protein was demonstrated in the early days of May 1985. It was totally specific for the ω^- site, and absent in the ω^d construct. The ω^n sites were not cleavable. The "mobility" of the ω group I intron in crosses was explained by the enzymatic activity of the protein encoded by that intron, and this protein, now called I-SceI, was the first characterized "homing endonuclease" (Colleaux et al. 1986).

7 The Unusual Enzymatic Properties of the Group I Intron-Encoded Homing Endonuclease I-SceI

I knew that I-SceI could not be similar to the bacterial restriction enzymes for two reasons. First, *E. coli* survived my plasmid expression experiments. Second, more importantly, during mitochondrial crosses, the novel intron copy was always inserted at the appropriate location, not elsewhere in the mitochondrial genome. Therefore, my enzyme had to be specific enough to cleave only one site in the mitochondrial genome. The parallel with HO was also suggestive (Kostriken and Heffron 1984). My priority was, therefore, to map the recognition site of I-SceI and estimate its recognition specificity. As this had to be done in vitro, not in vivo, purification of I-SceI was needed.

I engaged in this task using laborious classical protein purification methods, without any knowledge of the proper assay conditions that we determined much later (Monteilhet et al. 1990). None led to a completely pure fraction but some preparations were sufficiently depleted of contaminating nucleases for in vitro assays. Using these fractions, my first idea was to determine the mode of cleavage of I-SceI. This was done using terminally labeled DNA fragments, and the enzyme was shown to generate a 4 bp staggered cut with 3′ overhangs (Colleaux et al. 1988). On each DNA strand, hydrolysis occurs immediately upstream of the scissile phosphate groups, leaving 3′ OH and 5′ phosphate ends that can be religated directly with DNA ligase and ATP (Fig. 2). This mode of cleavage has now been shown as a general characteristic of the LAGLIDADG subclass of intron-encoded homing endonucleases (see Chevalier et al., this Vol.). It is also the mode of cleavage of most inteins. We determined much later (Perrin et al. 1993) that the double-strand cleavage activity of I-SceI results from two successive single-strand nicks during the same reaction, the bottom strand being cleaved first. It took 10 more years before this cleavage preference could be explained in atomic detail (Moure et al. 2003). In 1993, we had also determined that the turnover number of I-SceI reactions is very low, the enzyme being inhibited by one of the two cleaved products.

In 1987, determination of its recognition site was the next step to understanding the specificity of I-SceI. This was done using in vitro cleavage assays on mutant sites generated by oligonucleotides synthesized with one degenerate position at a time (Colleaux et al. 1988). The surprise was that the recognition site of I-SceI extended over 18 nucleotides, from position –7 to position +11 relative to the intron insertion site (Fig. 2). This was unprecedented. Even if the specificity of recognition of I-SceI is less than 4^{18} (because some nucleotide substitutions are tolerated), this was much higher than anything known before. Only HO had a similarly long recognition site.

With I-SceI, back in 1988, the route was opened to cleave entire genomes at a single, predetermined site (see Sect. 9, below). However, before going into that, the important point for the time was that the recognition site of I-SceI was extending on both sides of the intron insertion site, explaining why the intron-plus DNA is not cleaved, and consequently how ω^+ mitochondria escape self-destruction (Fig. 2). The mechanism of intron homing was now clear (Fig. 1b), but there was no explanation of its very existence. A unique self-splicing group I intron of yeast mitochondria, conferring no phenotype to the cell, was encoding an extremely specific DNA endonuclease only for the sake of its own propagation.

8 The First Additional Homing Endonucleases Discovered After I-SceI

In 1987, mobile introns with properties in crosses similar to the yeast ω intron were described in the chloroplast of *Chlamydomonas eugametos* (Lemieux and Lee 1987) and in the mitochondria of *C. smithii* (Boynton et al. 1987), two unicellular green algae. The first gave rise to the novel homing endonuclease I-CeuI (Marshall and Lemieux 1992).

During the same year, Pedersen-Lane and Belfort (1987), studying a bacteriophage intron previously identified from its self-splicing ability, showed that it had the signatures of a mobile intron. By analogy to the yeast ω intron, they subsequently demonstrated that this mobility was under the control of the intron open reading frame (Quirk et al. 1989). The fact that group I introns were also found in *E. coli* bacteriophages was obviously important for our understanding of intron evolution, but has also played a significant role in the characterization of another class of novel homing endonucleases, the first of which was I-TevI (Bell-Pedersen et al. 1991).

Shortly thereafter, a mobile group I intron was discovered in the nuclear DNA of a particular isolate of the ciliate *Physarum polycephalum* (Muscarella and Vogt 1989). Ironically, this intron is inserted exactly at the same site as the intron on which RNA catalysis was discovered in *Tetrahymena*, another ciliate. The *P. polycephalum* intron has led to the discovery of the homing endonuclease I-PpoI (Muscarella et al. 1990). By the end of 1989, half a dozen examples of mobile introns and homing endonucleases had been reported. And many other cases were suspected, some of which were not conclusively characterized but had been reported in a variety of organisms in the early 1980s (reviewed in Dujon 1989).

Historically, the second case of a mobile intron in yeast mitochondria was the fourth intron of the cytochrome oxidase subunit I gene. Wenzlau et al. (1989) had observed that it was absent in *Saccharomyces norbensis* and, after crosses with *S. cerevisiae*, they found a "polarity" phenomenon (co-conversion of flanking exons) characteristic of mobile introns. The important point here was not that the intron-encoded protein was a putative endonuclease, but that this protein was a latent maturase that could be reactivated by a single base substitution (Dujardin et al. 1982). Using in vitro replacement of non-universal codons, Delahodde et al. (1989) showed that this protein was an endonuclease that, subsequently, became known as I-SceII. Other cases of maturases with endonuclease function have been reported since (Seraphin et al. 1992; Schapira et al. 1993; Pellenz et al. 2002), and slowly the exact relationship between these two distinct properties of the same proteins is being clarified (see Belfort 2003; Caprara and Waring, this Vol.).

In 1989, I wrote a short review on group I intron homing (Dujon 1989) and the present nomenclature was adopted for the intron-encoded homing endonucleases (Dujon et al. 1989). The review mentioned six homing endonucleases. REBASE (http://rebase.neb.com/rebase) now lists 93 such enzymes, more than two-thirds of which are encoded by introns, the complement being made of inteins and proteins encoded by free-standing genes such as HO (F-SceII). This list is far from complete. For the introns of rDNA alone (the gene where I-SceI was found), 87 different putative homing endonucleases have been recently identified by the now fashionable in silico approach (Haugen and Bhattacharya 2004).

9 Use of the I-SceI Endonuclease in Heterologous Systems

At the end of 1987, I moved my laboratory from the CNRS at Gif-sur-Yvette to the Institut Pasteur in Paris. Given the exceptionally long recognition sequence of I-SceI, I was foreseeing a tool to cleave chromosomes, and possibly entire genomes, at a single predetermined site. Given the then emerging genome programs, especially the yeast sequencing project, YAC cloning and the first gene replacements in mammalian cells, this sounded to me like an attractive idea. One difficulty was the partial cleavage reactions produced when the universal code equivalent was expressed in heterologous systems. With Agnès Thierry, a young technician in my lab, I constructed a novel, totally synthetic gene for I-SceI, which has proven more efficient for expression, its codons being optimized for *E. coli* and yeast. This artificial gene has been distributed to nearly 150 laboratories in the world for academic research, and is also used for commercial production under agreement with Institut Pasteur. The sequence of the artificial gene is unpublished but can be obtained upon request. This gene has been successfully adapted with different promoters for in vivo expression in a very large variety of organisms including yeasts, fungi, insects, mammals, fishes, plants, ascidia, protozoa, bacteria, and viruses. A simple query of PubMed with the term "I-SceI" returns nearly 150 publications relating its use as a site-specific endonuclease for gene targeting and for studies on double-strand break repair, non-homologous end joining and other recombination mechanisms.

However, before these applications could start, a few technical details had to be worked out to demonstrate that I-Sce I could actually cleave only a single site in an entire eukaryotic genome. Yeast was instrumental for this demonstration (Thierry et al. 1991). To do this, artificial I-SceI sites were introduced at desired locations in its chromosomes by transformation with properly designed gene cassettes. Yeast chromosomes were then separated by the novel pulsed field gel electrophoresis technique. Treatment with I-SceI cleaved the yeast chromosome at the artificially inserted cognate site, leaving other chromosomes intact. This strategy has been very useful for the yeast sequencing program. Having inserted I-SceI sites at a large variety of chromosomal locations in different transformed clones, we could map the yeast genome at high resolution (Thierry and Dujon 1992; Tettelin et al. 1995, 1998; Thierry et al. 1995). Similar experiments were done on YAC clones to map mouse genes (Colleaux et al. 1993; Rougeulle et al. 1994; Fairhead et al. 1995).

But the major interest was in the in vivo experiments. There was no indication that the protein, which naturally acts in mitochondria where it is synthesized, could ever enter the nucleus if synthesized on cytoplasmic ribosomes. For me, this was a good enough reason to try. The experiment was immediately successful. Induction of I-SceI by activation of an artificial promoter was generating specific double-strand breaks at desired locations in artificial yeast plasmids and in chromosomes (Plessis et al. 1992; Fairhead and Dujon 1993). This offered us a very powerful novel tool that we, and others, are now extensively using to study double-strand break repair mechanisms and other genome instability phenomena in yeast (Ricchetti et al. 1999, 2003; Richard et al. 1999, 2003), or to engineer the yeast genome (Fairhead et al. 1996, 1998).

But yeast was not enough; the real challenge was higher eukaryotes. A young student in Jean-François Nicolas' lab, André Choulika, enthusiastically engaged in mouse cell experiments. Using specifically engineered retroviral vectors to insert I-SceI sites in the mouse genome, he rapidly demonstrated the power of this homing endonuclease to induce site-specific homologous recombination, opening the way for efficient gene replacement and targeting. Compared to controls, homologous recombination was stimulated almost 1000-fold (Choulika et al. 1994, 1995). Many others have now done similar experiments in a variety of organisms (above). For myself, I was pleased to collaborate with Holger Puchta and Barbara Hohn in plant genetics (Puchta et al. 1993, 1996) where the I-SceI homing endonuclease has great potential.

10 Epilogue

Thirty-five years have now passed since the presence of the first homing endonuclease was revealed by its consequences on the mitochondrial markers flanking the mobile intron. Even the most perfect molecular parasites have their weaknesses. The complicated story that led to this discovery is one that cannot be repeated. It has never been related before and I am pleased to do so in this book. If the yeast ω intron had not been studied, other routes would have probably eventually led to these enzymes, given their widespread distribution in nature and modern molecular techniques and genomic strategies. However, it is unclear whether they would have been recognized as being as special as they are. We still do not exactly understand the origin and evolution of intron homing, but the prospect of applications of homing endonucleases for gene targeting and genome engineering has now raised considerable interest in them. It is quite possible that other biological phenomena as specific as intron homing and its nucleases exist in nature. Detecting them is the challenge for modern biology. Creating the environment for this to happen is the challenge for policy-makers.

Acknowledgements. During the long story that I have summarized in this chapter with my own personal bias, I had the chance to meet and interact with numerous people without whom the history of intron-encoded homing endonucleases would not have been the same. It is impossible for me to recall all of them. May the absent forgive me. Beside those personally cited in the text, I should mention particularly: P. Avner, B. Backhaus, H. Baranowska, A. Beauvais, L. Belcour, M. Bolotin-Fukuhara, H. Blanc, J. Boyer, J. Brosius, G. Burger, M.-P. Carlotti, G. Church, M. Claisse, D. Coen, A.M. Colson, G. Cottarel, F. Denis, J. Deutsch, G. Dujardin, M. Eck, E. Fabre, C. Fairhead, G. Faye, G. Fischer, F. Foury, B. Frey, L. Gaillon, N.W. Gilham, J.E. Haber, A. Harington, E. Heard, C. Jacq, N. Jacquesson-Breuleux, Z. Kotylak, A. Kruszewska, F. Lang, J. Lazowska, B. Llorente, E. Luzi, G. Michaelis, C. Monteilhet, R. Morimoto, P. Netter, O. Ozier-Kalogeropoulos, S. Pellenz, A. Perrin, E. Petrochilo, A. Plessis, E. Pratje, G.-F. Richard, M. Ricchetti, C. Rougeulle, G. Schmitz, D. Schwartz, R. Schweyen, M. Soler, M. Somlo, A. Spassky, I. Stroke, H. Tettelin, M. Turmel, C. Vahrenholz, L. Weill, K. Wolf, and D. Worth. Some are no longer among us. They have all contributed to the saga of the first homing endonuclease, either directly by their experiments and efficient collaborations, or indirectly by their friendly discussions.

References

Avner P, Coen D, Dujon B, Slonimski PP (1973) Mitochondrial Genetics. IV. Allelism and mapping studies of oligomycin resistant mutants in *S. cerevisiae*. Mol Gen Genet 125:9–52

Barrell BG, Bankier AT, Drouin J (1979) A different genetic code in human mitochondria. Nature 282:189–194

Belfort M (2003) Two for the price of one: a bifunctional intron-encoded DNA endonuclease-RNA maturase. Genes Dev 17:2860–2863

Bell-Pedersen D, Quirk SM, Bryk M, Belfort M (1991) I-Tev I, the endonuclease encoded by the mobile td intron, recognizes binding and cleavage domains on its DNA target. Proc Natl Acad Sci USA 88:7719–7723

Bolotin M, Coen D, Deutsch J, Dujon B, Netter P, Petrochilo E, Slonimski PP (1971) La recombinaison des mitochondries chez *Saccharomyces cerevisiae*. Bull Inst Pasteur Paris 69:215–239

Bonitz SG, Berlani R, Coruzzi G, Li M, Macino G, Nobrega F G, Nobrega MP, Thalenfeld BE, Tzagoloff A (1980) Codon recognition rules in yeast mitochondria. Proc Natl Acad Sci USA 77:3167–3170

Bos JL, Heyting C, Borst P (1978) An insert in the single gene for the large ribosomal RNA in yeast mitochondrial DNA. Nature 275:336–338

Boynton JE, Harris EH, Burkhart BB, Lamerson PM, Gilham NW, (1987) Transmission of mitochondrial and chloroplast genome in crosses of *Chlamydomonas*. Proc Natl Acad Sci USA 84:2391–2395

Choulika A, Perrin A, Dujon B, Nicolas J-F (1994) The yeast I-SceI meganuclease induces site-directed chromosomal recombination in mammalian cells. CR Acad Sci Paris 317:1013–1019

Choulika A, Perrin A, Dujon B, Nicolas J-F (1995) Induction of homologous recombination in mammalian chromosomes by using the I-SceI system of *Saccharomyces cerevisiae*. Mol Cell Biol 15:1968–1973

Church GM, Slonimski PP, Gilbert W (1979) Pleiotropic mutations within two yeast mitochondrial cytochrome genes block mRNA processing. Cell 18:1209–1215

Coen D, Deutsch J, Netter P, Petrochilo E, Slonimski PP (1970) Mitochondrial genetics in yeast. I. Methodology and phenomenology. Symp Soc Exp Biol 24. Cambridge Univ Press, Cambridge, pp 449–496

Colleaux L, d'Auriol L, Betermier M, Cottarel G, Jacquier A, Galibert F, Dujon B (1986) Universal code equivalent of a yeast mitochondrial intron reading frame is expressed into *E. coli* as a specific double strand endonuclease. Cell 44:521–533

Colleaux L, d'Auriol L, Galibert F, Dujon B (1988) Recognition and cleavage site of the intron encoded omega transposase. Proc Natl Acad Sci USA 85:6022–6026

Colleaux L, Rougeulle P, Avner P, Dujon B (1993) Rapid physical mapping of YAC inserts by random integration of I-SceI sites. Hum Mol Genet 2:265–271

Davies RW, Waring RB, Ray JA, Brown TA, Scazzocchio C (1982) Making ends meet: a model for RNA splicing in fungal mitochondria. Nature 300:719–724

Delahodde A, Goguel V, Becam AM,Creusot F, Perea J, Banroques J, Jacq C (1989) Site-specific DNA endonuclease and RNA maturase activities of two homologous intron-encoded proteins from yeast mitochondria. Cell 56:431–441

Dorner S, Polacek N, Schulmeister U, Panuschka C, Barta A (2002) Molecular aspects of the ribosomal peptidyl transferase. Biochem Soc Trans 30:1131–1136

Dujardin G, Jacq C, Slonimski PP (1982) Single base substitution in an intron of oxidase gene compensates splicing defects of the cytochrome b gene. Nature 298:628–632

Dujon B (1979) Mutants in a mosaic gene reveal function for introns. Nature 282:777–778

Dujon B (1980) Sequence of the intron and flanking exons of the mitochondrial 21S rRNA gene of yeast strains having different alleles at the ω and RIB 1 loci. Cell 20:185–197

Dujon B (1981) Mitochondrial genetics and Functions. In: Strathern JN, Jones JR, Broach EW (eds) Molecular biology of the yeast Saccharomyces: life cycle and inheritance. Cold Spring Harbor Laboratory Press, Cold Spring Harbor, pp 505–635

Dujon B (1989) Group I introns as mobile genetic elements: facts and mechanistic speculations – a review. Gene 82:91–114

Dujon B, Jacquier A (1983) Organization of the mitochondrial 21S rRNA gene in *Saccharomyces cerevisiae*: mutants of the peptidyl transferase centre and nature of the ω locus. In: Schweyen RJ, Wolf K, Kaudewitz F (eds) Mitochondria 1983. De Gruyter, Berlin, pp 389–403

Dujon B, Slonimski PP (1976) Mechanisms and rules for transmission, recombination and segregation of mitochondrial genes in *Saccharomyces cerevisiae*. In: Bücher T, Neupert W, Sebald W, Werner S (eds) Genetics and biogenesis of chloroplasts and mitochondria. Elsevier North Holland Biomed Press, Amsterdam, pp 393–403

Dujon B, Slonimski PP, Weill L (1974) Mitochondrial genetics. IX. A model for recombination and segregation of mitochondrial genomes in *Saccharomyces cerevisiae*. Genetics 78:415–437

Dujon B, Kruszewska A, Slonimski PP, Bolotin-Fukuhara M, Coen D, Deutsch J, Netter P, Weill L (1975) Mitochondrial genetics. X. Effects of UV irradiation on transmission and recombination of mitochondrial genes in *Saccharomyces cerevisiae*. Mol Gen Genet 137:29–72

Dujon B, Bolotin-Fukuhara M, Coen D, Deutsch J, Netter P, Slonimski PP, Weill L (1976) Mitochondrial genetics. XI. Mutations at the mitochondrial locus ω affecting the recombination of mitochondrial genes in *Saccharomyces cerevisiae*. Mol Gen Genet 143:131–165

Dujon B, Cottarel G, Colleaux L, Betermier M, Jacquier A, d'Auriol L, Galibert F (1985) Mechanism of integration of an intron within a mitochondrial gene: a double strand break and the transposase function of an intron encoded protein as revealed by *in vivo* and *in vitro* assays. In: Quagliariello E, Slater EC, Palmieri F et al. (eds) Achievements and perspectives of mitochondrial research, vol II. Biogenesis. Elsevier, Amsterdam, pp 215–225

Dujon B, Colleaux L, Jacquier A, Michel F, Monteilhet C (1986) Mitochondrial introns as mobile genetic elements: the role of intron-encoded proteins. In: Wickner R, Hinnebusch A, et al (eds) Extrachromosomal elements in lower eucaryotes. Plenum, New York, pp 5–27

Dujon B, Belfort M, Butow RA, Jacq C, Lemieux C, Perlman PS, Vogt VM (1989) Mobile introns: definition of terms and recommended nomenclature. Gene 82:115–118

Dujon B, Sherman D, Fischer G, Durrens P, Casaregola S, Lafontaine I et al (2004) Genome evolution in yeast. Nature 430:35–44

Fairhead C, Dujon B (1993) Consequences of a unique double strand break in yeast chromosomes: death or homozygosis. Mol Gen Genet 240:170–180

Fairhead C, Heard E, Arnaud D, Avner P, Dujon B (1995) Insertion of unique sites into YAC arms for rapid physical analysis following YAC transfer into mammalian cells. Nucleic Acids Res 23:4011–4012

Fairhead C, Llorente B, Denis F, Soler M, Dujon B (1996) New vectors for combinatorial deletions in yeast chromosomes and for gap-repair cloning using "split-marker" recombination. Yeast 12:1439–1457

Fairhead C, Thierry A, Denis F, Eck M, Dujon B (1998) Mass-murder of ORFs from three regions of chromosome XI from *S. cerevisiae*. Gene 223:33–46

Faye G, Dennebouy N, Kujawa C, Jacq C (1979) Inserted sequence in the mitochondrial 23S ribosomal RNA gene of the yeast *Saccharomyces cerevisiae*. Mol Gen Genet 168:101–109

Haugen P, Bhattacharaya D (2004) The spread of LAGLIDADG homing endonuclease genes in rDNA. Nucleic Acids Res 32:2049–2057

Hensgens LAM, Bonen L, de Haan M, van der Horts G, Grivell LA (1983) Two intron sequences in yeast mitochondrial *COXI* gene: homology among URF-containing introns and strain-dependent variation in flanking exons. Cell 32:379–389

Jacq C, Lazowska J, Slonimski PP (1980) Sur un nouveau mécanisme de la régulation de l'expression génétique. CR Acad Sci Paris 290:1–4
Jacquier A, Dujon B (1983) The intron of the mitochondrial 21S rRNA: distribution in different yeast species and sequence comparison between *Kluyveromyces thermotolerans* and *Saccharomyces cerevisiae*. Mol Gen Genet 192:487–499
Jacquier A, Dujon B (1985) An intron encoded protein is active in a gene conversion process that spreads an intron into a mitochondrial gene. Cell 41:383–394
Kostriken R, Heffron F (1984) The product of the *HO* gene is a nuclease: purification and characterization of the enzyme. Cold Spring Harbor Symp Quant Biol 49:89–97
Kruger K, Grabowski PJ, Zaug AJ, Sands J, Gottschling DE, Cech TR (1982) Self-splicing RNA: autoexcision and autocyclization of the ribosomal RNA intervening sequence of *Tetrahymena*. Cell 31:147–157
Lambowitz AM, Perlman PS (1990) Involvement of aminoacyl-tRNA synthethases and other proteins in group I and group II intron splicing. Trends Biochem Sci 15:440–444
Lazowska J, Jacq C, Slonsimski PP (1980) Sequence of introns and flanking exons in wild-type and box3 mutants of cytochrome b reveals an interlaced splicing protein coded by an intron. Cell 22:333–348
Lemieux C, Lee RW (1987) Non-reciprocal recombination between alleles of the chloroplast 23S rRNA gene in interspecific *Chlamydomonas* crosses. Proc Natl Acad Sci USA 84:4166–4170
Macreadie IG, Scott RM, Zinn AR, Butow RA (1985) Transposition of an intron in yeast mitochondria requires a protein encoded by that intron. Cell 41:395–402
Macino G, Coruzzi G, Nobrega FG, Li M, Tzagoloff A (1979) Use of the UGA terminator as a tryptophan codon in yeast mitochondria. Proc Natl Acad Sci USA 76:3784–3785
Marshall P, Lemieux C (1992) The I-*Ceu* I endonuclease recognizes a sequence of 19 base pairs and preferentially cleaves the coding strand of the *Chlamydomonas moewusii* chloroplast large subunit rRNA gene. Nucleic Acids Res 20:6401–6407
Meselson MS, Radding CM (1975) A general model for genetic recombination. Proc Natl Acad Sci USA 72:358–361
Michel F, Dujon B (1983) Conservation of RNA secondary structures in two intron families including mitochondrial-, chloroplast- and nuclear-encoded members. EMBO J 2:33–38
Michel F, Jacquier A, Dujon B (1982) Comparison of fungal mitochondrial introns reveals extensive homologies in RNA secondary structure. Biochimie 64:867–881
Monteilhet C, Perrin A, Thierry A, Colleaux L, Dujon B (1990) Purification and characterization of the *in vitro* activity of I-SceI, a novel and highly specific endonuclease encoded by a group I intron. Nucleic Acids Res 18:1407–1413
Moure CM, Gimble FS, Quiocho FA (2003) The crystal structure of the gene targeting homing endonuclease I-SceI reveals the origins of its target site specificity. J Mol Biol 334:685–695
Muscarella DE, Vogt VM (1989) A mobile group I intron in the nuclaer rDNA of *Physarum polycephalum*. Cell 56:443–454
Muscarella DE, Ellison EL, Ruoff BM, Vogt VM (1990) Characterization of I-*Ppo*, an intron-encoded endonuclease that mediates homing of a group I intron in the ribosomal DNA of *Physarum polycephalum*. Mol Cell Biol 10:3386–3396
Netter P, Petrochilo E, Slonimski PP, Bolotin-Fukuhara M, Coen D, Deutsch J, Dujon B (1974) Mitochondrial genetics. VII. Allelism and mapping studies of ribosomal mutants resistant to chloramphenicol, erythromycin and spiramycin in *S. cerevisiae*. Genetics 78:1063–1100
Pedersen-Lane J, Belfort M (1987) Variable occurrence of the *nrdB* intron in T-even phages suggests intron mobility. Science 237:182–184

Pellenz S, Harington A, Dujon B, Wolf K, Schäfer B (2002) Characterization of the I-*Spom* I endonuclease from fission yeast: insights into the evolution of a group I intron-encoded homing endonuclease. J Mol Evol 55:302–313

Perrin A, Buckle M, Dujon B (1993) Asymmetrical recognition and activity of the I-SceI endonuclease on its site and on intron-exon junctions. EMBO J 12:2939–2947

Plessis A, Perrin A, Haber JE, Dujon B (1992) Site-specific recombination determined by I-SceI, a mitochondrial group I intron-encoded endonuclease expressed in the yeast nucleus. Genetics 130:451–460

Puchta H, Dujon B, Hohn B (1993) Homologous recombination in plant cells is enhanced by in vivo induction of double strand breaks into DNA by a site-specific endonuclease. Nucleic Acids Res 21:5034–5040

Puchta H, Dujon B, Hohn B (1996) Two different but related mechanisms are used in plants for the repair of genomic double-strand breaks by homologous recombination. Proc Natl Acad Sci USA 93:5055–5060

Quirk SM, Bell-Pedersen D, Belfort M (1989) Intron mobility in the T-even phages: high frequency inheritance of group I introns promoted by intron open reading frames. Cell 56:455–465

Ricchetti M, Fairhead C, Dujon B (1999) Mitochondrial DNA repairs double strand breaks in yeast chromosomes. Nature 402:96–100

Ricchetti M, Dujon B, Fairhead C (2003) Distance from the chromosome end determines the efficiency of double strand break repair in subtelomeres of haploid yeast. J Mol Biol 328:847–862

Richard G-F, Dujon B, Haber J (1999) Double-strand break repair can lead to high frequencies of deletions within short CAG/GTG trinucleotide repeats. Mol Gen Genet 261:871–882

Richard G-F, Cyncynatus C, Dujon B (2003) Contractions and expansions of CAG/CTG trinucleotide repeats occur during ectopic conversion in yeast, by a MUS81-independent mechanism. J Mol Biol 326:769–782

Rougeulle C, Colleaux L, Dujon B, Avner P (1994) Generation and characterization of an ordered lambda clone array for the 460 kb region surrounding the murine *Xist* sequence. Mammal Genome 5:416–423

Seraphin B, Faye G, Hatat D, Jacq C (1992) The yeast mitochondrial intron ai5alpha: associated endonuclease activity and in vivo mobility. Gene 113:1–8

Schapira M, Desdouets C, Jacq C, Perea J (1993) I-SceIII an intron-encoded DNA endonucleases from yeast mitochondria. Asymmetrical DNA binding properties and cleavage reaction. Nucleic Acids Res 21:3683

Slonimski PP, Tzagoloff A (1976) Localization in yeast mitochondrial DNA of mutations expressed in a deficiency of cytochrome oxidase and/or coenzyme QH2-cytochrome c reductase. Eur J Biochem 61:27–41

Slonimski PP, Claisse ML, Foucher M, Jacq C, Kochko A, Lamouroux A, Pajot P, Perrodin G, Spyridakis A, Wambier-Kluppel ML (1978) Mosaic organization and expression of the mitochondrial DNA region controlling cytochrome c reductase and oxidase. III. A model of structure and function. In: Bacila M (ed) Biochemistry and genetics of yeasts: pure and applied aspects. Academic Press, New York, pp 391–401

Sor F, Fukuhara H (1982) Identification of two erythromycin resistance mutations in the mitochondrial gene coding for the large ribosomal RNA in yeast. Nucleic Acids Res 10:6571–6577

Strathern JN, Klar AJ, Hicks JB, Abraham JA, Ivy JM, Nasmyth KA, McGill C (1982) Homothallic switching of yeast mating type cassettes is initiated by a double-strand cut in the MAT locus. Cell 31:183–192

Subik J, Kovacova V, Takacsova G (1977) Mucidin resistance in yeast: isolation, characterization and genetic analysis of nuclear and mitochondrial mucidin-resistant mutants of *Saccharomyces cerevisiae*. Eur J Biochem 73:275–286

Tabak HF, van der Horst G, Osinga KA, Arnberg AC (1984) Splicing of large ribosomal precursor RNA and processing of intron RNA in yeast mitochondria. Cell 39:623–629

Tettelin H, Thierry A, Fairhead C, Perrin A, Dujon B (1995) In vitro fragmentation of yeast chromosomes and yeast artificial chromosomes at artificially inserted sites and applications to genome mapping. Methods Mol Genet Microb Gene Technol 6:81–107

Tettelin H, Thierry A, Goffeau A, Dujon B (1998) Physical mapping of chromosomes VII and XV of *Saccharomyces cerevisiae* at 3.5 kb average resolution to allow their complete sequencing. Yeast 14:601–616

Thierry A, Dujon B (1992) Nested chromosomal fragmentation in yeast using the meganuclease I-SceI: a new method for physical mapping of eukaryotic genomes. Nucleic Acids Res 20:5625–5631

Thierry A, Perrin A, Boyer J, Fairhead C, Dujon B, Frey B, Schmitz G (1991) Cleavage of yeast and bacteriophage T7 genomes at a single site using the rare cutter endonuclease I-SceI. Nucleic Acids Res 19:189–190

Thierry A, Gaillon L, Galibert F, Dujon B (1995) Construction of a complete genomic library of *Saccharomyces cerevisiae* and physical mapping of chromosome XI at 3.7 kb resolution. Yeast 11:121–135

Wenzlau JM, Saldanha RJ, Butow RA, Perlman PS (1989) A latent intron-encoded maturase is also an endonuclease needed for intron mobility. Cell 56:421–430

Wolf K, Dujon B, Slonimski PP (1973) Mitochondrial genetics. V. Multifactorial mitochondrial crosses involving a mutation conferring paromomycin-resistance in *S. cerevisiae*. Mol Gen Genet 125:53–90

Zinn AR, Butow RA (1985) Nonreciprocal exchange between alleles of the yeast mitochondrial 21S rRNA gene: kinetics and the involvement of a double-strand break. Cell 40:887–895

The LAGLIDADG Homing Endonuclease Family

BRETT CHEVALIER, RAYMOND J. MONNAT, JR., BARRY L. STODDARD

1 Introduction

The LAGLIDADG protein family includes the first identified and biochemically characterized intron-encoded proteins (Dujon 1980; Lazowska et al. 1980; Jacquier and Dujon 1985), as described in this volume by Dujon. It has been variously termed the 'DOD', 'dodecapeptide', 'dodecamer', and 'decapeptide' endonuclease family, based on the conservation of a ten-residue sequence motif (Dujon 1989; Dujon et al. 1989; Belfort et al. 1995; Belfort and Roberts 1997; Dalgaard et al. 1997; Chevalier and Stoddard 2001). The LAGLIDADG endonucleases are the most diverse of the homing endonuclease families. Their host range includes the genomes of plant and algal chloroplasts, fungal and protozoan mitochondria, bacteria and *Archaea* (Dalgaard et al. 1997). One reason for the wide phylogenetic distribution of LAGLIDADG genes appears to be their remarkable ability to invade unrelated types of intervening sequences, including group I introns, archaeal introns and inteins (Belfort and Roberts 1997; Chevalier and Stoddard 2001). Descendents of LAGLIDADG homing endonucleases also include the yeast HO mating type switch endonuclease (Jin et al. 1997), which is encoded by an independent reading frame rather than within an intron, but does carry remnants of an inactive intein domain (Haber and Wolfe, this Vol.), and maturases that assist in RNA splicing (Delahodde et al. 1989; Lazowska et al. 1989; Schafer et al. 1994; Geese and Waring 2001; Caprara and Waring, this Vol.).

B. Chevalier, B.L. Stoddard (e-mail: bstoddar@fhcrc.org)
Division of Basic Sciences, Fred Hutchinson Cancer Research Center, 1100 Fairview Ave. N. A3-025, Seattle, Washington 98109, USA

R.J. Monnat, Jr.
Departments of Pathology and Genome Sciences, Box 357470, University of Washington, Seattle, Washington 98195, USA

Nucleic Acids and Molecular Biology, Vol. 16
Marlene Belfort et al. (Eds.)
Homing Endonucleases and Inteins

Members of the LAGLIDADG family are segregated into groups that possess either one or two copies of the conserved LAGLIDADG motif. Enzymes that contain a single copy of this motif, such as I-CreI (Thompson et al. 1992; Wang et al. 1997) and I-CeuI (Turmel et al. 1997), act as homodimers and recognize consensus DNA target sites that are constrained to palindromic or near-palindromic symmetry. Enzymes that have two copies of the LAGLIDADG motif (such as I-SceI, the first LAGLIDADG enzyme to be discovered) act as monomers, possess a pair of structurally similar nuclease domains on a single peptide chain, and are not constrained to symmetric DNA targets (Agaard et al. 1997; Dalgaard et al. 1997; Lucas et al. 2001). In both subfamilies, the LAGLIDADG motif residues play both structural and catalytic roles (see below).

Free-standing LAGLIDADG endonucleases (i.e., those that are not covalently associated with intein domains) recognize DNA sites that typically range from 18 to 22 base pairs. They cleave both DNA strands across the minor groove, to generate mutually cohesive four base 3′ overhangs (Chevalier and Stoddard 2001). Like most, if not all nucleases, LAGLIDADG homing endonucleases require divalent cations for activity.

Upon invasion of a novel biological target site, homing endonucleases and their associated mobile introns or inteins can persist, diversify and spread to similar sites of related hosts. The evolution of related homing endonucleases subsequent to a founding intron invasion event has been elegantly described for at least one LAGLIDADG endonuclease branch, which contains the I-CreI enzyme (Lemieux et al. 1988; Turmel et al. 1995; Chevalier et al. 2003). (A similar study has more recently been reported for the intein-associated PI-SceI lineage; Posey et al. 2004.) I-CreI is encoded within a group I intron present in the chloroplast large subunit (LSU) rDNA of the green alga *Chlamydomonas reinhardtii*; the insertion site of this intron corresponds to position 2593 in the *Escherichia* coli 23S rDNA (Turmel et al. 1995). Sequence analysis of chloroplast and mitochondrial LSU rDNAs from numerous other green algae have disclosed 15 similar open-reading frames (ORFs) within identically positioned introns. Three of these genes were shown to encode active endonucleases that are isoschizomers of I-CreI, including I-MsoI from *Monomastix* (Lucas et al. 2001). Although the native target sites of I-CreI and I-MsoI differ at 2 out of 22 base pair positions, each endonuclease efficiently cleaves both target sites. Threading the I-MsoI sequence onto the I-CreI structure suggests significant protein sequence divergence, especially at residues involved in DNA binding (Lucas et al. 2001); this observation has been confirmed by an X-ray crystal structure of I-MsoI (Chevalier et al. 2003). The structure also implies that the I-MsoI enzyme might recognize a larger number of sites (i.e. is more promiscuous in its DNA recognition profile) than does I-CreI.

2 Structures of LAGLIDADG Homing Endonucleases

The structures of six LAGLIDADG enzymes bound to their DNA targets have been determined. These include two isoschizomeric homodimers (I-CreI: Heath et al. 1997; Jurica et al. 1998; Chevalier et al. 2001, 2003 and I-MsoI: Chevalier et al. 2003), which are both encoded within group I introns in the 23S rDNA of the green algae *Chlamydomonas reinhardtii* and *Monomastix*; two pseudo-symmetric monomers (I-AniI: Bolduc et al. 2003 and I-SceI: Moure et al. 2003), which are encoded in mitochondrial introns of the fungi *Aspergillus nidulans* and *Saccharomyces cerevisiae*; one artificially engineered chimera (H-DreI: Chevalier et al. 2002, which is composed of a domain of the monomeric archaeal enzyme I-DmoI fused to a subunit of I-CreI); and an intein-associated endonuclease from yeast (PI-SceI: Moure et al. 2002). Structures of two additional enzymes have also been determined in the absence of DNA: the archaeal intron-encoded I-DmoI (encoded within an intron in the 23S rRNA gene of *Desulfurococcus mobilis*; Silva et al. 1999), and the archaeal intein-encoded PI-PfuI (found in the ribonucleotide reductase gene of *Pyrococcus furiosus*; Ichiyanagi et al. 2000). These crystallographic structures illustrate the structural and functional significance of the LAGLIDADG motif, the mechanism of DNA recognition and binding, and the structure and likely mechanism of their active sites.

LAGLIDADG enzyme domains form an elongated protein fold that consists of a core fold with mixed α/β topology (α-β-β-α-β-β-α). The overall shape of this domain is a half-cylindrical "saddle" that averages approximately 25×25×35 Å, with the longest dimension along a groove formed by the underside of the saddle. The surface of the groove is formed by an antiparallel, four-stranded β-sheet that presents a large number of exposed basic and polar residues for DNA contacts and binding. Each individual β-strand crosses the groove axis at an angle of ~45° and displays a continuous N- to C-terminal bend. The length of the core protein domain is often increased by extended loops connecting the β-strands at the periphery of the β-sheet structure. The β-sheets are stabilized by hydrophobic packing between the tops of the sheets and the α-helices of the core enzyme fold.

In the case of homodimeric enzymes, the full endonuclease structure is generated by a two-fold symmetry axis located at the N-termini of the individual subunits. For monomeric LAGLIDADG enzymes, a pseudo dyad symmetry axis at the same position arranges individual domains from a single peptide chain into similar relative positions (Fig. 1). For the monomeric enzymes the C- and N-terminal helices of the two related core domains (Dalgaard et al. 1997) are connected by flexible linker peptides with lengths between 3 residues to over 100 residues. In either enzyme subfamily, the complete DNA-binding surfaces of the full-length enzymes are 70–85 Å long, and thus can accommodate DNA targets of up to 24 base pairs.

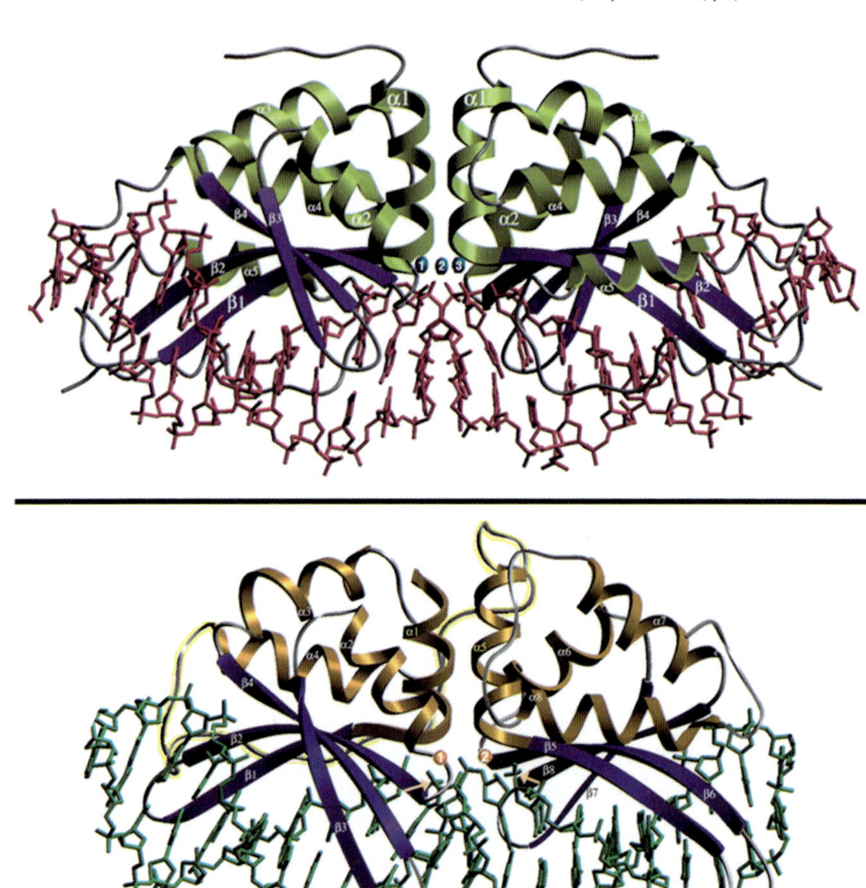

Fig. 1. Ribbon diagrams of homodimeric I-CreI (*top*) and asymmetric, monomeric I-AniI (*bottom*) endonucleases. In the latter, the core α-ββ-α-ββ-α domain fold is duplicated within the single polypeptide chain, and a long flexible linker (*highlighted in yellow*) connects their N- and C-termini, respectively. In the two structures, a perfect dyad symmetry axis, or a pseudo-symmetry axis, extends vertically in the plane of the page between the central two helices at the domain interface. The DNA target of I-CreI is 22 base pairs long and is a pseudo-palindrome; the DNA target of I-AniI is 19 base pairs long and asymmetric. Both enzymes use a four-stranded, antiparallel β-sheet to contact individual base pairs in the major groove of each DNA half-site. The DNA is in a very similar, slightly bent conformation in both structures. In I-CreI, there are three bound metal ions visible in the presence of manganese or magnesium. In the I-AniI structure, only two bound metal ions are visible, as discussed in the text.

The LAGLIDADG motif plays three distinct, but interrelated, roles in the structure and function of this enzyme family (Fig. 2). The first seven amino acid residues of each conserved motif form the last two turns of the N-terminal helices in each folded domain, which are packed against one another. Individual side chains from these helices participate either in core packing within individual domains or in contacts across the interdomain interface. The final three conserved residues (typically a Gly-Asp/Glu-Gly sequence) facilitate a tight turn from the N-terminal α-helix into the first β-strand of each DNA-binding surface. The conserved acidic residues of these sequences are positioned in the active sites and bind divalent cations that are essential for catalytic activity.

The structure and packing of the parallel, two-helix bundle in the domain interface of the LAGLIDADG enzymes are strongly conserved among the otherwise highly diverged members of this enzyme family. Helix packing at this interface is not mediated by a classic "ridges into grooves" strategy, but rather by small residues such as glycine and alanine that allow van der Waals contacts between backbone atoms along the helix–helix interface. The first two glycine and/or alanine residues in the LAGLIDADG motif participate directly in the dimer interface and allow tight packing of the helices. The close packing of the interface helices in these enzymes reflects the need to pack two symmetry-related endonuclease active sites less than 10 Å apart, to facilitate cleavage of homing site DNA across the narrow minor groove.

Despite little primary sequence homology among the LAGLIDADG homing endonucleases outside of the motif itself, the topologies of the endonuclease domains of the enzymes visualized to date, and the shape of their DNA-

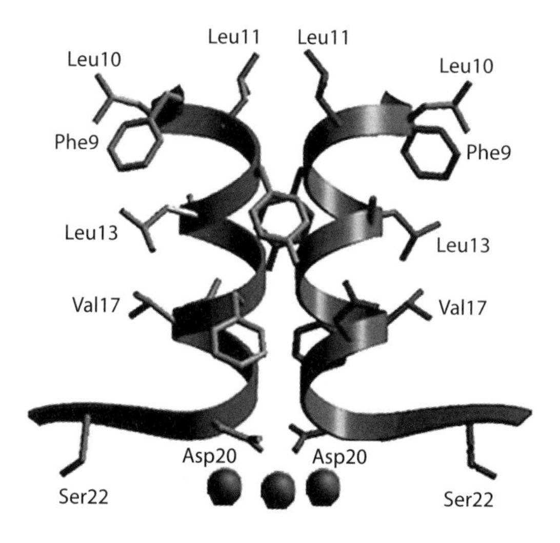

Fig. 2. The LAGLIDADG helices at the domain interface of I-CreI. Other enzymes in the family that have been visualized crystallographically have very similar motifs and structures. Note that a series of hydrophobic residues (Phe 9, Leu 10, 11 and 13, and Val 17) pack into the core of the individual enzyme domains, while other aromatic residues and small residues are involved in packing around and between the helices, respectively. The conserved acidic residues (Asp 20 and 20′) are contributed to the active sites where they participate in metal binding

bound β-sheets, are remarkably similar. A structural alignment of endonuclease domains and subunits in their DNA-bound conformation indicates that the structure of the central core of the β-sheets is well conserved (Bolduc et al. 2003). At least 12 Cα positions within these β-sheets are in close juxtaposition and have a Cα root-mean-square deviation (RMSD) of approximately 1 Å. These positions correspond to residues that make contacts to base pairs ±1 to 6 in each DNA half-site (see below). The conformations of the more distant ends of the β-strands and connecting turns are more poorly conserved, displaying RMSD values of over 3 Å for DNA-contacting residues. Similar alignments of intein-associated endonuclease domains indicate a more diverged structure of the β-sheet motifs.

In contrast to the LAGLIDADG enzymes, which contain a relatively compact structure in which DNA-binding and catalytic activities are intimately connected, the HNH and GIY-YIG homing endonuclease families have been shown by sequence analyses and by structural comparisons to display bipartite structures with separable catalytic and DNA-binding domains (Dalgaard et al. 1997; Derbyshire et al. 1997; VanRoey et al. 2001, 2002; Sitbon and Pietrokovski 2003; Shen et al. 2004). These enzymes often share common DNA-binding domain structures, which may indicate a common ancestral origin for a useful and reuseable binding domain. For example, both the GIY-YIG enzyme I-TevI and the HNH enzyme I-HmuI share a common helix-turn-helix motif at their C-termini that is critical for DNA recognition and binding (VanRoey et al. 2001; Shen et al. 2004). This pattern of swapping structural domains (which are usually part of tandemly arranged functional regions) is generally not observed for the LAGLIDADG family. However, recent analyses of homing endonuclease sequence alignments indicate that, in rare cases, the core fold of LAGLIDADG enzymes can be tethered to additional functional domains involved in DNA binding, usually termed NUMODS (nuclease-associated modular DNA-binding domains; Sitbon and Pietrokovski 2003). For example, a single copy of a canonical NUMOD1 region is found downstream (C-terminal) from the LAGLIDADG core of the intron-associated gene product of ORF Q0255 in yeast. This motif is similar to a conserved region of the bacterial sigma54-activator DNA-binding protein, and its C-terminal 15 amino acids are also similar to the N-terminal helix of typical helix-turn-helix (HTH) DNA-binding domains (Wintjens and Rooman 1996). In HTH domains, this helix is responsible for sequence-specific interactions with DNA.

3 Mechanisms of DNA Target Site Recognition and Specificity

As catalysts of the genetic mobility of introns and inteins, LAGLIDADG homing endonucleases (as well as other enzymes with the same biological function) must balance two somewhat contradictory requirements: they need to be highly sequence-specific, in order to promote precise intron transfer in their host genomes which are most often a chloroplast or mitochondrial genome, and yet must retain sufficient site recognition flexibility to allow successful lateral transfer in the face of sequence variation in genetically divergent hosts. LAGLIDADG homing endonucleases appear to solve these apparently contradictory problems by using a flexible homing site recognition strategy in which a well-defined, but limited, number of individual polymorphisms are tolerated by the enzyme without significant loss of binding affinity or cleavage efficiency. The biochemical basis of this flexible recognition strategy is to make phased, undersaturating DNA–protein contacts across long DNA target sites (Moure et al. 2002, 2003; Chevalier et al. 2003). The length of the interface provides overall high specificity, while formation of a broadly distributed set of phased, subsaturating contacts across the interface facilitates the recognition and accommodation of specific polymorphisms at individual target site positions (Fig. 3). The overall specificity of the LAGLIDADG endonucleases is not well established, but is generally thought to range from 1 in 10^8 to 10^9 random sequences for an average length of 20–22 base pairs (Chevalier et al. 2003).

In the protein–DNA interfaces visualized at high resolution (2.5–1.9 Å) for the LAGLIDADG family (I-CreI, I-MsoI, I-SceI and H-DreI), a set of four antiparallel β-strands in each enzyme domain provide direct and water-mediated contacts between residue side chains and nucleotide atoms in the major groove of each DNA half-site (Fig. 3). These contacts extend from base pairs ±3 to base pairs ±11 (the central four base pairs from −2 to +2, which are flanked by the scissile phosphate groups, are not in contact with the protein). Typically, strands β1 and β2 extend the entire length of this interface in each half-site, while strands β3 and β4 provide additional contacts to base pairs ±3, 4 and 5 in each complex. The LAGLIDADG endonucleases typically make contacts to approximately 65–75% of possible hydrogen-bond donors and acceptors of the base pairs in the major groove, make few or no additional contacts in the minor groove, and also contact approximately one-third of the backbone phosphate groups across the homing site sequence. These contacts are split evenly between direct and water-mediated interactions. A schematic of contacts formed by I-CreI to its pseudo-palindromic target site is shown in Fig. 3.

In the structures listed above, the DNA target is gradually bent around the endonuclease binding surface, giving an overall curvature across the entire length of the site of approximately 45°. In the homodimeric enzyme–DNA complexes with I-CreI and I-MsoI, the DNA is locally overwound between bases −3 to +3 (twist rising to ~50°), with a corresponding deformation in the base pair propeller twist and buckle angles for those same bases, leading to

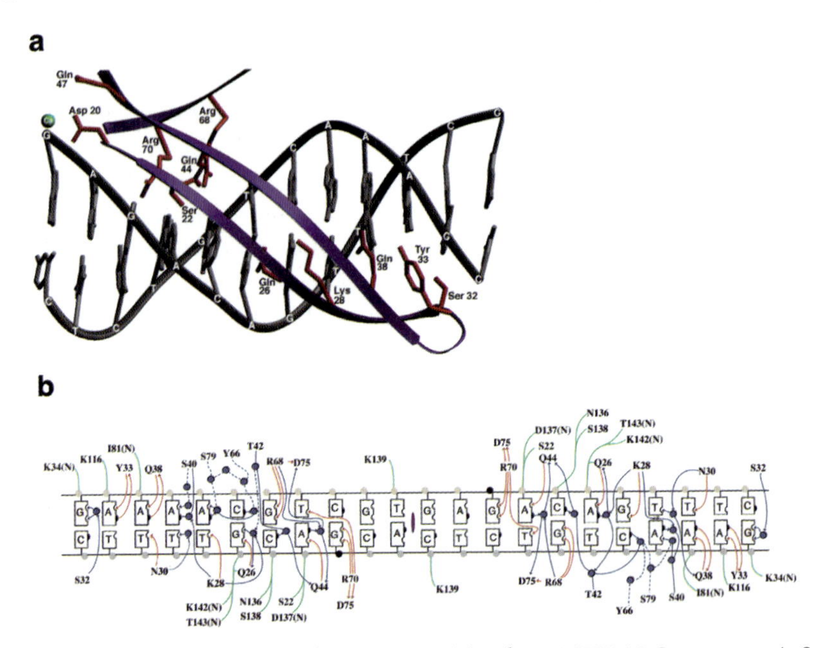

Fig. 3. Structural mechanism of DNA recognition by LAGLIDADG enzymes. A Structure of the β-sheet from a subunit of I-CreI in complex with its corresponding DNA target half-site. Note that every other side chain from a β-strand is pointed into the DNA major groove, and that the residues from adjacent β-strands are staggered in their positions to permit contact to several sequential bases. B Schematic of all observed contacts (both direct and water-mediated) between the I-CreI subunits and both DNA target half-sites, which differ in sequence at several positions (the full-length site is a pseudo-palindrome). The *blue circles* represent ordered water molecules. *Indentations* on bases represent H-bond acceptor groups; *bulges* on the bases represent H-bond donors. *Red lines* are direct contacts, *blue lines* are water-mediated, and *green lines* are contacts to backbone atoms of the DNA. *Dashed lines* represent 'double indirect' contacts to bases via two sequential bridging water molecules

narrowing of the minor groove at the site of DNA cleavage. The bending of the DNA is symmetric (Jurica et al. 1998). In the DNA complex with the monomeric enzymes, the central four base pairs of the cleavage sites generally display negative roll values, which translate into a similar narrowing of the minor groove. As a result, in all of these structures, the scissile phosphates are positioned approximately 5–8 Å apart and are located near bound metal ions in the active sites.

The distributions of related target site sequences that are recognized and cleaved by individual LAGLIDADG enzymes have been previously described using a variety of site preference screens (Argast et al. 1998; Gimble et al. 2003). In those experiments, target site variants that are recognized by the native enzyme are recovered from a randomized homing site library and sequenced. Using these data, the information content (specificity) at each base pair of the target site can be calculated using a computational method that accounts for the probability of each possible base being found at each position across the site (Schneider et al. 1986). The determination of crystallographic structures of the corresponding enzyme–DNA complexes, with the explicit visualization of direct and water-mediated contacts, facilitates an analysis of the correlation between the number and type of intermolecular contacts made to each base pair with the information content at each of these positions. Three general conclusions from these analyses are: (i) the specificity of base pair recognition to structurally unperturbed DNA sequence is proportional to the number of H-bond contacts to each base pair; (ii) the degree of specificity is not significantly attenuated by the use of solvent molecules as chemical bridges between nucleotide atoms and protein side chains; and (iii) information content is increased at individual base pairs, particularly near the center of the cleavage site, by indirect recognition of DNA conformational preferences.

4 Mechanism of DNA Cleavage

The kinetics and mechanism of catalysis have been particularly well studied for the I-CreI enzyme (Chevalier et al. 2004); many of these results appear to be generalizable to the LAGLIDADG family. The measured single-turnover kinetic rate constants, k^*_{max} and K_m^*, of the wild-type I-CreI enzyme are 0.03 $^{min^{-1}}$ and 1.0×10^{-4} nM, respectively, giving a value for catalytic efficiency (k^*_{max}/K_m^*) of 0.3 $nM^{-1}min^{-1}$. This enzyme and its relatives are all dependent on divalent cations for activity, similar to most if not all known endonucleases. A wide variety of divalent metal ions have been assayed for cleavage activity with I-CreI and display a wide range of effects (Chevalier et al. 2004). Two metals (calcium and copper) fail to support cleavage, two (nickel and zinc) display reduced cleavage activity, and three (magnesium, cobalt and manganese) display full activity under the conditions tested. The use of manganese in place of magnesium allows recognition and cleavage of a broader repertoire of DNA target sequences than is observed with magnesium, as is seen for a variety of endonuclease catalysts such as restriction enzymes.

The structures of the four endonuclease–DNA complexes that have been solved at relatively high resolution (I-CreI, I-MsoI, I-SceI and H-DreI) all indicate the presence of three bound divalent metal ions coordinated by a pair of overlapping active sites, with one shared metal participating in both cleavage reactions by virtue of interacting with the scissile phosphates and 3′ hydroxyl leaving groups on both DNA strands. The structures of these four enzymes differ somewhat in the precise position and binding interactions of the metals, but point to similar mechanisms where each strand is cleaved using a canonical two-metal mechanism for phosphodiester hydrolysis (Fig. 4). Whether this

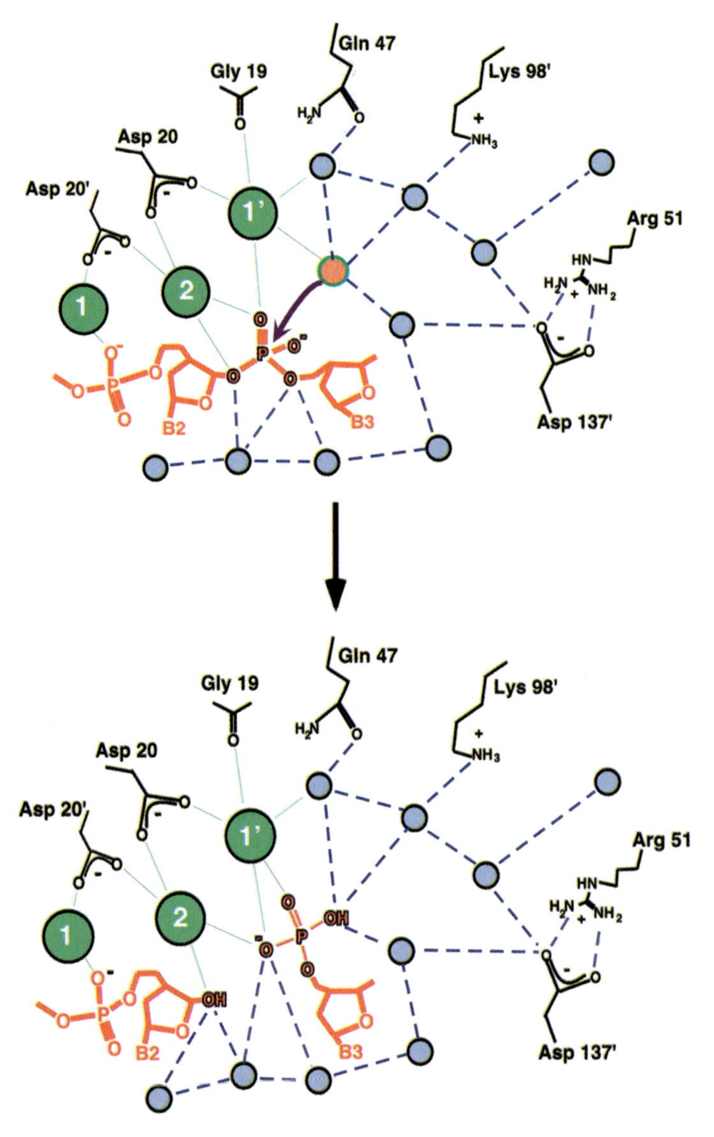

Fig. 4. Proposed mechanism of DNA hydrolysis for the I-CreI homing endonuclease. Other LAGLIDADG enzymes are also thought to follow a canonical two-metal phosphoryl hydrolysis pathway, but with significant variation in the positions and/or roles of basic residues and ordered water molecules. Those residues shown are all known to be essential or extremely important for DNA cleavage by I-CreI. Other LAGLIDADG enzymes display significant divergence at all positions except for the direct metal-binding residues (Asp 20 and 20′) from the LAGLIDADG motifs (Table 1)

unusual structural feature – a shared central divalent metal ion – imparts any particular kinetic order (or simultaneity) to the individual cleavage events is not known for the homodimeric enzymes. In contrast, the structure of the asymmetric I-SceI–DNA complex (Moure et al. 2003) clearly demonstrates that DNA cleavage must involve sequential cleavage of coding and non-coding DNA strands, with a significant conformational rearrangement of the active sites relative to DNA occurring between the two reactions.

In contrast, the structures of DNA complexes of one monomeric enzyme (I-AniI; Bolduc et al. 2003) and of the intein-associated PI-SceI, solved at lower resolution (~3 Å; Moure et al. 2002), have thus far revealed the presence of only two bound metal ions; a central, shared metal ion is not visible. It is unclear whether this reflects a significant difference in catalytic mechanism, reduced occupancy or poor structural ordering of the central metal ion, or simply a limitation of lower resolution crystallographic data.

In the high-resolution structures listed above, a single independently bound metal in each of the two endonuclease active sites coordinates a directly ligated water molecule, which is appropriately positioned for an in-line hydrolytic attack on a scissile phosphate group. The third, 'shared' central metal ion stabilizes the transition state phosphoanion and the 3′ hydroxylate leaving group for both strand cleavage events (Chevalier et al. 2001). In I-CreI, the central metal is jointly coordinated by one conserved acidic residue from each LAGLIDADG motif and by oxygen atoms from the scissile phosphates of each DNA strand. The unshared metals in each individual active site are also coordinated by a single LAGLIDADG carboxylate oxygen, as well as a non-bridging DNA oxygen atom and a well-ordered coordination shell of water molecules. One of the metal-bound water molecules in the active site is often in contact with a catalytically essential glutamine or asparagine residue. In addition to the attacking water molecule a well-ordered network of water molecules is distributed in a large pocket surrounding the DNA scissile phosphate group. These ordered solvent molecules extend from the metal-bound nucleophile to the leaving group 3′ oxygen and are themselves positioned or coordinated by several basic residues that line the solvent pocket.

At physiological pH, phosphate ester bonds have large barriers to cleavage even though they are thermodynamically unstable (Westheimer 1987). To efficiently catalyze the cleavage of phosphate esters, several chemical features are required, including a nucleophile, a basic moiety to activate and position that nucleophile, a general acid to protonate the leaving group, and the presence of one or more positively charged groups to stabilize the phosphoanion transition state (Galburt and Stoddard 2002). The diversity of chemical groups and metal ions available to proteins has made it possible for evolution to arrive at many diverse strategies that satisfy the above requirements. A common feature of many endonucleases (and other phosphoryl transfer enzymes) is the use of bound metal ions as cofactors, and a basic residue (such as a lysine) that directly activates the water molecule for nucleophilic attack.

The metal-dependent features of DNA hydrolysis described above are clearly imparted in the LAGLIDADG endonucleases by the conserved acidic residues of their namesake sequence motif, which directly coordinate divalent cations. However, the remaining residues in the active site are remarkable for their chemical and structural diversity (Chevalier and Stoddard 2001; Table 1). In fact, no enzyme in this family has an essential residue that has been unambiguously identified as a general base for activation of a water nucleophile. Indeed, these enzymes are unique compared to other hydrolytic endonucleases in that the basic residues in their active sites are <u>not</u> generally found in contact distance with metal-bound waters. Catalytically important basic residues, such as Lys 98 in I-CreI, which are involved in interactions with solvent molecules (including those in contact with the scissile phosphate), are poorly conserved, and in some cases absent. The only obvious common chemical feature of many of those residues is the capacity to either donate or accept one or more hydrogen bonds. It is possible that these peripheral active site residues are responsible for positioning and polarizing the solvent network in the active site to facilitate efficient proton transfer reactions to and from nucleophiles and 3′ leaving groups. Each branch of closely related enzymes may have adopted a unique active site solvent packing arrangement that is highly specialized. Furthermore, this rapidly diverging enzyme family

Table 1. Summary of conserved motif and active site residues for LAGLIDADG homing endonuclease structures

Enzyme	LAGLIDADG	Metal Binding		Basic Pocket	
I-CreI	LAGFVDGDG	D20	Q47	K98	R51
I-MsoI	IAGFLDGDG	D21	Q49	K104	K54
I-DmoI	LLGLIIGDG	D21	Q42	K120	K43
	IKGLYVAEG	E117	N129	–	K130
I-AniI	LVGLFEGDG	D15	L36	K94	D40
	LVGFIEAEG	E148	Q171	K227	G174
I-SceI	GIGLILGDA	D44	E61	K122	–
	LAYWFMDDG	D145	N192	K223	–
PI-SceI	LLGLWIGDG	D218	D229	K301	R231
	LAGLIDSDG	D326	T341	K403	H343
PI-PfuI	LAGFIAGDG	D149	D173	L220	–
	IAGLFDAEG	E250	M263	K322	–

may be broadly sampling and adopting significantly different combinations and configurations of chemical groups and associated water molecules to fulfill the catalytic roles described above.

References

Agaard C, Awayez MJ, Garrett RA (1997) Profile of the DNA recognition site of the archaeal homing endonuclease I-DmoI. Nucleic Acids Res 25:1523–1530

Argast GM, Stephens KM, Emond MJ, Monnat RJ (1998) I-PpoI and I-CreI homing site sequence degeneracy determined by random mutagenesis and sequential *in vitro* enrichment. J Mol Biol 280:345–353

Belfort M, Roberts RJ (1997) Homing endonucleases – keeping the house in order. Nucleic Acids Res 25:3379–3388

Belfort M, Reaban ME, Coetzee T, Dalgaard JZ (1995) Prokaryotic introns and inteins: a panoply of form and function. J Bacteriol 177:3897–3903

Bolduc JM, Spiegel PC, Chatterjee P, Brady KL, Downing ME, Caprara MG, Waring RB, Stoddard BL (2003) Structural and biochemical analyses of DNA and RNA binding by a bifunctional homing endonuclease and group I intron splicing cofactor. Genes Dev 17:2875–2888

Chevalier BS, Stoddard BL (2001) Homing endonucleases: structural and functional insight into the catalysts of intron/intein mobility (review; 144 refs). Nucleic Acids Res 29:3757–3774

Chevalier BS, Monnat RJ Jr, Stoddard BL (2001) The homing endonuclease I-CreI uses three metals, one of which is shared between the two active sites (see comments). Nature Struct Biol 8:312–316

Chevalier BS, Kortemme T, Chadsey MS, Baker D, Monnat RJ Jr, Stoddard BL (2002) Design, activity and structure of a highly specific artificial endonuclease. Mol Cell 10:895–905

Chevalier B, Turmel M, Lemieux C, Monnat RJ Jr, Stoddard BL (2003) Flexible DNA target site recognition by divergent homing endonuclease isoschizomers I-CreI and I-MsoI. J Mol Biol 329:253–269

Chevalier B, Sussman D, Otis C, Noel A-J, Turmel M, Lemieux C, Stephens K, Monnat RJ Jr, Stoddard BL (2004) Metal-dependent DNA cleavage mechanism of the I-CreI LAGLIDA-GA homing endonuclease. Biochemistry 43:14015–14026

Dalgaard JZ, Klar AJ, Moser MJ, Holley WR, Chatterjee A, Mian IS (1997) Statistical modeling and analysis of the LAGLIDADG family of site-specific endonucleases and identification of an intein that encodes a site-specific endonuclease of the HNH family. Nucleic Acids Res 25:4626–4638

Delahodde A, Goguel V, Becam AM, Creusot F, Perea J, Banroques J, Jacq C(1989) Site-specific DNA endonuclease and RNA maturase activities of two homologous intron-encoded proteins from yeast mitochondria. Cell 56:431–441

Derbyshire V, Kowalski JC, Dansereau JT, Hauer CR, Belfort M(1997) Two-domain structure of the td intron-encoded endonuclease I-TevI correlates with the two-domain configuration of the homing site. J Mol Biol 265:494–506

Dujon B (1980) Sequence of the intron and flanking exons of the mitochondrial 21S rRNA gene of yeast strains having different alleles at the omega and rib-1 loci. Cell 20:185–197

Dujon B (1989) Group I introns as mobile genetic elements: facts and mechanistic speculations – a review. Gene 82:91–114

Dujon B, Belfort M, Butow RA, Jacq C, Lemieux C, Perlman PS, Vogt VM (1989) Mobile introns: definition of terms and recommended nomenclature. Gene 82:115–118
Galburt E, Stoddard BL (2002) Catalytic mechanisms of restriction and homing endonucleases. Biochemistry 41:13851–13860
Geese WJ, Waring RB (2001) A comprehensive characterization of a group IB intron and its encoded maturase reveals that protein-assisted splicing requires an almost intact intron RNA. J Mol Biol 308:609–622
Gimble FS, Moure CM, Posey KL (2003) Assessing the plasticity of DNA target site recognition of the PI-SceI homing endonuclease using a bacterial two-hybrid selection system. J Mol Biol 334:993–1008
Heath PJ, Stephens KM, Monnat RJ, Stoddard BL (1997) The structure of I-CreI, a group I intron-encoded homing endonuclease. Nat Struct Biol 4:468–476
Ichiyanagi K, Ishino Y, Ariyoshi M, Komori K, Morikawa K (2000) Crystal structure of an archaeal intein-encoded homing endonuclease PI-PfuI. J Mol Biol 300:889–901
Jacquier A, Dujon B (1985) An intron-encoded protein is active in a gene conversion process that spreads an intron into a mitochondrial gene. Cell 41:383–394
Jin Y, Binkowski G, Simon LD, Norris D (1997) Ho endonuclease cleaves MAT DNA in vitro by an inefficient stoichiometric reaction mechanism. J Biol Chem 272:7352–7359
Jurica MS, Monnat RJ Jr, Stoddard BL (1998) DNA recognition and cleavage by the LAGLIDADG homing endonuclease I-CreI. Mol Cell 2:469–476
Lazowska J, Jacq C, Slonimski PP (1980) Sequence of introns and flanking exons in wild-type and box3 mutants of cytochrome b reveals an interlaced splicing protein coded by an intron. Cell 22:333–348
Lazowska J, Claisse M, Gargouri A, Kotylak Z, Spyridakis A, Slonimski PP (1989) Protein encoded by the third intron of cytochrome b gene in *Saccharomyces cerevisiae* is an mRNA maturase. Analysis of mitochondrial mutants, RNA transcripts proteins and evolutionary relationships. J Mol Biol 205:275–289
Lemieux B, Turmel M, Lemieux C (1988) Unidirectional gene conversions in the chloroplast of *Chlamydomonas* inter-specific hybrids. Mol Gen Genet 212:48–55
Lucas P, Otis C, Mercier JP, Turmel M, Lemieux C (2001) Rapid evolution of the DNA-binding site in LAGLIDADG homing endonucleases. Nucleic Acids Res 29:960–969
Moure CM, Gimble FS, Quiocho FA (2002) Crystal structure of the intein homing endonuclease PI-SceI bound to its recognition sequence. Nature Struct Biol 9:764–770
Moure CM, Gimble FS, Quiocho FA (2003) The crystal structure of the gene targeting homing endonuclease I-SceI reveals the origins of its target site specificity. J Mol Biol 334:685–696
Posey KL, Koufopanou V, Burt A, Gimble FS (2004) Evolution of divergent DNA recognition specificities in VDE homing endonucleases from two yeast species. Nucleic Acids Res 32:3947–3956
Schafer B, Wilde B, Massardo DR, Manna F, del Giudice L, Wolf K (1994) A mitochondrial group-I intron in fission yeast encodes a maturase and is mobile in crosses. Curr Genet 25:336–341
Schneider TD, Stormo GD, Gold L, Ehrenfeucht A (1986) Information content of binding sites on nucleotide sequences. J Mol Biol 188:415–431
Shen BW, Landthaler M, Shub DA, Stoddard BL (2004) DNA binding and cleavage by the HNH homing endonuclease I-HmuI. J Mol Biol 342:43–56
Silva GH, Dalgaard JZ, Belfort M, Roey PV (1999) Crystal structure of the thermostable archaeal intron-encoded endonuclease I-DmoI. J Mol Biol 286:1123–1136
Sitbon E, Pietrokovski S (2003) New types of conserved sequence domains in DNA-binding regions of homing endonucleases. Trends Biochem Sci 28:473–477

Thompson AJ, Yuan X, Kudlicki W, Herrin DL (1992) Cleavage and recognition pattern of a double-strand-specific endonuclease (I-CreI) encoded by the chloroplast 23S rRNA intron of *Chlamydomonas reinhardtii*. Gene 119:247–251

Turmel M, Cote V, Otis C, Mercier JP, Gray MW, Lonergan KM, Lemieux C (1995) Evolutionary transfer of ORF-containing group I introns between different subcellular compartments (chloroplast and mitochondrion) Mol Biol Evol 12:533–545

Turmel M, Otis C, Cote V, Lemieux C (1997) Evolutionarily conserved and functionally important residues in the I-CeuI homing endonuclease. Nucleic Acids Res 25:2610–2619

VanRoey P, Waddling CA, Fox KM, Belfort M, Derbyshire V (2001) Intertwined structure of the DNA-binding domain of intron endonuclease I-TevI with its substrate. EMBO J 20:3631–3637

VanRoey P, Meehan L, Kowalski JC, Belfort M, Derbyshire V (2002) Catalytic domain structure and hypothesis for function of GIY-YIG intron endonuclease I-TevI. Nat Struct Biol 9:806–811

Wang J, Kim H-H, Yuan X, Herrin DL (1997) Purification, biochemical characterization and protein-DNA interactions of the I-CreI endonuclease produced in *Escherichia coli*. Nucleic Acids Res 25:3767–3776

Westheimer FH (1987) Why nature chose phosphates. Science 235:1173–1178

Wintjens R, Rooman M (1996) Structural classification of HTH DNA-binding domains and protein-DNA interaction modes. J Mol Biol 262:294–313

HNH Endonucleases

Anthony H. Keeble, María J. Maté, Colin Kleanthous

1 Introduction

HNH endonucleases cleave phosphodiester bonds in many biological contexts, including intron homing, degradation and repair of genomic DNA, and restriction of viral DNA (Goodrich-Blair and Shub 1996; Malik and Henikoff 2000; Bujnicki et al. 2001; James et al. 2002; San Filippo and Lambowitz 2002; Walker et al. 2002). In the 10 years since the HNH motif was first reported by Shub et al. (1994) and Gorbalenya (1994), around 500 members have been identified, which have the following database identifiers: cd00085, SM00507 and pfam01844. HNH enzymes are found in all biological kingdoms, encoded by group I and group II introns, inteins, as well as free-standing open reading frames (Dalgaard et al. 1997).

In the first part of this chapter (Sect. 2), we describe the consensus sequence subsets that are associated with HNH enzymes and the proteins that accommodate them. We then present the biochemical properties of HNH enzymes as a group (Sect. 3), and finally describe recent structural data on HNH enzymes bound to DNA highlighting plausible cleavage mechanisms (Sect. 4). Throughout, we describe how HNH enzymes are part of a wider group of enzymes generally referred to as ββα-Me or His-Me endonucleases that also includes His-Cys homing endonucleases (Galburt and Jurica, this Vol.) and the eukaryotic apoptotic enzyme caspase-activated DNase (CAD).

A.H. Keeble, M.J. Maté, C. Kleanthous (e-mail: ck11@york.ac.uk)
Department of Biology, Area 10, University of York, Heslington YO10 5YW, UK

Nucleic Acids and Molecular Biology, Vol. 16
Marlene Belfort et al. (Eds.)
Homing Endonucleases and Inteins

2 The HNH Family – A Tree of Three Branches

2.1 The HNH Consensus Sequence

The ~34 amino acid HNH motif gained its name from the invariant histidine and asparagine residues that are characteristic of the motif. Since first described, many new sequences have been added to the group allowing a more

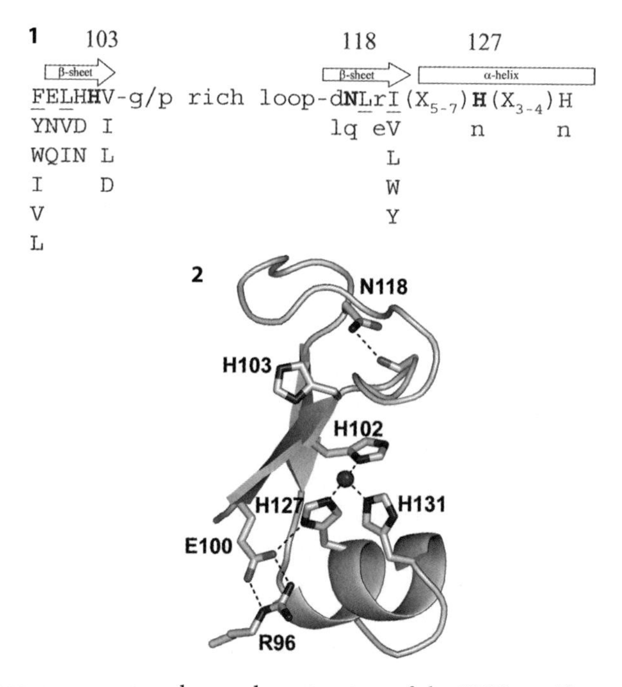

Fig. 1. Consensus sequence and secondary structure of the HNH motif. *1* Consensus sequence for the HNH motif. Residues in *bold* are those that denote the motif itself, *underlined* residues indicate positions where hydrophobic residues stabilize the fold in the context of the recipient protein scaffold (see Fig. 3). Residues are numbered according to the motif in the colicin E9 DNase domain. The secondary structure of the motif is indicated above the consensus sequence. *2* Structure of the histidine-rich HNH motif from the colicin E9 DNase bound to a single Zn^{2+} ion (1FSJ). The HNH motif is denoted by His103 (general base), Asn118 (secondary structure stabilization) and His127 (metal ion ligand), respectively. Glu100 and His131 form the semi-conserved Glu and His within the **E**X_1HH-HX_3**H** motif of Gorbalenya (1994). Glu100 is at the center of a hydrogen-bond network involving both His127 and Arg96, which are not strictly part of the HNH motif but used in the E9 DNase for binding and distorting the minor groove of DNA. His131 is only a metal ion ligand in the context of the Zn^{2+}-bound structure of the E9 DNase. For HNN enzymes, the metal ion ligands His102 and His127 are replaced by aspartate or asparagine and asparagine, respectively

refined consensus sequence to be established (Fig. 1.1). Described from the N-terminus, the HNH motif comprises five elements: (1) an absolutely conserved histidine (**H**NH), usually part of a His-His or Asp-His dyad, with this dyad flanked by hydrophobic residues; (2) a Gly/Pro loop of varying length; (3) an invariant asparagine (H**N**H) followed by a leucine residue; (4) a region of variable length (X_{5-7}) and irregular structure that contains one or more hydrophobic residues involved in stabilizing the motif; and (5) a histidine (HN**H**). Alternative residues at key positions are also shown in Fig. 1, notable amongst these are a Glu/Gln/Asn residue immediately N-terminal to the hydrophobic residue flanking the His-His dyad, and an additional histidine (or glutamine) 3–4 residues C-terminal to the HN**H** residue. It was in this form that the HNH motif (EX_1HH-HX_3H) was described by Gorbalenya (1994). Mutagenesis experiments have confirmed the importance of many of these residues for catalytic activity toward DNA, including the His-His dyad, the HN**H** residue, and the glutamate and C-terminal histidine residues (Walker et al. 2002).

Most of the available crystal structures of HNH-containing enzymes are of DNA-degrading bacterial toxins called colicins. These are plasmid-encoded toxins produced by *Escherichia coli* during times of stress to kill competing strains (James et al. 2002). Crystal structures of the endonuclease domains of colicin E7 (Ko et al. 1999) and E9 (Kleanthous et al. 1999) show that the HNH motif is composed of two β-strands and an α-helix with a metal ion sandwiched between the structural elements (Fig. 1.2), an arrangement resembling a Zn^{2+} finger (Grishin 2001). A single transition metal ion can bind within the histidine-rich HNH motif adopting tetrahedral geometry (Kleanthous et al. 1999; Ko et al. 1999), although the physiological relevance of this has proven controversial (see Sects. 3 and 4). The transition metal ligating residues project from the secondary structural elements themselves (in contrast to conventional Zn^{2+} fingers where the metal ion is coordinated by residues presented on loops) and comprise the first histidine of the His-His dyad, the HN**H** histidine, and the C-terminal histidine residue (Fig. 1). The asparagine of the motif (H**N**H) serves a structural role forming a backbone hydrogen bond across the motif (Kleanthous et al. 1999; Ko et al. 1999).

2.2 The Cysteine-Containing HNH Motif (cysHNH)

A large number of HNH enzymes, including the nrdB intron-encoded endonuclease I-TevIII and 5-methylcytosine-dependent restriction endonuclease McrA (Eddy and Gold 1991; Shub et al. 1994), have additional consensus sequences that take the form of two $CX_{2-4}C$ motifs, one 8–12 residues N-terminal to the His-His dyad and the other adjacent to the HN**H** residue (not shown). Modelling studies suggest that these cysteines point away from the active site and coordinate structural Zn^{2+} ions (Bujnicki et al. 2000), a feature that is shared with His-Cys enzymes such as I-PpoI (see Galburt and Jurica, this Vol.). We suggest such motifs are denoted as cysHNH motifs to distinguish them from the general HNH group of enzymes.

2.3 The HNN Variant

A further variation of the HNH motif exists that is modified in two respects: (1) the initial His-His dyad is replaced by an Asp-His or Asn-His pair; and (2) the HN**H** residue is replaced by an asparagine to form an HNN motif. These substitutions are highly relevant since the two changed residues ligate the active site metal ion (Maté and Kleanthous 2004; Fig. 1.2). The HNN variant can also be found in combination with two stabilizing $CX_{2-4}C$ couplets (cysHNN).

2.4 The HNH Motif Is Part of the Wider ββα-Me Superfamily of Endonucleases

Conservation of the HNH motif fold extends across a wider range of endonucleases, alternatively termed the ββα-Me or His-Me finger endonucleases (Kühlmann et al. 1999; Aravind et al. 2000). This structural homology was first identified by Kühlmann et al. (1999), who reported that the active sites of the His-Cys homing endonuclease I-PpoI (Fig. 2) and the non-specific nuclease from *Serratia* superimposed with that of the E9 DNase with a rmsd for main chain atoms of 1.2 and 1.5 Å, respectively. The central Asn-Leu pair of the HNH motif is also present in *Serratia* but replaced by a His-Leu pair in I-PpoI, in each case the polar residue stabilizing the motif fold. I-PpoI is a cysHNN variant while Serratia nuclease is an HNN motif enzyme. Many ββα-Me motif enzymes belong to the HNN group, including the Holliday junction resolving enzyme T4 endo VII (Aravind et al. 2000).

Other notable HNH/ββα-Me endonucleases are the apoptotic enzyme CAD, the periplasmic nuclease Vvn from *Vibrio vulnificus* and the homing endonuclease I-HmuI (Li et al. 2003; Shen et al. 2004; Woo et al. 2004). Figure 2 shows a comparison of the structures of CAD and Vvn alongside those of I-PpoI and the E9 DNase. It illustrates the similarity of the ββα-Me motif in enzymes that are otherwise structurally unrelated. The individual motifs are accommodated on the different enzyme scaffolds through the burial of 4–6 conserved (or conservatively substituted) hydrophobic amino acids. These

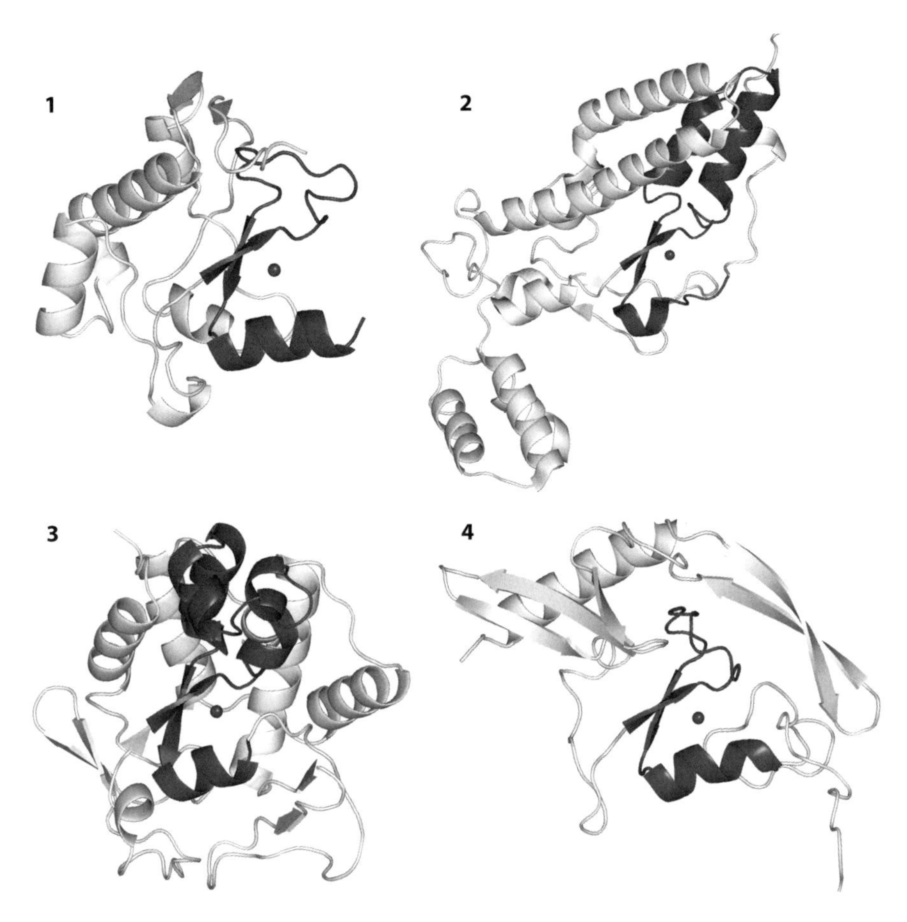

Fig. 2. The HNH/ββα-Me motif in different endonucleases. The figure shows the crystal structures of a variety of enzymes containing the motif (*black*). *1* Colicin E9 DNase (1FSJ); *2* CAD (1V0D); *3* Vvn (1OUO); *4* I-PpoI (1CYQ). The figure also highlights how a large section of non-conserved sequence connecting the β-strands of the motif adopts a variety of folds, forming an irregular structure in the E9 DNase and I-PpoI but folding into α-helices in CAD and Vvn

are shown in Fig. 3.1 for the colicin E9 DNase, with a corresponding structural superposition for the E9 DNase and CAD presented in Fig. 3.2. The hydrophobic residues dock into pockets in the recipient protein, essentially bolting the motif to the protein. The ability to dock the motif onto differing scaffolds explains its wide distribution in evolutionarily unrelated enzymes.

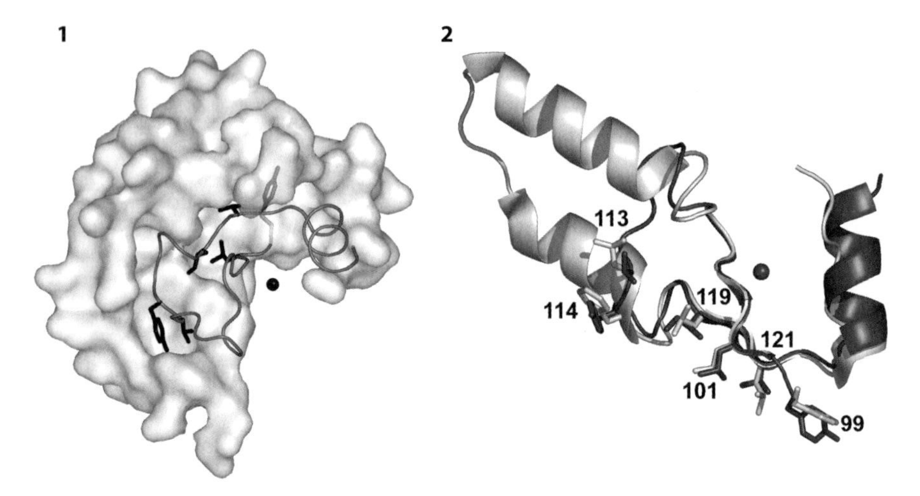

Fig. 3. The HNH/ββα-Me motif is bolted onto different protein scaffolds by hydrophobic residues that dock into pockets in the host protein. *1* The HNH motif of the colicin E9 DNase is shown in ribbon while the remainder of the molecule is shown as a molecular surface. The motif is accommodated through the insertion of (mostly conserved) hydrophobic residues into corresponding pockets in the E9 DNase. *2* Structural overlay of the HNH/ββα-Me motifs of the colicin E9 DNase (*black*; 1EMV) and CAD (*gray*, 1V0D) showing the positions of the key hydrophobic residues (numbered according to the E9 DNase) that enable the motif to be bolted onto different protein scaffolds

3 Enzymology of HNH/ββα-Me Motif Endonucleases

3.1 The HNH/ββα-Me Motif Is Functionally Adaptable

Enzymes that contain the HNH/ββα-Me motif and no other DNA recognition domains, as is the case for the bacterial colicins (Fig. 4.1), CAD and *Serratia* nuclease, show little cleavage specificity. This is consistent with the biological roles for these proteins whose function it is to degrade intracellular DNA (colicins, CAD), or, as in the case of *Serratia* nuclease, catabolize extracellular nucleic acids for nutrient uptake (Friedhoff et al. 1999; Widlak et al. 2000; James et al. 2002). In some cases (*Serratia*, colicins), this low specificity extends to cleaving ssDNA and RNA (Friedhoff et al. 1999; Pommer et al. 2001). DNA recognition specificity is essential for the homing function of intron-encoded endonucleases, with this specificity emanating from DNA recognition domains. For example, in addition to its HNH motif, I-HmuI contains three such domains (Fig. 4.2): (1) an N-terminal NUMOD4 (Nuclease-associated MODule, type 4) domain comprised of anti-parallel β-sheets that inserts into the major groove in a similar manner to the DNA recognition domain of I-PpoI (see Galburt and Jurica, this Vol.); (2) a central NUMOD3 domain comprised of a

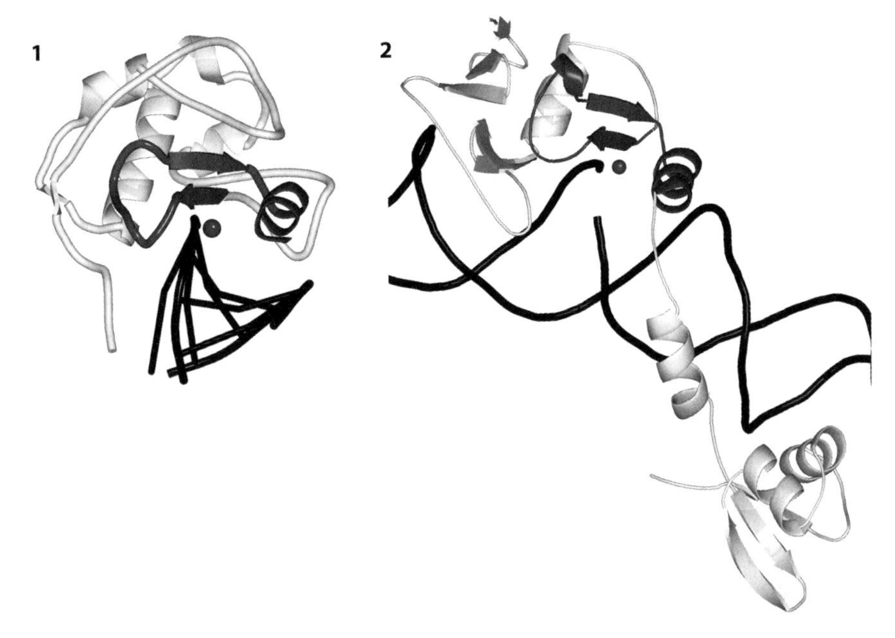

Fig. 4. Structures of HNH endonucleases bound to dsDNA. *1* Structure of the E9 DNase (2.9 Å) bound to double-stranded substrate DNA and Mg^{2+} (1V14). The E9 DNase HNH motif (*black*) binds to the minor groove of dsDNA (shown as *stick model*), inserting the α-helix of the motif and an adjoining loop into the groove and inducing a large structural distortion. This distortion is needed in order for the cleavable phosphodiester bond to be presented to the catalytic metal ion (*sphere*). *2* Structure of I-HmuI (2.9 Å) bound to double-stranded DNA and Mn^{2+} in the cleaved, product complex (Shen et al. 2004). A 40° bend is produced in the DNA by the insertion of a DNA recognition domain into the minor groove. The Mn^{2+} ion is adjacent to the cleaved bond. The HNN motif (*black*) is part of a larger protein with additional domains that contact DNA through contacts to the major and minor grooves (see text for details)

pair of α-helices that insert into the minor groove; (3) a C-terminal 'IENR1' domain containing a helix-turn-helix that inserts into the major groove (Sitbon and Pietrokovski 2003; Shen et al. 2004). The last two domains are similar to those found within the GIY-YIG homing endonuclease I-TevI (see van Roey and Derbyshire, this Vol.). This sharing of DNA recognition modules between HNH, His-Cys and GIY-YIG endonucleases indicates that 'domain exchange' has occurred during the evolution of these families (Shen et al. 2004). Interestingly, I-HmuI makes fewer sequence-specific contacts with DNA than either LAGLIDADG or His-Cys box homing enzymes and accordingly shows weaker sequence specificity, illustrated by its ability to cleave intron-plus and intron-minus genes (Shen et al. 2004).

Members of other homing endonuclease groups (see Chevalier et al., this Vol.) function as dimers with each active site cleaving one of the DNA strands. In cases where the enzyme is monomeric the protein contains two active sites and resembles a fused dimer (e.g. I-DmoI; Galburt and Stoddard 2002). The oligomeric status of HNH/ββα-Me enzymes can be quite variable; enzymes such as CAD are organized as higher order oligomers, sequence-specific enzymes such as I-PpoI are dimers, which permits cleavage of both DNA strands simultaneously, whereas I-HmuI, colicin DNases and I-CmoeI are monomers (Goodrich-Blair and Shub 1996; Galburt et al. 1999; Friedhoff et al. 1999; Drouin et al. 2000; Maté and Kleanthous 2004).

3.2 Biochemical Properties

Cleavage products of the different enzymes in the HNH/ββα-Me group vary from blunt end-products to overhangs of different sizes, with single-strand nicks produced in some cases such as I-HmuI, colicin E9 and the mitochondrial DNA repair protein MutS (Goodrich-Blair and Shub 1996; Pommer et al. 1998; Drouin et al. 2000; Malik and Henikoff 2000). Where single-strand nicking occurs this correlates with the enzyme being monomeric, although simply being monomeric is not sufficient as, for example, I-CmoeI is monomeric but still produces double-strand breaks (Drouin et al. 2000). Nevertheless, the products of phosphodiester hydrolysis by HNH/ββα-Me enzymes are always the same, 5′-phosphates and 3′-hydroxyls (Friedhoff et al. 1999; Widlak et al. 2000; Pommer et al. 2001). Whilst these are probably the most common cleavage products for endonucleases as a whole, formation of the 3′-hydroxyl is essential for the mechanism of homing/insertion of the genetic element encoding the enzyme.

In most cases, HNH/ββα-Me endonucleases have alkaline pH optima (pH 7.5–9), which is suggestive of one or more HNH histidine(s) being in the unprotonated form for catalysis to proceed (Friedhoff et al. 1999; Mannino et al. 1999; Drouin et al. 2000; Pommer et al. 2001; Widlak and Garrard 2001). This is consistent with the absolutely conserved **H**NH residue acting as a general base, as proposed for I-PpoI (Galburt et al. 1999). Monovalent salts such as NaCl generally inhibit HNH/ββα-Me endonucleases. This may, in part, be due to electrostatic screening of protein–DNA contacts. In the case of I-PpoI, Na^+ ions inhibit catalysis by displacing the Mg^{2+} to form a non-productive I-PpoI·Na^+·dsDNA complex (Galburt et al. 1999).

3.3 HNH/ββα-Me Enzymes Require Single Divalent Cations for Activity

HNH/ββα-Me enzymes in common with most endonucleases require divalent cations for catalysis. However, in contrast to the two-cation mechanisms of most restriction enzymes and some homing endonucleases (see Chevalier et al., this Vol.), only one metal ion is required for catalysis (Galburt et al. 1999; Miller et al. 1999; Maté and Kleanthous 2004). The most commonly used metal ion is Mg^{2+}, although many other divalent cations can also substitute, including Ca^{2+} and various transition metal ions such as Mn^{2+}, Co^{2+}, Cu^{2+} and Ni^{2+} (Wittmayer and Raines 1996; Friedhoff et al. 1999; Pommer et al. 1999; Widlak and Garrard 2001). The ability to utilize a broad range of metal ions is in contrast to most restriction endonucleases and the LAGLIDADG homing endonucleases (see Chevalier et al., this Vol.). The role of the metal ion (see below and Chevalier et al., this Vol.) is to stabilize the phosphoanion transition state and the 3′OH leaving group of the scissile bond, the latter via an activated water molecule, as proposed for I-PpoI by Galburt et al. (1999).

Some variation exists as to which metal ions will support catalysis in different enzymes of this group. Zn^{2+}, for example, can support catalysis by I-PpoI and the *Serratia* nuclease but not colicins or I-HmuI (Wittmayer and Raines 1996; Friedhoff et al. 1999; Pommer et al. 1999, 2001; Shen et al. 2004). There have been reports of activity with Zn^{2+} for colicin E7 DNase (Ku et al. 2002) although, in our hands, no colicin DNases exhibit Zn-dependent activity (Pommer et al. 2001; Keeble et al. 2002).

Given that a variety of metal ions support catalysis by HNH/ββα-Me enzymes the question arises to which is the physiological cofactor. Mg^{2+} is usually considered to be the in vivo metal ion for endonucleases due to its high natural abundance (~0.5 mM) and highest activity with endonucleases (Cowan 1998). This is in contrast to the very low concentrations of available transition metal ions (Changela et al. 2003). The importance of Mg^{2+} for cleavage of DNA by HNH enzymes, rather than transition metal ions, in vivo has been demonstrated for the colicin DNases. Walker et al. (2002) identified a group of seven active site residues that inactivated Mg^{2+} activity in vitro of which a subset were still active in the presence of transition metals. However, all mutants were biologically inactive (Walker et al. 2002).

Although Mg^{2+} is the likely metal ion in vivo for most HNH/ββα-Me endonucleases, the ability of the single Mg^{2+} ion to bind the enzyme in the absence of DNA varies. For example, Mg^{2+} can bind to *Serratia*, T4 endo VII and LtrA in the absence of DNA (Miller et al. 1999; Raaijmakers et al. 2001; San Filippo and Lambowitz 2002) but not to the colicin E9 DNase or I-PpoI (Pommer et al. 1999; Galburt et al. 2000). The metal ion is also required for binding the DNA in I-CmoeI (Drouin et al. 2000).

4 Structural Analysis of HNH Endonuclease Mediated dsDNA Cleavage

Focusing on the catalytic domains of HNH endonucleases, we describe the structural basis for dsDNA cleavage using the recently solved structures of the substrate (H103A E9·DNase·Mg^{2+}·dsDNA; Fig. 4.1) and product (wild-type I-HmuI·Mn^{2+}·dsDNA; Fig. 4.2) complexes (Maté and Kleanthous 2004; Shen et al. 2004). We also account for the apparent inactivity with Zn^{2+} ions within the HNH motif.

4.1 Relaxation of Substrate Strain Is Conserved in the Cleavage Mechanisms of HNH/ββα-Me Enzymes

A Mg^{2+}-bound substrate complex structure was obtained for the E9 DNase by mutation of the putative general base H**N**H residue (H103A), thereby inactivating the enzyme (Walker et al. 2002). A total of ~700 $Å^2$ of the E9 DNase surface area is buried on binding the dsDNA, with the HNH motif inserting into the minor groove (Maté and Kleanthous 2004). Strikingly, the V-shaped ββα-fold of the active site remains undistorted in the complex. Conversely, in order for the scissile phosphodiester bond of substrate DNA to reach the metal center at the base of the HNH motif, the DNA duplex is distorted (Fig. 4.1). The C-terminal α-helix (containing the HN**H** residue) lies parallel to the helical axis of the DNA minor groove, with residues preceding the motif and the metal ion inserted into the groove itself. The contacts to the DNA predominantly involve the phosphate backbone and cause a significant bend toward the major groove and concomitant widening of the minor groove (from 5.9 to ~9 Å; Fig. 4.1). In the case of apoE7 bound to dsDNA (Hsia et al. 2004) a similar central distortion is also seen, demonstrating that it is an intrinsic property of the HNH motif. Distortions are also seen for the substrate complexes of I-PpoI (Galburt et al. 1999) and Vvn (Li et al. 2003). In all cases distortion induces strain around the scissile bond. In the complex of the E9 DNase the distortion is brought about by the insertion of the Arg96-Glu100 salt bridge into the minor groove, with Arg96 making a hydrogen bond to the hydroxyl of a thymine on the opposing strand (Maté and Kleanthous 2004). An equivalent salt bridge is observed in Vvn causing similar distortion to the DNA minor groove (Li et al. 2003). The conformation of the DNA in the E9 DNase complex is stabilized by a network of hydrogen bonds between the nucleic acid and three charged residues Arg5, Arg54 and Asp51 that straddle the scissile bond. Consistent with the important roles of Arg5, Arg54, Arg96 and Glu100 in deforming the substrate, alanine mutants at these positions abolish dsDNA cleavage activity (Walker et al. 2002; A.H. Keeble and C. Kleanthous, unpubl. results).

Structural distortions are retained within the product complex of I-HmuI bound to Mn^{2+} and dsDNA although, in part, this is due to a 40° bend 4–5 base pairs downstream of the cleaved bond (Fig. 4.2; Shen et al. 2004). However, the DNA around the cleaved bond itself is no longer strained with the 5′ phosphate moving away from the 3′OH and adopting a relaxed, tetrahedral geometry (Shen et al. 2004). Relaxation of induced strain in the scissile phosphodiester upon cleavage has been proposed to be part of the mechanism of DNA cleavage by I-PpoI by Galburt et al. (1999; see also Galburt and Jurica, this Vol.) and has also been reported for Vvn (Li et al. 2003). Therefore, it is likely to be a conserved feature of the cleavage mechanism of all HNH/ββα-Me enzymes.

4.2 Metal Ion Coordination in the DNA-Bound Complex

In the structure of H103A E9 DNase bound to dsDNA and Zn^{2+}, the Zn^{2+} ion exhibits tetrahedral geometry, made up of three protein ligands, His102, His127 and His131, and a phosphate oxygen atom from the scissile bond (Fig. 5.1). This arrangement is also seen for H103A E9 DNase bound to Zn^{2+}

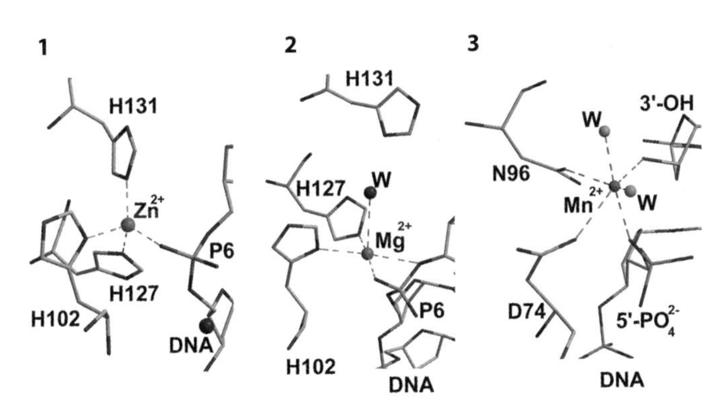

Fig. 5. The HNH/ββα-Me motif is an adaptable center able to bind both alkaline earth and transition metal ions. *1* H103A E9 DNase·Zn^{2+}·dsDNA complex showing tetrahedral geometry around the metal ion (1V15). The Zn^{2+} ion is coordinated by three motif histidine residues and a phosphate oxygen atom from the DNA; *2* H103A E9 DNase·Mg^{2+}·dsDNA complex showing distorted octahedral geometry (1V14). The metal ion is coordinated by two motif histidines and two phosphate oxygen atoms from the DNA. His131 disengages from the metal ion and is replaced by a water molecule. The sixth coordination site is presumed to be another water molecule. *3* Wild-type I-HmuI·Mn^{2+}·dsDNA product complex showing distorted octahedral geometry to the metal ion (Shen et al. 2004). The two metal-ligating residues from the enzyme are asparagine and aspartic acid in this HNN motif endonuclease, equivalent to the two histidines of the colicin E9 DNase

in the absence of DNA where the phosphate oxygen is replaced by water (Maté and Kleanthous 2004). In each complex the imidazole nitrogen–Zn^{2+} distances (2.0–2.1 Å) are similar to those of other tetrahedral transition metal sites within proteins. By contrast, the distance between the Zn^{2+} and the phosphate oxygen is shorter, at 1.9 Å, with similar Zn^{2+}–O bond distances also seen in the *E. coli* repair enzyme endo IV (Hosfield et al. 1999).

Two of the four histidines within the HNH motif (His102 and His127) are also coopted to bind the Mg^{2+} ion in the H103A·E9·Mg^{2+}·dsDNA complex (Fig. 5.2). The octahedral coordination of the Mg^{2+} ion is clearly very different from the tetrahedral geometry of the Zn^{2+} ion (Fig. 5.1). Furthermore, the metal ions are displaced relative to each other by ~1 Å (Maté and Kleanthous 2004). At the current resolution of the H103A E9 DNase·Mg^{2+}·dsDNA complex (2.9 Å) only 5/6 ligands to the Mg^{2+} can be identified. Two oxygen atoms of the scissile phosphate (O1P and O3′), together with nitrogen atoms from the His102 and His127, form the equatorial ligands, with one of the axial positions taken by a water molecule (Fig. 5, 2). The sixth position is assumed to be another water molecule (Maté and Kleanthous 2004). Octahedral coordination is also observed for the Mn^{2+} ion within the I-HmuI·dsDNA complex (Fig. 5. 3) but here two DNA oxygen atoms contact to the metal ion as equatorial and axial ligands (Shen et al. 2004).

This H103A E9 DNase·Mg^{2+}·dsDNA complex represents one of only two examples in the literature where the coordination shell of a Mg^{2+} ion in an enzyme active has multiple histidine residues, the other being the F plasmid TraI relaxase (Datta et al. 2003). In this case the Mg^{2+}–N bond distances are in the range 2.3–2.6 Å which are comparable to the distances for His102 and His127 to the Mg^{2+} ion (2.5 and 2.2 Å, respectively). The octahedral coordination of the metal center is distorted, similar to that seen in the I-PpoI·Mg^{2+}·DNA substrate complex (Galburt et al. 1999), indicating further similarities in the cleavage mechanism between these two ββα-Me endonucleases. In contrast to I-PpoI and Serratia nuclease that contact the bound Mg^{2+} ion through a single protein side chain, E9 DNase, CAD and Vvn endonucleases use two protein ligands that are superimposable with each other; Asp262 and His308 in CAD, Glu79 and Asn127 in Vvn, and His102 and His127 in the E9 DNase (Li et al. 2003; Maté and Kleanthous 2004; Woo et al. 2004).

4.3 Mechanism of Mg^{2+}-Dependent Cleavage by HNH Endonucleases and Why Zn^{2+} Does Not Support Catalytic Activity

Many mechanistic similarities exist between the E9 DNase and I-PpoI including minor groove binding by the ββα-Me motif, a distorted substrate configuration in the DNA-bound complex, a single Mg^{2+} ion coordinated to the motif, and a strained octahedral geometry around the catalytic metal ion. It is likely therefore that the two enzymes cleave the dsDNA by similar mechanisms. In this mechanism, the HNH histidine residue acts as a general base that activates a water molecule for nucleophilic attack of the scissile phosphodiester bond (this role is taken by His98 in I-PpoI). The Mg^{2+} ion is coordinated by His102 and His127 in the E9 DNase whereas Asn119 is the only protein coordination site in I-PpoI (equivalent to His127). The Mg^{2+} ion nevertheless likely serves similar functions in both, acting to stabilize the phosphoanion transition state and activating a water molecule to protonate the 3′OH leaving group (Galburt et al. 1999). A similar mechanism has been proposed for the *Serratia* nuclease where His89 acts as the general base and Asn119 stabilizes the phosphoanion transition state and activates a water molecule for protonating the leaving group (Friedhoff et al. 1999). In *Serratia*, Glu127 makes a water-mediated interaction with the Mg^{2+} whereas this role is taken by His131 in the E9 DNase (shown in Fig. 5, 2 for E9). The roles of these residues are still not clear although they are required for activity since mutations at both reduce activity (Miller et al. 1999; Walker et al. 2002). An equivalent residue is found within the active site of T4 endo VII (Raaijmakers et al. 2001) and I-PpoI (Asn123). Most HNH/ββα-Me endonucleases use arginines to bind and/or distort substrate DNA, although details differ between enzymes. For example, Arg5 in the E9 DNase binds ground-state DNA but the equivalent residue in I-PpoI (Arg61) binds only within the product complex.

We have been unable to detect Zn^{2+}-catalyzed endonuclease activity for any colicin DNase, which has been difficult to rationalize given that the HNH motif resembles a zinc finger and binds Zn^{2+} ions with the greatest affinity of all transition metals (Pommer et al. 1999). Moreover, other transition metals such as Ni^{2+} and Co^{2+} are active in the motif (Pommer et al. 1998, 1999). The current structures provide an explanation for these observations. His131, one of the coordinating histidines to the Zn^{2+} ion, is displaced by more than 2 Å in the Mg^{2+}-bound structure (Fig. 5, 2). We have proposed previously that transition metal mediated activity with metals such as Ni^{2+} and Co^{2+} requires two non-protein-bound coordination sites to the tetrahedral metal ion, one for ligation of the scissile phosphate oxygen, the other for the activation of a water molecule for leaving group stabilization (Pommer et al. 2001). Although no structure for Ni^{2+} bound to an E9 DNase·dsDNA complex is yet available, an earlier structure for the E9 DNase HNH motif bound with Ni^{2+} and a phosphate molecule is informative (Kleanthous et al. 1999). Here, again, His131 disengages from the metal ion, in this case taking a position midway between that seen in the Mg^{2+} and Zn^{2+} structures. This may allow a water molecule to replace the histidine and so serve as a protonating device for DNA hydrolysis, by a mechanism similar to that proposed for Mg^{2+}-based activity. Consistent with the proposal, rapid reaction kinetics suggests that Zn^{2+} binding to the E9

DNase induces an additional conformational change relative to Ni^{2+} binding, which is most likely attributable to His131 binding to the Zn^{2+} ion, generating a catalytically inert complex (Keeble et al. 2002).

5 Conclusions

HNH endonucleases are part of a widespread group of enzymes whose active sites are built around an evolutionarily conserved ββα-Me structural motif. First described in homing enzymes, HNH endonucleases cleave DNA in a range of biological contexts. Biochemical and structural analyses suggest that HNH/ββα-Me enzymes share a conserved, single metal ion (Mg^{2+} or Ca^{2+}) cleavage mechanism, with the HN**H** residue (or an equivalent asparagine in HN**N** enzymes) acting as a metal ligand. The other conserved histidine of the motif (**H**NH) is a general base that activates a hydrolytic water molecule, which then attacks the metal-coordinated scissile phosphodiester bond. By analogy with the mechanism for the His-Cys enzyme I-PpoI (also a ββα-Me endonuclease), the Mg^{2+} ion in HNH enzymes is proposed to stabilize the pentavalent phosphoanion transition state and activate a water molecule to act as a catalytic acid, enabling protonation of the 3′-hydroxyl leaving group. Notwithstanding our current level of understanding of HNH enzymes and their relationship to the broader group of ββα-Me endonucleases, a number of key questions remain to be resolved: (1) To what extent does substrate strain play a role in catalysis since structures of DNA bound to ββα-Me active sites show highly distorted minor grooves and distorted octahedral geometry to the bound Mg^{2+} ion? (2) The current structures for DNA and metal ions bound to the substrate complex of the colicin E9 DNase and the product complex of I-HmuI show subtly different DNA coordination chemistries to the metal ion. Are these mechanistically relevant? (3) What are the pK_a's of the different active site residues in their liganded states and are they appropriate for their proposed roles? (4) Do all the residues that contact the phosphate backbone and scissile bond play similar catalytic roles in different HNH/ββα-Me enzymes? For example, Arg5 of the E9 DNase is structurally equivalent to Arg61 of I-PpoI yet appears to contact the ground-state substrate, whereas Arg61 only contacts the product phosphate. (5) Is the mechanism of phosphodiester hydrolysis by HNH/ββα-Me enzymes different when the motif is occupied by a transition metal ion such as Ni^{2+} or Co^{2+}, as suggested by work on the colicin E9 DNase? (6) Which step (binding and distortion, phosphodiester hydrolysis or product release) is rate-limiting for catalysis by HNH/ββα-Me enzymes? The next few years should resolve many of these issues by which time we will be able to more fully appreciate how adaptable the motif is as a catalytic center for the hydrolysis of phosphodiester bonds.

Acknowledgements. Work on HNH enzymes in C.K.'s lab has been supported by funding from the Biotechnology and Biological Sciences Research Council of the UK and the Wellcome Trust. We would like to thank Charles Heise, Nicholas Housden, Lorna Lancaster and Daniel Walker for critical reading of the manuscript and helpful suggestions. We also acknowledge the work of our collaborators Geoffrey Moore (Norwich) and Richard James (Nottingham) for their valuable contributions to nuclease colicin research.

References

Aravind L, Makarova KS, Koonin EV (2000) Holliday junction resolvases and related nucleases: identification of new families, phyletic distribution and evolutionary trajectories. Nucleic Acids Res 28:3417–3432

Bujnicki JM, Radlinska M, Rychlewski L (2000) Atomic model of the 5-methylcytosine-specific restriction enzyme McrA reveals an atypical zinc finger and structural similarity to ββαMe endonucleases. Mol Microbiol 37:1280–1281

Bujnicki JM, Radlinska M, Rychlewski L (2001) Polyphyletic evolution of type II restriction enzymes revisited: two independent sources of second-hand folds revealed. Trends Biochem Sci 26:9–11

Changela A, Chen K, Xue Y, Holschen J, Outten CE, O'Halloran TV, Mondragon A (2003) Molecular basis of metal-ion selectivity and zeptomolar sensitivity by CueR. Science 301:1383–1387

Cowan JA (1998) Metal activation of enzymes in nucleic acid biochemistry. Chem Rev 98:1067–1088

Dalgaard JZ, Klar AJ, Moser MJ, Holley WR, Chatterjee A, Mian IS (1997) Statistical modeling and analysis of the LAGLIDADG family of site-specific endonucleases and identification of an intein that encodes a site-specific endonuclease of the HNH family. Nucleic Acids Res 25:4626–4638

Datta S, Larkin C, Schildbach JF (2003) Structural insights into single-stranded DNA binding and cleavage by F factor TraI. Structure 11:1369–1379

Drouin M, Lucas P, Otis C, Lemieux C, Turmel M (2000) Biochemical characterization of I-CmoeI reveals that this H-N-H homing endonuclease shares functional similarities with H-N-H colicins. Nucleic Acids Res 28:4566–4872

Eddy SR, Gold L (1991) The phage T4 nrdB intron: a deletion mutant of a version found in the wild. Genes Dev 5:1032–1041

Friedhoff P, Franke I, Krause KL, Pingoud A (1999) Cleavage experiments with deoxythymidine 3′,5′-bis-(p-nitrophenyl phosphate) suggest that the homing endonuclease I-PpoI follows the same mechanism of phosphodiester bond hydrolysis as the non-specific Serratia nuclease. FEBS Lett 443:209–214

Galburt EA, Chevalier B, Tang W, Jurica MS, Flick KE, Monnat RJ Jr, Stoddard BL (1999) A novel endonuclease mechanism directly visualized for I-PpoI. Nat Struct Biol 6:1096–1099

Galburt EA, Stoddard BL (2002) Catalytic mechanisms of restriction and homing endonucleases. Biochemistry 41:13851–13860

Galburt EA, Chadsey MS, Jurica MS, Chevalier BS, Erho D, Tang W, Monnat RJ Jr, Stoddard BL (2000) Conformational changes and cleavage by the homing endonuclease I-PpoI: a critical role for a leucine residue in the active site. J Mol Biol 300:877–887

Goodrich-Blair H, Shub DA (1996) Beyond homing: competition between intron endonucleases confers a selective advantage on flanking genetic markers. Cell 84:211–221

Gorbalenya AE (1994) Self-splicing group I and group II introns encode homologous (putative) DNA endonucleases of a new family. Protein Sci 3:1117–1120

Grishin NV (2001) Treble clef finger – a functionally diverse zinc-binding structural motif. Nucleic Acids Res 29:1703–1714

Hosfield DJ, Guan Y, Haas BJ, Cunningham RP, Tainer JA (1999) Structure of the DNA repair enzyme endonuclease IV and its DNA complex: double-nucleotide flipping at abasic sites and three-metal-ion catalysis. Cell 98:397–408

Hsia KC, Chak KF, Liang PH, Cheng YS, Ku WY, Yuan HS (2004) DNA binding and degradation by the HNH protein ColE7. Structure 12:205–214

James R, Penfold CN, Moore GR, Kleanthous C (2002) Killing of E. coli cells by E group nuclease colicins. Biochimie 84:381–389

Keeble AH, Hemmings AM, James R, Moore GR, Kleanthous C (2002) Multistep binding of transition metals to the H-N-H endonuclease toxin colicin E9. Biochemistry 41:10234–10244

Kleanthous C, Kuhlmann UC, Pommer AJ, Ferguson N, Radford SE, Moore GR, James R, Hemmings AM (1999) Structural and mechanistic basis of immunity toward endonuclease colicins. Nat Struct Biol 6:243–252

Ko TP, Liao CC, Ku WY, Chak KF, Yuan HS (1999) The crystal structure of the DNase domain of colicin E7 in complex with its inhibitor Im7 protein. Structure 7:91–102

Ku WY, Liu YW, Hsu YC, Liao CC, Liang PH, Yuan HS, Chak KF (2002) The zinc ion in the HNH motif of the endonuclease domain of colicin E7 is not required for DNA binding but is essential for DNA hydrolysis. Nucleic Acids Res 30:1670–1678

Kühlmann UC, Moore GR, James R, Kleanthous C, Hemmings AM (1999) Structural parsimony in endonuclease active sites: should the number of homing endonuclease families be redefined? FEBS Lett 463:1–2

Li CL, Hor LI, Chang ZF, Tsai LC, Yang WZ, Yuan HS (2003) DNA binding and cleavage by the periplasmic nuclease Vvn: a novel structure with a known active site. EMBO J 22:4014–4025

Malik HS, Henikoff S (2000) Dual recognition-incision enzymes might be involved in mismatch repair and meiosis. Trends Biochem Sci 25:414–418

Mannino SJ, Jenkins CL, Raines RT (1999) Chemical mechanism of DNA cleavage by the homing endonuclease I-PpoI. Biochemistry 36:16178–16186

Maté MJ, Kleanthous C (2004) Structure-based analysis of the metal-dependent mechanism of H-N-H endonucleases. J Biol Chem 279:34763–34769

Miller MD, Cai J, Krause KL (1999) The active site of Serratia endonuclease contains a conserved magnesium-water cluster. J Mol Biol 288:975–987

Pommer AJ, Wallis R, Moore GR, James R, Kleanthous C (1998) Enzymological characterization of the nuclease domain from the bacterial toxin colicin E9 from *Escherichia coli.* Biochem J 334:387–392

Pommer AJ, Kuhlmann UC, Cooper A, Hemmings AM, Moore GR, James R, Kleanthous C (1999) Homing in on the role of transition metals in the HNH motif of colicin endonucleases. J Biol Chem 274:27153–27160

Pommer AJ, Cal S, Keeble AH, Walker D, Evans SJ, Kuhlmann UC, Cooper A, Connolly BA, Hemmings AM, Moore GR, James R, Kleanthous C (2001) Mechanism and cleavage specificity of the HNH endonuclease colicin E9. J Mol Biol 314:735–749

Raaijmakers H, Toro I, Birkenbihl R, Kemper B, Suck D (2001) Conformational flexibility in T4 endonuclease VII revealed by crystallography: implications for substrate binding and cleavage. J Mol Biol 308:311–323

San Filippo J, Lambowitz AM (2002) Characterization of the C-terminal DNA-binding/DNA endonuclease region of a group II intron-encoded protein. J Mol Biol 324:933–951

Shen BW, Landthaler M, Shub DA, Stoddard BL (2004) DNA binding and cleavage by the HNH homing endonuclease I-HmuI. J Mol Biol 342:43–56

Shub DA, Goodrich-Blair H, Eddy SR (1994) Amino acid sequence motif of group I intron endonucleases is conserved in open reading frames of group II introns. Trends Biochem Sci 19:402–404

Sitbon E, Pietrokovski S (2003) New types of conserved sequence domains in DNA-binding regions of homing endonucleases. Trends Biochem Sci 28:473–477

Walker DC, Georgiou T, Pommer AJ, Walker D, Moore GR, Kleanthous C, James R (2002) Mutagenic scan of the H-N-H motif of colicin E9: implications for the mechanistic enzymology of colicins, homing enzymes and apoptotic endonucleases. Nucleic Acids Res 30:3225–3234

Widlak P, Garrard WT (2001) Ionic and cofactor requirements for the activity of the apoptotic endonuclease DFF40/CAD. Mol Cell Biochem 218:125–130

Widlak P, Li P, Wang X, Garrard WT (2000) Cleavage preferences of the apoptotic endonuclease DFF40 (caspase-activated DNase or nuclease) on naked DNA and chromatin substrates. J Biol Chem 275:8226–8232

Wittmayer PK, Raines RT (1996) Substrate binding and turnover by the highly specific I-PpoI endonuclease. Biochemistry 35:1076–1083

Woo EJ, Kim YG, Kim MS, Han WD, Shin S, Robinson H, Park SY, Oh BH (2004) Structural mechanism for inactivation and activation of CAD/DFF40 in the apoptotic pathway. Mol Cell 14:531–539

GIY-YIG Homing Endonucleases – Beads on a String

PATRICK VAN ROEY, VICTORIA DERBYSHIRE

1 Introduction

Homing endonucleases are intron- and intein-encoded proteins that initiate the mobility of their particular host elements. They recognize an intron- or intein-less version of their host gene and introduce a double-strand break in the DNA. This break is then repaired by copying the intron- or intein-plus allele, using the hosts cellular machinery. This results in a gene-conversion event whereby both alleles become intron- or intein-plus (Belfort et al. 2002).

Homing endonucleases can be classified into four distinct families, based on the presence of conserved sequence elements: the LAGLIDADG, GIY-YIG, His-Cys box, and HNH families (Belfort et al. 2002). However, the His-Cys box and HNH families have been hypothesized to constitute a single ββα-Me family, and recent structural data support this classification (Kuhlmann et al. 1999; Shen et al. 2004).

The GIY-YIG endonucleases, the second most numerous family and the focus of this chapter, were first identified through the presence of the sequence GIY–$X_{10/11}$–YIG in intron-encoded proteins of filamentous fungi and bacteriophage T4 (Michel and Dujon 1986; Cummings et al. 1989). As additional sequences became available, it became clear that these motifs were shared with other proteins, including intergenic endonucleases (Belfort and Perlman 1995). A detailed computational analysis of proteins that contain the GIY-YIG motif showed that family members share several additional sequence elements (Kowalski et al. 1999). Together, these elements form a conserved

V. Derbyshire (e-mail: vicky.derbyshire@wadsworth.org)
P. Van Roey (e-mail: vanroey@wadsworth.org)
Wadsworth Center, New York State Department of Health, Center for Medical Sciences, 150 New Scotland Avenue, Albany, New York 12208, USA

Nucleic Acids and Molecular Biology, Vol. 16
Marlene Belfort et al. (Eds.)
Homing Endonucleases and Inteins

module of 70–100 amino acids, containing up to five distinct sequence motifs (Fig. 1A). These motifs range from 7 to 17 amino acids in length and all but one include at least one highly conserved residue. Motif A contains the signature GIY and YIG elements, motif B an absolutely conserved Arg residue, motif D a conserved Glu, and motif E a highly conserved Asn. Motif C is less well

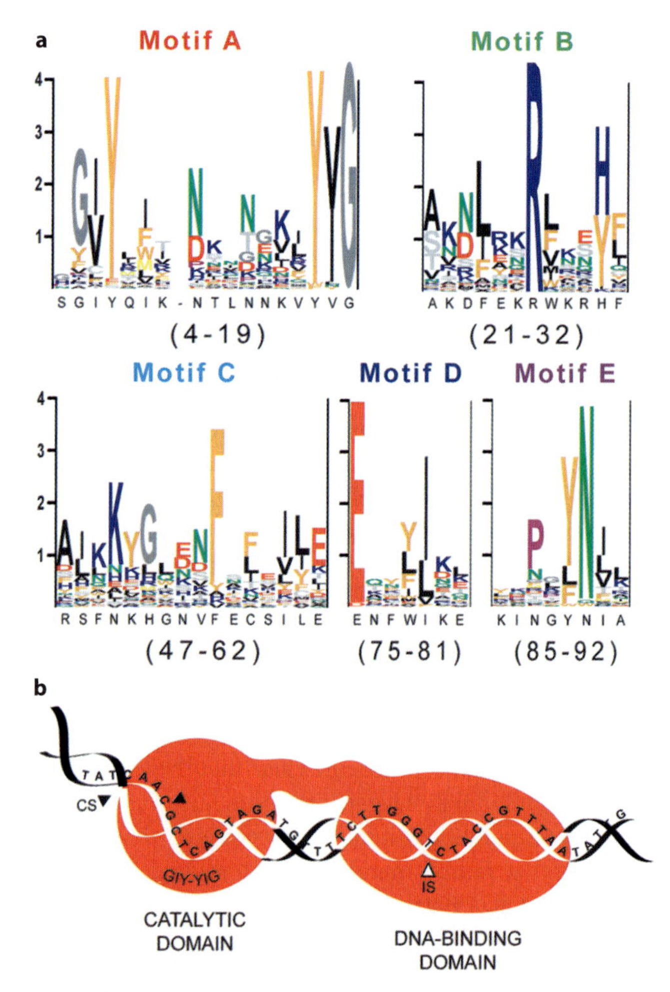

Fig. 1. **a** LOGOS (Schneider and Stephens 1990; Henikoff et al. 1995) representation of the GIY-YIG sequence module, based on those presented by Kowalski et al. (1999), but updated with sequences from the literature and sequence databases. Beneath each logo motif is the sequence of I-TevI, with its amino acid position in the protein. **b** Cartoon presentation of GIY-YIG endonuclease I-TevI and its homing site. The enzyme is represented as a *red dumbbell.* The DNA shown to be contacting the protein in footprinting experiments is *white. CS* Cleavage site; *IS* intron insertion site

conserved and is actually absent in approximately one-third of the proteins. Work on I-TevI as a model GIY-YIG enzyme has shown that the conserved Arg and Glu residues (Arg 27 and Glu 75) are essential for catalytic activity (Derbyshire et al. 1997; Kowalski et al. 1999). These and other data have led to the idea that the GIY-YIG module forms a "catalytic cartridge" that associates with a variety of DNA-binding domains, giving the individual GIY-YIG endonucleases their sequence specificity (Derbyshire et al. 1997). This concept has been borne out by structural and functional studies that show that these enzymes are assembled of individual modules similar to beads on a string.

2 Occurrence and Functions of GIY-YIG Enzymes

The GIY-YIG module has now been found in at least 60 proteins, representing a diverse range of family members. They include proteins encoded by group I introns in bacteriophage and in the mitochondrial and chloroplast genomes of fungi, algae and liverworts; in intergenic proteins of phage, viruses, and bacteria; and the UvrC subunits of archaeal and bacterial (A)BC DNA excision repair complexes (Kowalski et al. 1999).

Approximately one-third of the GIY-YIG family members are intron-encoded proteins, several of which have been shown to have endonuclease activity and therefore most likely function in the mobility of their host intron in a classic homing pathway (see Belfort, this Vol.; Dujon, this Vol.). Another one-third are intergenic proteins, most of unknown function. However, a set of seven related intergenic family members in the T-even phages have been characterized (Sharma et al. 1992; Sharma and Hinton 1994; Kadyrov et al. 1997; Belle et al. 2002; Liu et al. 2003). These Seg proteins (similarity to endonucleases of group I introns) are not present in all of the related T-even phages. SegF and SegG are present in T4 but not in T2, and, in mixed infections, both proteins have been shown to play a role in marker exclusion, resulting in preferential inheritance of T4 markers over T2 markers (Belle et al. 2002; Liu et al. 2003).

UvrC is an integral part of the UvrABC complex that carries out nucleotide excision repair (Sancar 1996). DNA containing a damaged base is recognized, and the phosphodiester backbone is nicked both 3′ and 5′ of the site of damage. UvrC contains two distinct active sites responsible for each endonucleolytic cleavage. The GIY-YIG module, found in the N-terminal portion of the protein, is responsible for nicking of the DNA at the fourth or fifth phosphodiester bond 3′ to the site of DNA damage (Verhoeven et al. 2000). The module is entirely separate from the active site responsible for 5′ cleavage, which is located in the C-terminal half of the protein (Lin and Sancar 1992; Moolenaar et al. 1998).

Additional comparative analyses have defined a similar conserved module, the Uri domain (UvrC and intron-encoded endonucleases; Aravind et al. 1999), which is found in proteins in the UvrC endonuclease superfamily, as well as in a number of small uncharacterized proteins from bacteria, eukaryotes, and viruses, that are potentially also involved in DNA repair. Among these, a protein called Cho, the product of the *ydjC* gene, has been identified in *E. coli* (Moolenaar et al. 2002). Interestingly, Cho, like UvrC, can incise DNA 3′ of a lesion, but at a site that is four nucleotides further away from the lesion than the site where UvrC cleaves. Cho is postulated to play a role in excision repair as a "back-up" to UvrC, in situations where the latter might be impeded from acting by the presence of adducts that are simply too large to permit its access to the DNA. Additionally, a Uri/GIY-YIG domain has been identified in retrotransposable elements in pufferfish and *Drosophila* (Volff et al. 2001).

Finally, sequence database searches with type II restriction endonucleases as queries have shown that some of these enzymes may be related to homing endonucleases (Bujnicki et al. 2001). Nine appear to be related to the HNH family, and three, Eco29kI, MraI, and NgoMIII, are similar to GIY-YIG proteins. While the similarities are only marginally significant, the alignments nevertheless appear to be valid, since all of the most conserved and catalytically important residues are present in these enzymes.

The diversity of occurrence and location of the GIY-YIG endonuclease domain are remarkable and the lack of conservation of sequences in other modules of the proteins is entirely consistent with the concept that the GIY-YIG module evolved separately from the associated domains that have additional distinct functions, such as DNA-binding domains.

3 I-TevI as the Model GIY-YIG Enzyme: Structure and Function

I-TevI is the most extensively studied GIY-YIG endonuclease. It is encoded by a group I intron found in the thymidylate synthase (*td*) gene of bacteriophage T4. As described in other chapters in this volume, homing endonucleases are site-specific enzymes that recognize fairly lengthy DNA targets with some degree of sequence tolerance. Even given this description, I-TevI is extremely unusual in that the 28-kDa protein recognizes a 37-bp target site as a monomer (Mueller et al. 1995) and that it can tolerate mutations at any base pair in its substrate and still effect cleavage (Bryk et al. 1993). The protein interacts with its *td* homing site by binding to the DNA at two sites that are separated by a 5-bp segment. The primary binding site is approximately 20 bp long and spans the intron insertion site (IS). The second site is about 12 bp long and includes the cleavage site (CS), which is 23–25 bp upstream of the IS (Fig. 1B). In addition to tolerating nucleotide substitutions, the enzyme can cope with larger perturbations in its recognition site. It can bind and cleave substrates that have large insertions or deletions between the IS and CS, in many cases reaching out or pulling back to cleave at the natural CS (Bryk et al. 1995). This flexibility in recognition of a two-domain DNA target is achieved through the flexible organization of the two-domain protein. Limited proteolysis and mutagenesis experiments defined I-TevI as having two stable functional domains, an N-terminal catalytic domain and a C-terminal DNA-binding domain, separated by a proteolytically sensitive linker (Derbyshire et al. 1997; Fig. 1B). The isolated C-terminal domain binds to the homing site with the same affinity as does the full-length protein, suggesting that most, if not all, binding energy originates from this region of the protein. Although the isolated N-terminal domain is unable to bind to the homing site, the full-length enzyme has preferred cleavage sites and the catalytic domain must play a role in selecting these sites. In particular, the enzyme prefers sites with a G-C bp at position –23 relative to the IS and a C-G at position –27 (Bryk et al. 1995; Edgell et al. 2004a,b). This selectivity has important implications for I-TevI function (see Sect. 3.4).

While the three-dimensional structure of full-length I-TevI bound to its substrate has not yet been reported, crystal structures of the catalytic domain (Van Roey et al. 2002) and of the DNA-binding domain in complex with the primary binding site (Van Roey et al. 2001) allow interpretation of the biochemical data to give a detailed image of the recognition properties of the enzyme.

3.1 Catalytic Domain

The crystal structures of two derivatives of the N-terminal catalytic domain of I-TevI, corresponding to the first 97 residues of the protein, have been determined (Van Roey et al. 2002). The two distinct catalytic mutants (R27A and E75A) adopt the same molecular conformation, and their structures allowed the construction of a composite wild-type molecule. The first 92 amino acids, corresponding precisely to the GIY-YIG module (Fig. 1A), form a globular domain with a unique fold that consists of a twisted, three-stranded, antiparallel β-sheet, flanked by two α-helices on one side, and by the third α-helix on the other (Fig. 2A). The GIY and YIG sequences are located on the first and second β-strands, respectively. The role of the GIY sequence appears to be strictly structural, as the segment is found at the core of the molecule, whereas the YIG segment (YVG, residues 17–19, in the case of I-TevI) is both structurally and functionally important. The side chain of the hydrophobic residue Val 18 is part of the hydrophobic core of the molecule, while Tyr 17 and Gly 19 are external and centrally located in a shallow concave surface of the molecule

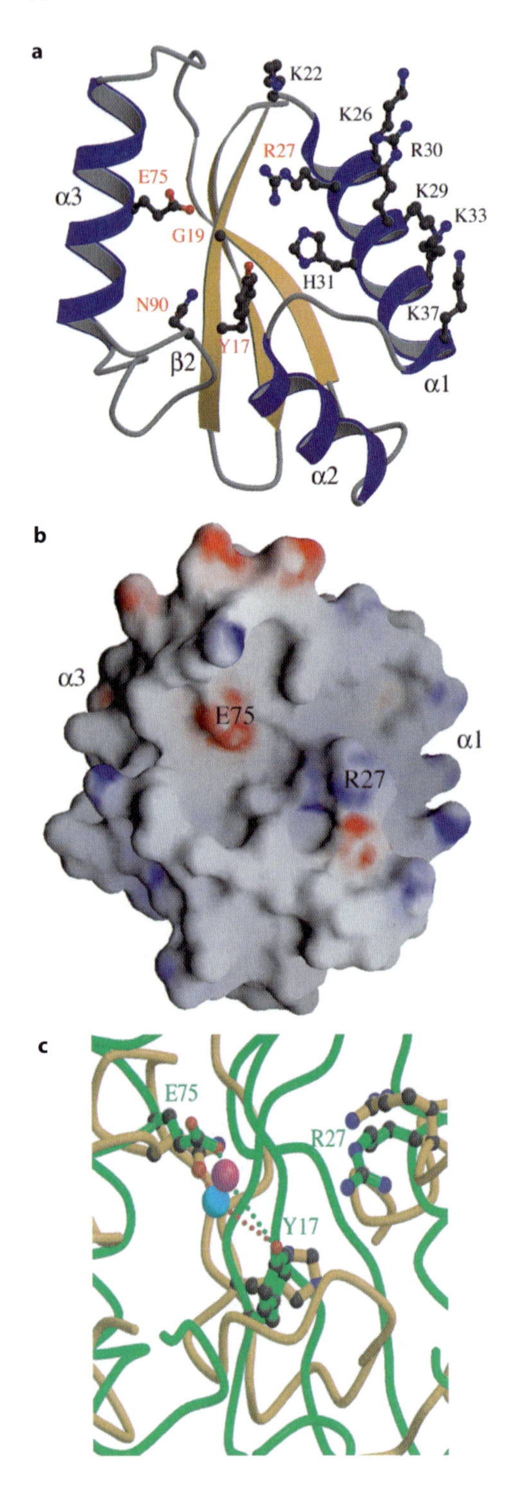

Fig. 2. Structure of the catalytic domain of I-TevI. **a** Ribbon diagram showing secondary structure elements, conserved residues (*red*), and positively charged residues of α-helix 1, which is proposed to interact with the DNA substrate. **b** Electrostatic potential map of the substrate–interaction surface. α-helix 1 is on the *right* with catalytic residues Glu 75 and Arg 27 in the center of the shallow concave surface. **c** Overlap of key catalytic residues of I-TevI (*green*) and I-PpoI (*orange*). The labels refer to the I-TevI residue numbers. The figure illustrates the coincidence of the divalent cation binding sites in the two enzymes (I-TevI, *magenta*; I-PpoI, *cyan*)

that is formed primarily by β-strand 2 and α-helices 1 and 3 (Fig. 2A). This surface also includes sections of all four of the other conserved motifs (B, C, D, and E) as well as the most important conserved residues Tyr 17, Gly 19, Arg 27, Glu 75, and Asn 90. Accordingly, this surface was proposed to be the site of substrate interaction. α-helix 1, which forms one edge of this surface, is highly positively charged. The side chains of Arg 30, Arg 27, and His 31 stack together above the surface while the side chains of Lys 22, Lys 26, Lys 29, Lys 33, and Lys 37 align along the outer rim (Fig. 2A, B). This pattern of positively charged residues suggests a role for α-helix 1 in interaction with the DNA through phosphate backbone contacts.

The importance of Arg 27 and Glu 75 in the catalytic mechanism has been established through mutagenesis studies (Derbyshire et al. 1997; Kowalski et al. 1999). Furthermore, the role of Glu 75 in a metal-ion-based catalytic mechanism was supported by structural data, from crystals soaked in $MnCl_2$, that revealed the presence of the cation bound to this residue (Van Roey et al. 2002). Additionally, the location of the divalent cation, 3.8 Å from the Cα of Gly 19, clarifies the necessity of the Gly at this position, since any side chain would sterically hinder cation binding.

Comparisons with other nucleases revealed a level of structural correspondence between key residues of I-TevI and those of I-PpoI, an unrelated homing endonuclease in the His-Cys box family (Galburt et al. 1999). Like I-TevI, I-PpoI contains a single residue (Asn 119) that binds the divalent cation responsible for DNA cleavage and an arginine (Arg 61) that is important in the mechanism. A third residue of I-PpoI (His 98) is required for activation of a water molecule that serves as the nucleophile in the reaction mechanism. Superposition of Asn 119, Arg 61, and His 98 of I-PpoI onto Glu 75, Arg 27 and Tyr 17 of I-TevI, respectively, reveals strong similarity in the three-dimensional arrangement of these residues (Fig. 2C). In addition, Glu 75 of I-TevI and Asn 119 of I-PpoI are similarly located on helices at one edge of the molecular surface, and the Mn^{2+} site of I-TevI is nearly identical to the site of the metal ion in I-PpoI ternary complexes. This analysis suggests similar functions for the matched residues, especially Glu 75/Asn 119 and Tyr 17/His 98, in a related catalytic mechanism. However, it should be noted that I-PpoI effects double-stranded DNA cleavage as a dimer, with the two active sites each being responsible for cleavage of one strand. At this time it is unclear how I-TevI, as a monomer, is able to cleave both strands. Also, while there is coincidence of key catalytic residues between I-PpoI and I-TevI, the local fold of I-TevI does not correspond to the ββα-Me structural motif that is shared by the His-Cys box and HNH homing endonucleases. This divergence is consistent with the the idea that the GIY-YIG enzymes form an independent family.

3.2 DNA-Binding Domain

The structure of the C-terminal DNA-binding domain of I-TevI, in complex with a 20-bp duplex DNA that contains the primary binding region of the homing site, has been determined (Van Roey et al. 2001). The protein used for these crystallization experiments consisted of amino acids 130–245, based on limited proteolysis experiments that defined this as a stable domain (Derbyshire et al. 1997). However, only the residues beyond position 149 were visible in the structure (Fig. 3A). This domain adopts an extended conforma-

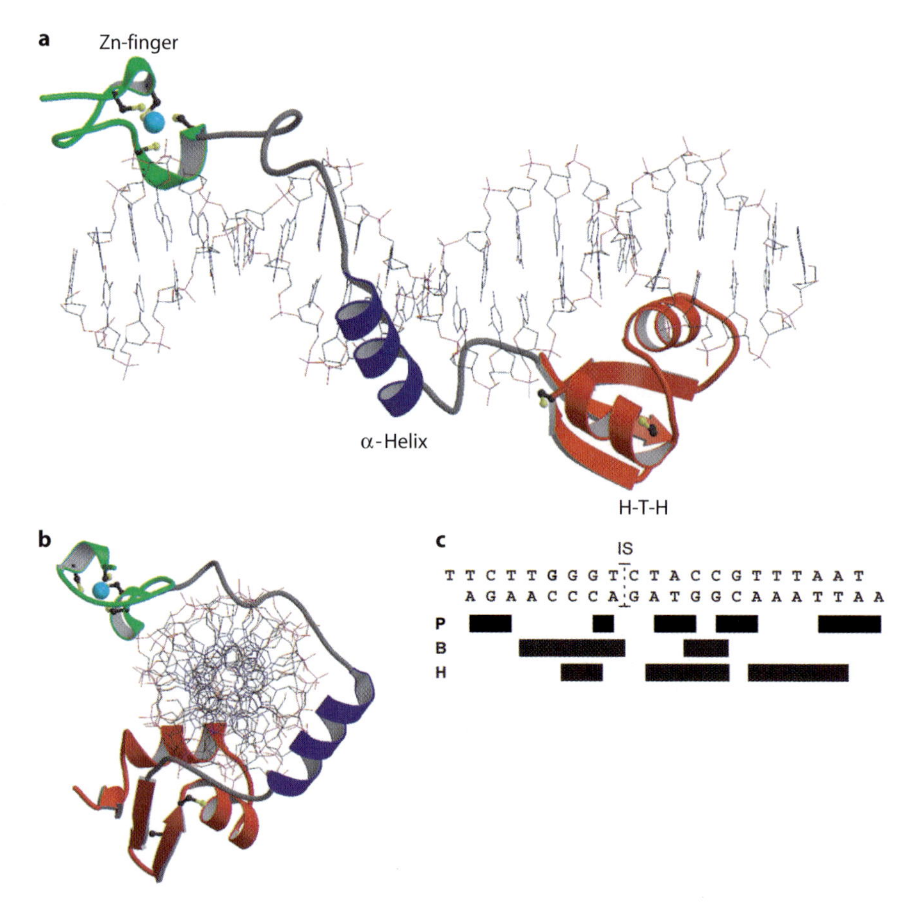

Fig. 3. Structure of the DNA-binding domain of I-TevI in complex with its primary binding site. **a** Ribbon diagram showing arrangement of subdomains. **b** As in **a**, but viewed along the DNA axis. **c** Summary of protein–DNA contacts along the DNA. *P* Phosphate-backbone contacts; *H* hydrophobic contacts; *B* hydrogen-bonding contacts to the bases; *IS* intron insertion site

tion, winding along the full length of the substrate and contacting the phosphate backbone throughout. Unlike the single-domain structure of the catalytic domain, the DNA-binding domain is composed of three subdomains that are connected by segments that lack defined secondary structure. The subdomains are a Zn-finger, a minor groove-binding α-helix, and an unusually small helix-turn-helix (H-T-H) domain. With the exception of the H-T-H domain, which places a helix in the major groove, the protein occupies the minor groove and does not greatly perturb the DNA conformation (Fig. 3B). The extended nature of the molecule helps to account for how a small enzyme recognizes such a long substrate.

The H-T-H subdomain sits in the major groove but forms no hydrogen bonds with the bases. Instead, the domain makes extensive contacts with the phosphate backbone, and its second helix is inserted into the major groove, where it makes hydrophobic contacts with a series of thymine bases, thereby imparting considerable specificity to this sequence-tolerant enzyme (Fig. 3C). The α-helix inserted in the minor groove also has a hydrophobic surface contacting the bases, and it only forms hydrogen bonds with the phosphate backbone. Contacts between the Zn-finger and the DNA are limited to two hydrogen bonds with the phosphate backbone. The few base contacts that are actually made are with residues in the joining segments between the subdomains (Fig. 3C). These hydrogen-bonding contacts provide the only, albeit limited, specificity of the interaction between the protein and DNA, apart from the specificity resulting from the selectivity for an AT-rich region by the H-T-H domain. This is entirely consistent with the observed sequence tolerance of I-TevI. Clearly, many base substitutions can be tolerated because so few bases in the homing site are actually contacted directly. In addition, it appears that many contacts are redundant. Thus, the loss of one or two does not significantly affect the enzyme's ability to bind to its target.

It is perhaps not surprising that I-TevI contacts the DNA largely in the minor groove, given that T4 phage DNA is highly modified, containing glucosylated 5-hydroxymethyl cytosine residues, with the bulky adduct occupying the major groove. Logically, the region of the DNA that is contacted by the H-T-H subdomain in the major groove is devoid of cytosines.

3.3 Flexible Linker and Distance Determination

The crystal structures define I-TevI as comprising a catalytic domain and a tripartite DNA-binding domain connected by a long linker. These data are also consistent with evidence that a large region in the center of the enzyme is sensitive to protease digestion and that deletions of two to five amino acids can be tolerated by the enzyme. This suggests that this portion of the molecule acts as a flexible spacer between the two functional domains (Kowalski et al. 1999). However, it is now clear that the Zn-finger and parts of the linker actually play a more complex role in I-TevI cleavage activity (Dean et al. 2002 and unpubl. results).

In order to determine the role in binding affinity for each of the subdomains in the DNA-binding domain, a number of derivatives were made by systematically removing segments of the protein (Dean et al. 2002). Deletion of the H-T-H domain at the C-terminus was immediately deleterious, but removal of residues from the N-terminus was well tolerated. Most remarkably, mutants missing the Zn-finger were able to bind to the homing site as well as (if not better than) the full domain, leading to the conclusion that the Zn-finger does not contribute significantly to DNA binding.

The Zn-finger of I-TevI is highly unusual in that it is extremely small and has a non-canonical $CXC(X)_{10}CX_2C$ sequence. To date, no other GIY-YIG protein has been identified that contains a Zn-finger motif. Similar sequences have only been found in the β′-subunits of bacterial RNA polymerases, but those Zn-fingers are as yet structurally and functionally uncharacterized (Campbell et al. 2001). Interestingly, Zn-finger mutants in the context of full-length I-TevI were only minimally compromised in their ability to bind and cleave homing-site substrates (Dean et al. 2002). However, experiments carried out to map the exact CS for each protein on wild-type and mutant substrates, with insertions or deletions between the IS and CS, highlighted the function of the Zn-finger (Fig. 4A). The wild-type protein cleaves a wild-type substrate at the CS, 23–25 bp upstream of the IS. For substrates having modest insertions and deletions it can reach forward or pull back to find the natural cleavage sequence, but for larger insertions and deletions it prefers to cleave at the natural distance, rather than at the natural sequence. For the Zn-finger mutants, the picture is quite different. These proteins prefer to cleave all forms of the substrate at the natural sequence. Even a derivative in which the full Zn-finger (18 amino acids) is deleted reaches out to find the natural cleavage sequence in DNA substrates containing insertions.

These results indicate that the loss of the Zn-finger correlates with the loss of a distance determinant for cleavage. Two distinct models for this role of the Zn-finger, both based on protein–protein interactions with another part of the enzyme, have been proposed (Dean et al. 2002). Such interactions could involve either the Zn-finger as a "catalytic clamp" locking down the catalytic domain, and reducing its ability to stray too far, or, alternatively, as an "organizer" for the extremely long linker so that, once again, the catalytic domain is constrained and thus cleaves at the appropriate distance. More recent data suggest that parts of the linker also contribute to distance determination. Consequently, a combination of the two models may be correct (Fig. 4B; Liu et al., unpubl.).

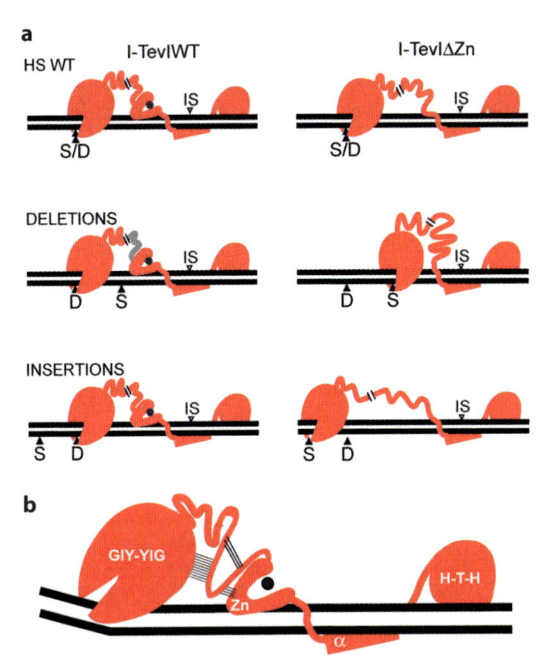

Fig. 4. The Zn-finger as a distance determinant. **a** Cartoon representation of I-TevI showing cleavage activity of wild-type and Zn-finger mutants (I-TevIΔZn) on wild-type and mutant homing-site substrates. **b** Model for Zn-finger function, showing putative protein–protein interactions among the Zn-finger, the linker, and the catalytic domain

3.4 I-TevI Endonuclease Is Bifunctional, Also Serving As a Transcriptional Autorepressor

Expression of I-TevI in phage is tightly regulated. I-TevI's host gene, thymidylate synthase, is expressed from a T4 middle promoter. I-TevI itself, however, is not expressed from that transcript, because its ribosome-binding site and start codon are sequestered in a stem-loop structure (Gott et al. 1988). This results in tight translational control, presumably to protect the host and phage from this somewhat promiscuous enzyme. I-TevI is expressed later in infection from a T4 late promoter without translational repression, because the stem-loop cannot be formed in that transcript.

There is sequence similarity between a region that overlaps the I-TevI promoter and the homing site of the enzyme (Fig. 5A). Notably, in vivo experiments have shown that I-TevI can repress its own expression, representing a second level of control, now at the transcriptional level (Edgell et al. 2004a). In vitro experiments subsequently showed that the DNA-binding domain of I-TevI binds to this operator sequence with the same affinity as to the homing site, even though there are 6-bp substitutions in the 20-bp primary recognition sequence. In contrast, I-TevI cleaves the operator site very poorly (approximately 100-fold less efficiently), because there is no sequence similarity upstream of the primary binding site. In particular, there is no sequence that corresponds to the CS. The low-level cleavage that does occur is at a site 14 bp upstream from the sequence analogous to the intron IS, where a preferred cleavage context is fortuitously located.

To determine how the protein can bind such a variant sequence, the crystal structure of the I-TevI DNA-binding domain bound to a 20-bp duplex corresponding to the operator sequence was solved (Fig. 5B). The overall structure of this complex is very similar to that of the homing-site complex, with a root-mean-square deviation (RMSD) for the Cα of the protein of 0.67 Å. The DNA structure is largely unchanged, except for small differences in the regions where the DNA interacts with the extended regions between the subdomains.

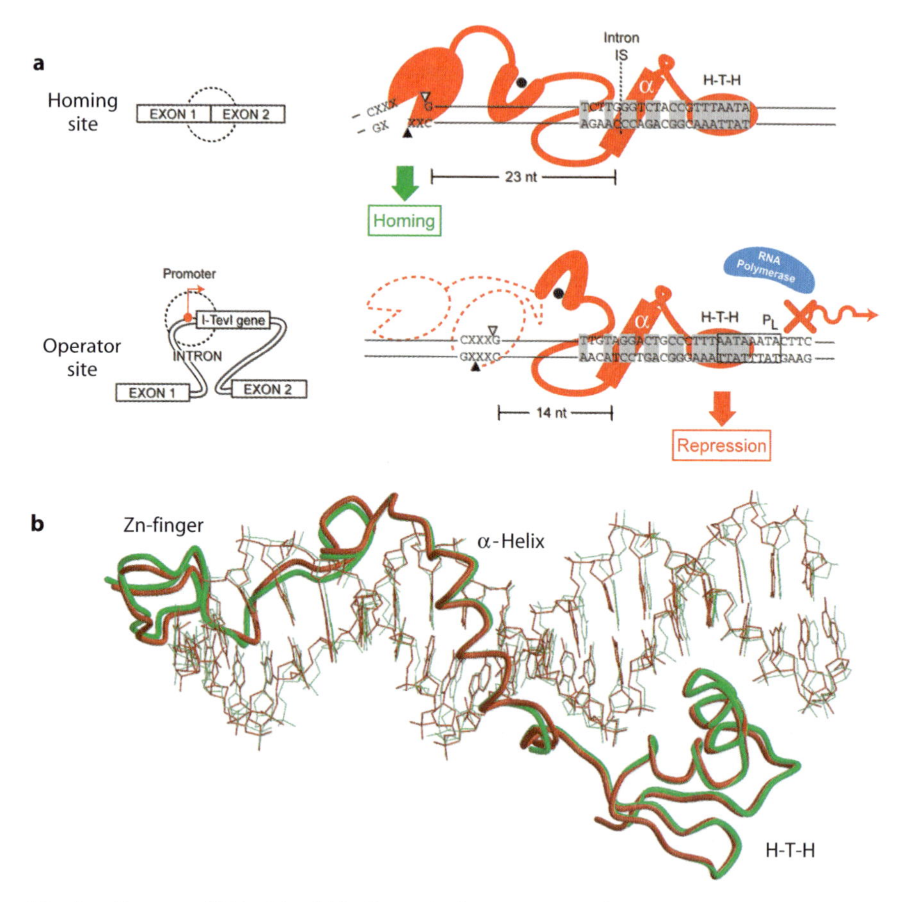

Fig. 5. **a** Cartoon of I-TevI (*red*) binding to its homing site and to its promoter site. Sequence identities between the two are shaded in *gray*. **b** Superposition of I-TevI DNA-binding domain complexes with homing-site (*red*) and operator-site (*green*) DNA duplexes

In general, nucleotide substitutions are accommodated by a small number of changes in the conformation of the side chains of amino acids, resulting in a small number of changes in contacts to the bases in the unstructured regions, thus yielding alternate hydrogen-bond interactions.

I-TevI has a great deal of flexibility in its ability to interact with DNA. This flexibility likely contributes to the enzyme's ability to home to new sites. In addition, because of its defined sequence preferences at the cleavage site, I-TevI can function as a repressor, binding to its operator site to control expression, but with a 100-fold reduction in cleavage. This endows the enzyme with a second distinct biological function (Fig. 5A).

4 DNA-Binding Domain Diversity and Conserved Modules

There appears to be very little sequence similarity outside of the GIY-YIG module among the members of the GIY-YIG family, except among proteins closely related to I-TevI, like I-BmoI, an intron-encoded enzyme from *Bacillus mojavensis* (Edgell and Shub 2001). I-TevI and I-BmoI can be aligned through their GIY-YIG domains and through parts of their DNA-binding domains. I-BmoI has a H-T-H domain analogous to that of I-TevI, but it lacks a Zn-finger domain. Instead, it appears to have three copies of the minor groove-binding α-helix present in I-TevI (Sitbon and Pietrokovski 2003).

However, a number of short stretches of conserved sequence have been identified, through multiple sequence alignments (Sitbon and Pietrokovski 2003), that are found in many GIY-YIG proteins and are also shared with certain HNH endonucleases and other proteins. In most cases, the structural and functional significance of these sequence segments is unclear, but one such conserved domain (NUMOD 3; Sitbon and Pietrokovski 2003) corresponds to the minor groove-binding α-helix present in the DNA-binding domain of I-TevI, suggesting that this might be conserved among many of the otherwise divergent proteins. Indeed, NUMOD 3 is predicted to be present in several GIY-YIG enzymes, including four of the Seg proteins. Interestingly, it is present in multiple copies in several other proteins, besides I-BmoI, suggesting the presence of multiple DNA-binding α-helices.

The modular nature of I-TevI has led us to propose that the GIY-YIG catalytic cartridge can acquire specificity by associating with various DNA-binding domains (Derbyshire et al. 1997). The structure of the DNA-binding domain of I-TevI, and comparisons with the sequences of other family members, show this to be the case. This modular organization that separates cleavage and DNA binding into distinct domains is a feature shared with the HNH proteins but not with the LAGLIDADG and His-Cys box endonucleases, which have their catalytic and DNA-binding functions within a single subunit or protein domain. The structure of I-HmuI, an HNH homing endonuclease encoded by a group I intron in the *B. subtilis* bacteriophage SPO1, has recently been solved (Shen et al. 2004). What is most remarkable about this structure is that, despite the fact that I-HmuI and I-TevI are from unrelated endonuclease families, their DNA-binding domains have very similar structures (Fig. 6). I-HmuI has a C-terminal H-T-H domain and two copies of a minor groove-binding α-helix. The structures of the H-T-H domain and of the second α-helix resemble those of I-TevI very closely. However, the protein–DNA interactions carried out by these subdomains are actually somewhat different in the two proteins. Consistent with its high sequence tolerance, I-TevI interacts primarily with the phosphate backbone and ribose moieties whereas I-HmuI, which is much more sequence specific, makes many contacts directly with the bases. The α-helix of I-HmuI is bound much more deeply into the minor groove of its target than that of I-TevI. This causes a significant distortion of the DNA, whereas the I-TevI homing site is largely unperturbed.

It would thus appear that the DNA-binding modules are shuffled with different endonuclease cartridges, and that each then adapts to the particular biology of the organism in which the enzyme resides. On one hand, the mosaicism of these endonucleases and their genes provides evidence for modular exchange and therefore rapid evolution. On the other hand, adaptations to

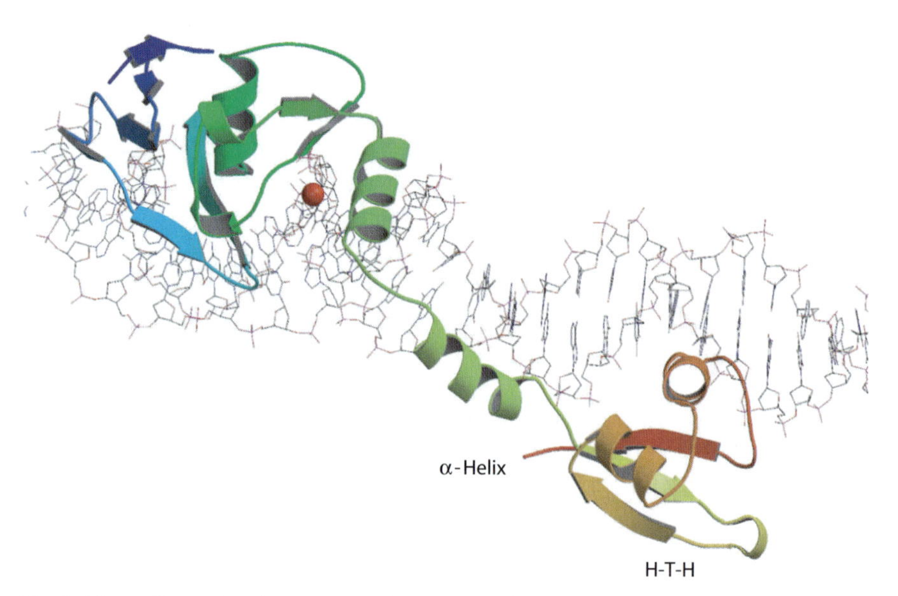

Fig. 6. Three-dimensional structure of I-HmuI with its substrate (Shen et al. 2004). The H-T-H domain and the second minor groove-binding α-helix are very similar, both in structure and relative orientation to the corresponding segments of I-TevI (see Fig. 3A)

their specific host ensure the enzyme's persistence. This combination of promiscuity and restraint is a hallmark of successful parasitic elements.

Acknowledgements. We thank Maryellen Carl for expert secretarial assistance, John Dansereau for preparing the figures, and Joe Kowalski for preparing the LOGOS figure. This work is supported by NIH grant GM44844 to Marlene Belfort.

References

Aravind L, Walker DR, Koonin EV (1999) Conserved domains in DNA repair proteins and evolution of repair systems. Nucleic Acids Res 27:1223–1242

Belfort M, Perlman PS (1995) Mechanisms of intron mobility. J Biol Chem 270:30237–30240

Belfort M, Derbyshire V, Cousineau B, Lambowitz A (2002) Mobile introns: pathways and proteins. In: Craig N, Craigie R, Gellert M, Lambowitz A (eds) Mobile DNA II. ASM Press, Washington, DC, pp 761–783

Belle A, Landthaler M, Shub DA (2002) Intronless homing: site-specific endonuclease SegF of bacteriophage T4 mediates localized marker exclusion analogous to homing endonucleases of group I introns. Genes Dev 16:351–362

Bryk M, Quirk SM, Mueller JE, Loizos N, Lawrence C, Belfort M (1993) The *td* intron endonuclease makes extensive sequence tolerant contacts across the minor groove of its DNA target. EMBO J 12:2141–2149

Bryk M, Belisle M, Mueller JE, Belfort M (1995) Selection of a remote cleavage site by I-*Tev*I, the *td* intron-encoded endonuclease. J Mol Biol 247:197–210

Bujnicki JM, Radlinska M, Rychlewski L (2001) Polyphyletic evolution of type II restriction enzymes revisited: two independent sources of second-hand folds revealed. Trends Biochem Sci 26:9–11

Campbell EA, Korzheva N, Mustaev A, Murakami K, Nair S, Goldfarb A, Darst SA (2001) Structural mechanism for rifamycin inhibition of bacterial DNA polymerase. Cell 104:901–912

Cummings DJ, Michel F, McNally KL (1989) DNA sequence analysis of the 24.5 kilobase pair cytochrome oxidase subunit I mitochondrial gene from *Podospora anserina*: a gene with sixteen introns. Curr Genet 16:381–406

Dean AB, Stanger MJ, Dansereau JT, Van Roey P, Derbyshire V, Belfort M (2002) Zinc finger as distance determinant in the flexible linker of intron endonuclease I-TevI. Proc Natl Acad Sci USA 99:8554–8561

Derbyshire V, Kowalski JC, Dansereau JT, Hauer CR, Belfort M (1997) Two-domain structure of the *td* intron-encoded endonuclease I-TevI correlates with the two-domain configuration of the homing site. J Mol Biol 265:494–506

Edgell DR, Shub DA (2001) Related homing endonucleases I-BmoI and I-TevI use different strategies to cleave homologous recognition sites. Proc Natl Acad Sci USA 98:7898–7903

Edgell DR, Derbyshire V, van Roey P, LaBonne S, Stanger MJ, Li Z, Boyd TM, Shub DA, Belfort M (2004a) Intron endonuclease I-TevI also functions as a transcriptional autorepressor. Nat Struct Mol Biol 11:936–944

Edgell DR, Stanger MJ, Belfort M (2004b) Coincidence of cleavage sites of intron endonuclease I-TevI and critical sequences of the host thymidylate synthase gene. J Mol Biol 343:1231–1241

Galburt EA, Chevalier B, Tang W, Jurica MS, Flick KE, Monnat RJ Jr, Stoddard BL (1999) A novel endonuclease mechanism directly visualized for I-PpoI. Nat Struct Biol 6:1096–1099

Gott JM, Zeeh A, Bell-Pedersen D, Ehrenman K, Belfort M, Shub DA (1988) Genes within genes: independent expression of phage T4 intron open reading frames and the genes in which they reside. Genes Dev 2:1791–1799

Henikoff S, Henikoff JG, Alford WJ, Pietrokovski S (1995) Automated construction and graphical presentation of protein blocks from unaligned sequences. Gene 163:17–26

Kadyrov FA, Shlyapnikov MG, Kryukov VM (1997) A phage T4 site-specific endonuclease, SegE, is responsible for a non-reciprocal genetic exchange between T-even-related phages. FEBS Lett 415:75–80

Kowalski JC, Belfort M, Stapleton MA, Holpert M, Dansereau JT, Pietrokovski S, Baxter SM, Derbyshire V (1999) Configuration of the catalytic domain of intron endonuclease I-TevI: coincidence of computational and molecular findings. Nucleic Acids Res 27:2115–2125

Kuhlmann UC, Moore GR, James R, Kleanthous C, Hemmings AM (1999) Structural parsimony in endonuclease active sites: should the number of homing endonuclease families be redefined? FEBS Lett 463:1–2

Lin J-J, Sancar A (1992) Active site of (A)BC excinuclease. I. Evidence for 5′ incision by UvrC through a catalytic site involving Asp^{399}, Asp^{438}, Asp^{466} and His^{538} residues. J Biol Chem 267:17688–17692

Liu Q, Belle A, Shub DA, Belfort M, Edgell DR (2003) SegG endonuclease promotes marker exclusion and mediates co-conversion from a distant cleavage site. J Mol Biol 334:13–23

Michel F, Dujon B (1986) Genetic exchanges between bacteriophage T4 and filamentous fungi? Cell 46:323

Moolenaar GF, Uiterkamp RS, Zwijnenburg DA, Goosen N (1998) The C-terminal region of the *Escherichia coli* UvrC protein, which is homologous to the C-terminal region of the human ERCC1 protien, is involved in DNA binding and 5′-incision. Nucleic Acids Res 26:462–468

Moolenaar GF, van Rossum-Fikkert S, van Kesteren M, Goosen N (2002) Cho, a second endonuclease involved in *Escherichia coli* nucleotide excision repair. Proc Natl Acad Sci USA 99:1467–1472

Mueller JE, Smith D, Bryk M, Belfort M (1995) Intron-encoded endonuclease I-TevI binds as a monomer to effect sequential cleavage via conformational changes in the *td* homing site. EMBO J 14:5724–5735

Sancar A (1996) DNA excision repair. Annu Rev Biochem 65:43–81

Schneider TD, Stephens RM (1990) Sequence logos a new way to display consensus sequences. Nucleic Acids Res 18:6097–6100

Sharma M, Hinton DM (1994) Purification and characterization of the SegA protein of bacteriophage T4, an endonuclease related to proteins encoded by group I introns. J Bacteriol 176:6439–6448

Sharma M, Ellis RL, Hinton DM (1992) Identification of a family of bacteriophage T4 genes encoding proteins similar to those present in group I introns of fungi and phage. Proc Natl Acad Sci USA 89:6658–6662

Shen BW, Landthaler M, Shub DA, Stoddard BL (2004) DNA binding and cleavage by the HNH homing endonuclease I-HmuI. J Mol Biol 342:43–56

Sitbon E, Pietrokovski S (2003) New types of conserved sequence domains in DNA-binding regions of homing endonucleases. Trends Biochem Sci 28:473–477

Van Roey P, Waddling CA, Fox KM, Belfort M, Derbyshire V (2001) Intertwined structure of the DNA-binding domain of intron endonuclease I-TevI with its substrate. EMBO J 20:3631–3637

Van Roey P, Meehan L, Kowalski J, Belfort M, Derbyshire V (2002) Catalytic domain structure and hypothesis for function of GIY-YIG intron endonuclease I-TevI. Nat Struct Biol 9:806–811

Verhoeven EEA, van Kesteren M, Moolenaar GF, Visse R, Goosen N (2000) Catalytic sites for 3′ and 5′ incision of *Escherichia coli* nucleotide excision repair are both located in UvrC. J Biol Chem 275:5120–5123

Volff J-N, Hornung U, Schartl M (2001) Fish retroposons related to the *Penelope* element of *Drosophila virilis* define a new group of retrotransposable elements. Mol Genet Genom 265:711–720

His-Cys Box Homing Endonucleases

Eric A. Galburt, Melissa S. Jurica

Abstract

Homing endonucleases are often grouped into four families based on distinct sequence motifs. One of these families is known as the His-Cys box homing endonucleases and contains two clusters of conserved histidine and cysteine residues over a central 100 amino acid region. At last count, 23 members of this family had been identified. The open reading frames (ORFs) of these proteins are contained within mobile group I introns found in nuclear rDNA genes of several protists. The nuclear location of these introns and ORFs is currently unique among the homing endonuclease families and poses an intriguing puzzle regarding their expression from non-coding rRNA transcripts.

The best-studied member of the His-Cys box homing endonucleases is I-PpoI from the myxomycete *Physarum polycephalum*. Following an introduction to all of the known members of the His-Cys box endonuclease family, much of the following chapter will outline the extensive characterization of I-PpoI structure and function. Although our understanding of how I-PpoI is expressed in cells is still not fully complete, the means by which I-PpoI specifically recognizes a single cleavage site in the host genome to mediate homing of its host intron is widely accepted. Details of DNA recognition and the catalytic mechanism of nucleolytic cleavage have been ascertained from both *in vivo* and *in vitro* activity assays as well as from extensive X-ray crystallographic structural analyses of the enzyme bound to its DNA substrate.

E.A. Galburt (e-mail: eagalburt@lbl.gov)
Physical Biosciences, Lawrence Berkeley National Laboratory, 1 Cyclotron Road Mail Stop 20A-355, Berkeley, California 94720, USA

M.S. Jurica (e-mail: jurica@biology.ucsc.edu)
Molecular, Cell and Developmental Biology, Center for Molecular Biology of RNA, UC Santa Cruz, 1156 High Street, Santa Cruz, California 95064, USA

Nucleic Acids and Molecular Biology, Vol. 16
Marlene Belfort et al. (Eds.)
Homing Endonucleases and Inteins

1 The His-Cys Box Family of Homing Endonucleases

Similar to members of other homing endonuclease families, the His-Cys box homing endonucleases are encoded by group I introns that exhibit mobility into intron-less alleles of their host genes. For all His-Cys box family members identified thus far, the host genes are nuclear rDNA loci from several species of protists including slime molds, fungi, algae, and amoebae. This endonuclease family is characterized by a series of conserved histidine and cysteine residues over ~100 amino acid stretch of the protein (Johansen et al. 1993). The family now includes 6 enzymes that have demonstrated activity and 17 putative members from full-length open-reading frames (ORFs; Table 1). The enzymes with demonstrated activity are I-PpoI from *Physarum polycephalum* (Muscarella et al. 1990; Wittmayer et al. 1998), I-DirI from *Didymium iridis* (Johansen et al. 1997), and a set of four highly related proteins that act as isoschizomers (I-NjaI, I-NanI, I-NitI, and I-NgrI) from several *Naegleria* species (Elde et al. 1999, 2000; Decatur et al. 2000). These enzymes recognize several variants of their respective homing sites which are ~18 nucleotide (nt) pseudopalindromic sequences. In general, the His-Cys box proteins are small, on the order of 160–290 amino acids in length. It has been suggested that their size may be limited by length restrictions of their encoding mobile introns.

In addition to the full-length His-Cys box ORFs, 20 pseudogenes have also been identified in group I introns found at the same position of rDNA sequences as those encoding active endonucleases. These pseudogene sequences contain nonsense codons that interrupt the putative ORF and therefore are not likely to encode an active endonuclease. Due to their high sequence divergence, many of the His-Cys box ORFs and pseudogenes were found by locating their host group I intron in nuclear ribosomal genes and correlating the observation of a larger than normal intron with the presence of an ORF-like sequence. The large number of pseudogenes is taken as evidence that homing endonuclease genes may be unstable and therefore lost during evolution (Cho et al. 1998; Goddard and Burt 1999; Haugen et al. 1999; Foley et al. 2000; Muller et al. 2001; Bhattacharya et al. 2002; Nozaki et al. 2002)

It is interesting to note that in addition to the mobility of the group I introns encoding the homing endonucleases, there is accumulating evidence that the enzyme ORFs are also mobile (Mota and Collins 1988; Lambowitz and Belfort 1993; Loizos et al. 1994; Ogawa et al. 1997; Pellenz et al. 2002; Haugen et al. 2004). This prospect complicates the matter of the inheritance of homing introns, which center primarily on questions of vertical versus lateral transfer of the sequences. Bhattacharya and coworkers performed phylogenetic analysis of the His-Cys box enzymes, examining their relationship to the encoding group I introns, as well as those introns' relationships to the host gene in terms of position of the ORF within the intron and position of the in-

Table 1. Members of the His-Cys box homing endonuclease family. *ACT* Enzyme with demonstrated activity; *FL* predicted full-length ORF, those marked with * contain predicted spliceosomal introns; *PG* pseudogene

HE	Organism	rDNA insertion site	Intron insertion site[a]	No. aa	Type	Reference
I-PpoI	*Physarum polycephalum*	L1925	s-P1	163	ACT	Muscarella et al. (1990)
I-NgrI	*Naegleria gruberi*	S516	s-P6	245	ACT	Einvik et al. (1997); Decatur et al. (2000)
I-NjaI	*Naegleria jamiesoni*	S516	s-P6	245	ACT	Elde et al. (1999)
I-NanI	*Naegleria andersoni*	S516	s-P6	245	ACT	Elde et al. (2000)
I-NitI	*Naegleria italica*	S516	s-P6	245	ACT	Elde et al. (2000)
I-DirI	*Didymium iridis*	S956	s-P2	261*	ACT	Decatur et al. (1995)
I-NaeIIP	*Naegleria sp. NG874*	L1926	s-P1	148	FL	Haugen et al. (2002)
I-NmoIP	*Naegleria morganensis (NG236)*	L2563	s-P1	175	FL	de Jonckheere and Brown (1998)
I-NcaIP	*Naegleria carteri*	S516	s-P6	245	FL	de Jonckheere and Brown (1998)
I-NclIP	*Naegleria clarki*	S516	s-P6	245	FL	de Jonckheere (1994)
I-NaeIP	*Naegleria sp. NG872*	S516	s-P6	244	FL	de Jonckheere and Brown (1998)
	Naegleria sp. NG597	S516	s-P6	245	FL	de Jonckheere (1994)
	Naegleria sp. NG434	S516	s-P6	245	FL	de Jonckheere (1994)
	Naegleria sp. NG560	S516	s-P6	245	FL	de Jonckheere (1994)
I-PteIP	*Porphyra tenera*	S516	s-P2	162	FL	Haugen et al. (1999)
I-PabIP	*Porphyra abbottae*	S516	s-P2		FL	Muller et al. (2001)
I-EmyIP	*Ericoid mycorrhizal*	S943	s-P8	294*	FL	Perotto et al. (2000)
I-MteIP	*Monoraphidium terrestre*	S943	a-P8	276	FL	Haugen et al. (2002)
I-CpiIP	*Capronia pilosella*	S943	s-P8	307	FL	Haugen et al. (2004)
I-CmoIP	*Coemansia mojavensis*	S943	a-P8	171	FL	Tanabe et al. (2002)
I-PchIP	*Pleopsidium chlorophanum*	S943	a-P8	256	FL	Haugen et al. (2004)
I-SdiIP	*Scytalidium dimidiatum*	S943	?	290	FL	P. Haugen (pers. comm.)
I-DirIIP	*Didymium iridis*	S952	a-P8	192*	FL	Vader (1998)
-	*Candida albicans*	L1923	a-P2.1		PG	Haugen et al. (2004)

Table 1. *(Continue)*

HE	Organism	rDNA insertion site	Intron insertion site[a]	No. aa	Type	Reference
-	*Candida dublineinsis*	L1923	a-P2.1		PG	Haugen et al. (2004)
-	*Cercomonas sp.*	S1190	P9		PG	P. Haugen (pers. comm.)
-	*Ericoid mycorrhizal*	S1199	a-P9		PG	Haugen et al. (2004)
-	*Nectria galligena*	S1199	a-P9		PG	Johansen and Haugen (1999)
-	*Protomyces pachydermus*	S1506	s-P9		PG	Haugen et al. (2004)
-	*Arthrobotrys superba*	S1506	s-P9		PG	Haugen et al. (2004)
-	*Tilletiopsis oryzicola*	S1506	a-P9		PG	Haugen et al. (2004)
-	*Porphyra spiralis*	S1506	a-P1		PG	Haugen et al. (1999)
-	*Bangia atropupurea*	S1506	a-P1		PG	Haugen et al. (1999)
-	*Bangia fuscopurpurea*	S1506	a-P1		PG	Muller et al. (2001)
-	*Porphyra tenera*	S1506	a-P1		PG	Haugen et al. (1999)
-	*Porphyra sp. 2*	S1506	a-P1		PG	Haugen et al. (1999)
-	*Porphyra umbilicus*	S1506	a-P1		PG	Muller et al. (2001)
-	*Bangia fuscopurpurea*	S516	s-P2		PG	Haugen et al. (1999)
-	*Acanthamoeba sp. KA/E4*	S516	s-P2		PG	Haugen et al. (2004)
	Porphyra kanakaensise	S516	s-P2		PG	Muller et al. (2001)
-	*Pseudohalonectria lignicola*	S943	s-P8		PG	Haugen et al. (2004)
-	*Cordyceps pseudomilitaris*	S943	a-P8		PG	Haugen et al. (2004)
-	*Beauveria bassiana*	S943	s-P8		PG	Yokoyama et al. (2002)

[a]s and a refer to ORF insertion into the sense or antisense orientation relative to the rDNA transcript.

tron in the rDNA gene (Haugen et al. 2004). They found that the *Naeglaria* family of enzymes, with the exceptions of I-NmoI and I-NaeII, appear to be examples of vertical inheritance. Here, highly related endonuclease ORFs situated in the same position (the P6 arm) of related group I introns are found in the same rDNA site (S516). However, examples of likely lateral transfer of endonuclease sequences independent of the encoding intron also exist. In one case, four endonuclease ORFs that group together phylogenetically (I-PpoI, I-NaeII, and two *Candida* pseudogenes) are found at different sites in two different subclasses of group I introns. The ORFs of I-PpoI and I-NaeII are found in the sense orientation in the P1 element of the IC1 subclass of group I introns, whereas the *Candida* pseudogenes are in the antisense orientation in the P2.1 element of the IE subclass. Interestingly, both of these introns are located in similar locations of the LSU gene (L1923 and L1926).

2 Expression of His-Cys Box Homing Endonucleases from Nuclear rDNA Transcripts

The position of the ORFs of the His-Cys box homing endonucleases in nuclear rDNA genes poses several questions regarding the expression of these proteins. In the nuclei of eukaryotes, RNAs destined to be translated into protein are transcribed by RNA polymerase II (Pol II) and subjected to extensive processing (5′ capping, splicing and polyadenylation). Transcription and processing appear to exhibit cross talk that can reciprocally increase the efficiency of each event (Maniatis and Reed 2002; Proudfoot et al. 2002; Le Hir et al. 2003). Furthermore, RNA processing events greatly affect the stability, export, localization, and translation of the messenger RNA. Conversely, rDNA is transcribed by RNA Pol I in the nucleolus, and rRNA transcripts undergo a very different processing (cleavage and modification) to ready them for assembly into the ribosomal subunits.

For the His-Cys box homing endonucleases found in these rRNA transcripts questions of which polymerase generates the translated message and what pre-mRNA processing is necessary for expression have been raised. This situation is in contrast to the other families of homing endonucleases that so far have been found exclusively in organelle genomes. In organelles, transcription and translation take place in the same compartment, and pre-mRNA processing appears to be minimal. Also, it is believed that a single RNA polymerase is responsible for all organelle transcription. In addition to transcription and processing of the His-Cys box endonuclease RNAs, there is also the matter of which portion of the transcript is exported to the cytoplasm and eventually translated (Fig. 1). It has been shown that group I introns are primarily spliced co-transcriptionally from nascent rRNA transcripts (Brehm

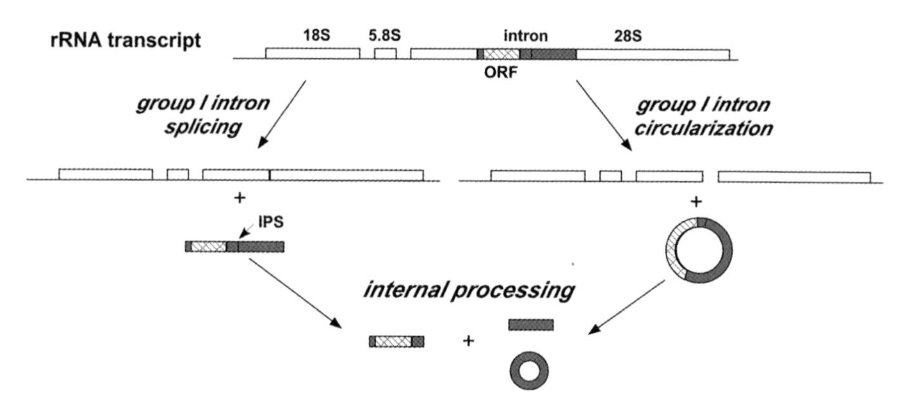

Fig. 1. Schematic of transcription and processing of RNA containing the His-Cys box homing endonuclease ORF. *IPS* refers to the internal processing site of nucleolytic cleavage by the (sometimes second) ribozyme

et al. 1983). In addition to the splicing reaction, the ribozyme can alternately catalyze the circularization of the full intron (leaving the flanking exons unjoined). Although this reaction is not beneficial to the host and appears at a low frequency in vivo, it may generate an RNA species resistant to degradation that plays a role in either endonuclease expression or intron mobility (Zaug et al. 1983; Nielsen et al. 2003). Another ribozyme-mediated processing event described for group I introns is hydrolysis at internal sites, resulting in a separation of the ribozyme from the ORF sequence. Internal processing appears to be carried out by the same ribozyme responsible for self-splicing in most cases. However, the *Didymium iridis* S956 intron encoding the I-DirI endonuclease and the *Naegleria* introns inserted at S516 both contain "twin" ribozymes: one that catalyzes intron excision and exon ligation and one that catalyzes hydrolysis at two internal processing sites to release itself along with the endonuclease ORF (Johansen and Vogt 1994; Decatur et al. 1995; Einvik et al. 1997).

Studies of I-PpoI expression from the PpLSU3 intron integrated into the *S. cerevisiae* genomic rDNA showed that endonuclease activity could not be detected when RNA Pol I was inactivated (Lin and Vogt 1998). This result indicates that I-PpoI is expressed from an RNA Pol I transcript. Additionally, expression was not dependent on internal processing of the intron. In fact, higher enzyme activity was observed when processing sites were mutated. This result and data showing that the full-length intron was found in the cytoplasm was taken as evidence that I-PpoI is translated from the full-length excised intron. Further studies in this system showed that a heterologous ORF (β-galactosidase) replacing the I-PpoI ORF could also be expressed from the RNA

Pol I transcript, although at lower efficiency (~3%) than a similar RNA Pol II transcript (Lin and Vogt 2000). The lower expression does not appear to be due to transcript levels, but is likely due to lower efficiency of nuclear export and translation, which are known to be stimulated by the 5′ cap and poly-A tail found on RNA Pol II transcripts and are presumably lacking from the excised intron.

It appears that the stories for other His-Cys box endonuclease expression might be more complicated. For example, activity of the second ribozyme responsible for internal processing in the twin introns of *Didymium* and *Naegleria* are necessary for expression of their homing endonucleases (Vader et al. 1999; Decatur et al. 2000). Intriguingly, the *Didymium* I-DirI ORF also contains what appear to be a small spliceosomal intron and a polyadenylation site, hallmarks of RNA Pol II transcripts (Vader et al. 1999). When the *Didymium* transcripts were examined, the majority species in the cytoplasm consisted of the ORF released by internal processing with the spliceosomal intron sequence absent and the 3′ end polyadenylated. This species also specifically associated with polysomes, indicating that it is the messenger RNA. Although production of this RNA from a cryptic RNA Pol II promoter has not been rigorously disproven, all current data point to this being a unique example of these selfish sequences somehow incorporating host organism pre-mRNA processing in a RNA Pol I transcript to improve expression.

3 Structure, DNA Binding, and Catalytic Mechanism

Much of our understanding of His-Cys box homing endonuclease structure and function is derived from studies of I-PpoI. Although very high sequence divergence within the family challenges generalization of the information garnered from these studies, it is likely that common themes of protein structure stabilization, DNA recognition and catalytic mechanism exist.

3.1 Structure of a His-Cys Box Endonuclease: I-PpoI

Currently, I-PpoI is the only member of the His-Cys box family of homing endonucleases with a known structure (Flick et al. 1998). This small enzyme (163 aa) forms a stable homodimer, and X-ray structures have been solved of the DNA-bound and apo forms of the protein. I-PpoI displays a fold of mixed α/β topology with the dimensions of 25×35×80 Å (Fig. 2A). The extended length of the dimer allows the protein to fully interact with its 18-bp DNA homing site. Each monomer of I-PpoI contains three antiparallel β-sheets flanked by two long α-helices and a long carboxy-terminal tail. The folded structure is stabilized by two zinc ions that are 15 Å apart in the monomer. The central interface of the enzyme dimer is small and highly solvated, with subunit contacts that bury only 700 Å^2 of surface area. The C-terminal tails (residues 146–

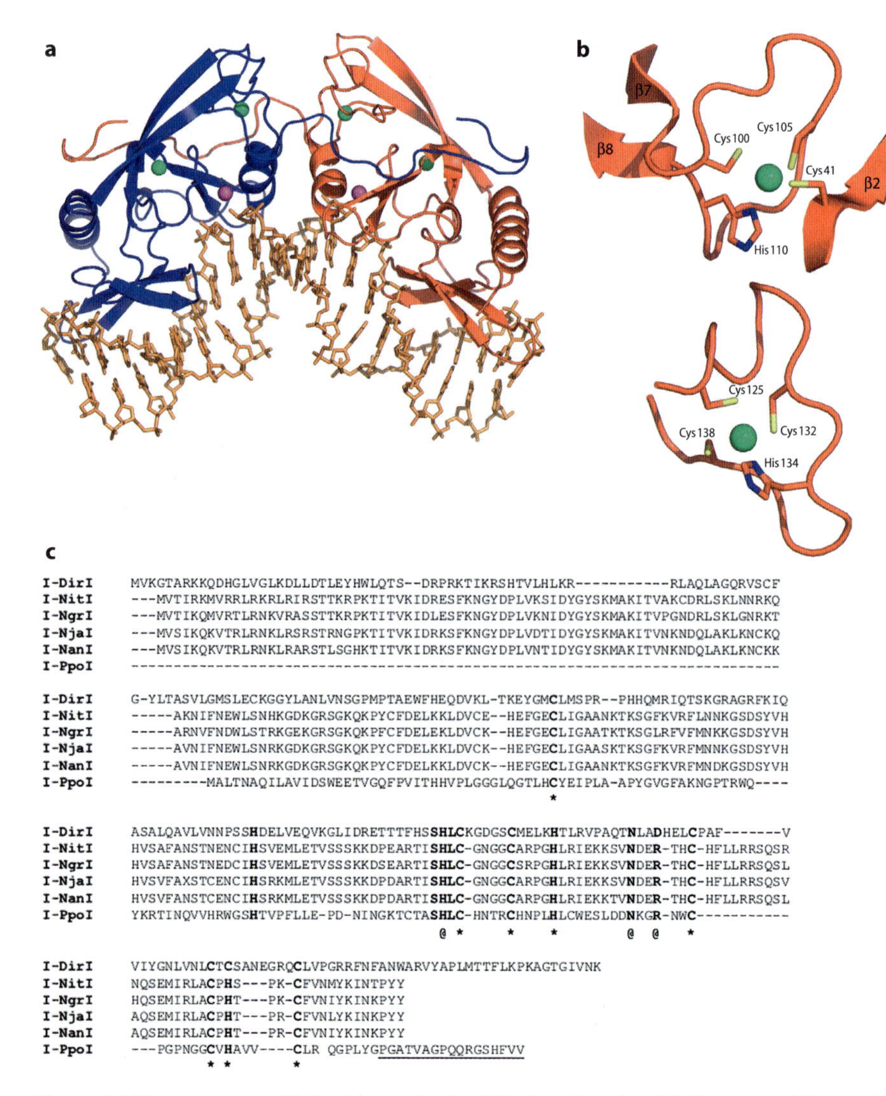

Fig. 2. (**a**) The structure of I-PpoI bound to its DNA homing site. (**b**) Close-up of the two I-PpoI zinc-binding sites created by conserved histidine and cysteine residues. (**c**) Sequence alignment of the known active His-Cys box homing endonucleases. Conserved residues are indicated in *bold.* * designates residues involved in zinc binding. @ designates active site residues. The C-terminal dimerization domain of I-PpoI is *underlined*

163) are domain-swapped and extend 34 Å across opposite monomers, burying an additional 900 $Å^2$ per subunit (Fig. 2C).

Eight of the residues conserved among the known His-Cys box homing endonucleases are involved in zinc coordination (Fig. 2B, C). In contrast to other zinc-binding motifs (e.g. RING finger) in DNA-binding proteins that are primarily associated with DNA recognition, the I-PpoI zinc-binding motifs play a central role in stabilizing the folded structure of the enzyme. In fact, the zinc ions appear to substitute for a tightly packed hydrophobic core, allowing this rather small protein to have an extended footprint for DNA binding. The first bound zinc ion is coordinated by a cluster of three cysteine and one histidine ligands. One side chain (Cys 41) is contributed by an amino-terminal β-strand (β2) and the remaining three side chains (Cys 100, Cys 105 and His 110) are donated from a short loop between β7 and β8. The second zinc ion is coordinated by a short cluster of four side chains (Cys 125, Cys 132, His 134 and Cys 138).

The remaining conserved residues of the His-Cys box homing endonucleases are positioned in the active site of the enzyme (Fig. 2C). A conserved asparagine coordinates a divalent metal ion that in turn contacts the 3′ hydroxyl of the cleaved DNA and four bound water molecules. The metal is positioned to interact with the scissile phosphate and cannot position a water molecule for in-line nucleophilic attack. Instead, a conserved histidine residue (His 98) is positioned to activate a water molecule.

As noted, the zinc binding and active site residues found in the I-PpoI structure are conserved among the His-Cys box family members. However, outside of these sequences, the family is highly divergent (Fig. 2C). Most of the other His-Cys box enzymes contain an additional long N-terminal extension that make them nearly 50% larger than I-PpoI. Also, the C-terminal tail of I-PpoI that serves as a dimerization motif does not appear to be conserved in most of the other His-Cys box enzymes. As all these enzymes cleave nearly symmetric homing sites (see below for details), it is likely they also dimerize but may have evolved alternate means for dimer stabilization.

3.2 DNA Binding and Recognition

The characterized His-Cys box endonucleases recognize extended (up to 20 nt) pseudopalindromic homing sites. Members of the family cleave to generate 4–5 nt 3′ overhangs, indicating cleavage across the minor groove. Unlike restriction sites, homing sites may vary in sequence at many of their individual nucleotide positions while still being recognized and cleaved. Site preference for I-PpoI was explored by an in vitro selection strategy using a plasmid library containing partially randomized cleavage sites (Argast 1998). Selected sites revealed that I-PpoI tolerates base-pair substitutions at several positions within the homing site with some positions being more stringently recognized.

I-PpoI was crystallized with a 20-nt DNA duplex containing a perfect palindrome variant of its recognition sequence (Flick et al. 1998). The twofold symmetry of the recognition site is mirrored in the homodimer structure of the enzyme with each monomer contacting either half of the palindrome. I-PpoI employs an antiparallel β-sheet motif that adopts a curvature and twist complementary to the DNA and forms an extended interface with the DNA major groove. The enzyme places five alternating side chains from the second β-sheet (β3–β4–β5) of each enzyme monomer in the major groove of each DNA half-site to contact base pairs 5–9 (Fig. 3). Additional DNA contacts are made in the center of the complex within the minor groove and with several phosphate groups in the cleavage site.

Unlike most restriction endonucleases, I-PpoI does not fully read the DNA by contacting all the possible hydrogen bond donors and acceptors present-

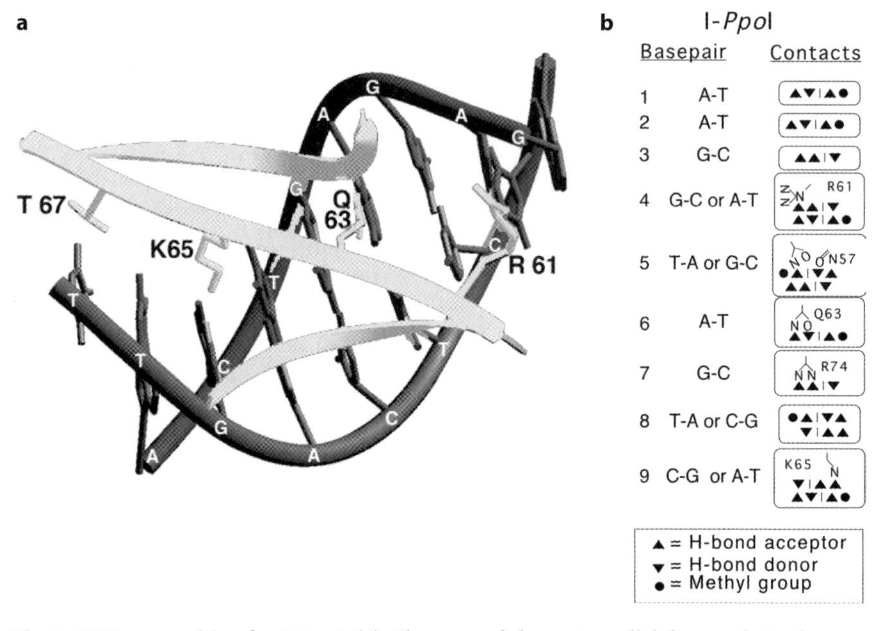

Fig. 3. DNA recognition by I-PpoI. (**a**) Close-up of the antiparallel β-strands in the major groove of the recognition half-site. (**b**) Schematic representation of contacts made by I-PpoI in a recognition half-site with the sequence-specific hydrogen bond donors and acceptors presented in the major groove. Where alternate base pairs can be recognized by I-PpoI both nucleotide identities are shown

ed by the homing sequence (Fig. 3). It directly satisfies only 8 of 24 hydrogen-bonding possibilities in the major groove (Jurica and Stoddard 1999). These limited contacts explain much of the variability allowed within I-PpoI's target site. Positions where I-PpoI makes multiple contacts with a given base pair were highly conserved for a single base-pair orientation in the homing site selection experiments (Argast 1998). For example, the bipartite contact between Gln 63 and Ade +6 observed in the structure nicely explains why an adenine was selected at this position. Conversely, positions with a single protein contact were found with two of the four possible base-pair identities.

The screen for homing site variants revealed an absolute requirement for A-T base pairs in the central four positions of the homing site (Argast 1998). However, no sequence-specific contacts between protein and DNA were observed in this region in the co-crystal structures (Flick et al. 1998; Galburt et al. 2000). The preference for these base pairs probably reflects the DNA sequence requirements of the severely bent conformation of the target across these base pairs. The distortion results in a compaction of the major groove and an expansion of the minor groove such that at the center of the homing site the minor groove is 5 Å wider than the major groove. The bent conformation increases the buried surface between protein and DNA, thus stabilizing the complex. At the same time, the bend results in the proper positioning of the substrate phosphodiester bonds relative to the two active sites. I-PpoI distorts the DNA substrate to widen the minor groove, thus separating the two scissile phosphates and facilitating the positioning of two sets of active site residues across the usually narrow minor groove.

Does I-PpoI induce bending or selectively bind bent DNA? Circular permutation experiments suggest that the homing site does not exhibit a stable kink; however, the structure of I-PpoI in the apo form is virtually unchanged from the DNA-bound form, indicating that the protein does not bind unbent B-form DNA (Galburt et al. 2000). One residue, Leu 116, is positioned near the DNA bend in the complex crystal structure and makes an edge-on contact to Ade +2 and a face-on contact to Gua +3. Substitution of an alanine at this position (L116A) has a dramatic effect on both binding and catalysis (Galburt et al. 2000). The structure of an L116A/DNA complex is similar to the wild-type complex except the DNA is less bent and the central AATT base pairs are disordered. This suggests that the surface area buried by Leu 116 and nucleotides bracketing the bend are crucial for stable binding of the deformed conformation of the DNA.

It is interesting to note that a subfamily of His-Cys box endonucleases, typified by I-NjaI, has been shown to generate 5-base, 3′ overhangs (Elde et al. 1999). In these cases, the homing site scissile phosphates would be further apart in unperturbed B-form DNA (15 Å instead of 10 Å), and their corresponding homing endonucleases do not contain residues that correspond to

Leu 116 in I-PpoI. This increased spacing of active sites and the corresponding lack of a leucine residue make it likely that the DNA-binding mode of these enzymes differs to the mode observed for I-PpoI. In conclusion, even homing endonucleases within the same family appear to have evolved a variety of DNA-binding modes and active site chemistries that accomplish the same biological function.

3.3 Catalytic Mechanism

Enzymatic cleavage by I-PpoI has been studied both biochemically and structurally. I-PpoI cleaves its target site with a k_{cat}/K_m of 10^8 M^{-1} s^{-1} and is activated by many divalent metal ions (in order of activity: Mg^{2+}>Mn^{2+}>Ca^{2+}=Co^{2+}>Ni^{2+}>Zn^{2+}; Lowery et al. 1992; Ellison and Vogt 1993; Wittmayer and Raines 1996). A series of I-PpoI–DNA complex structures have been solved along the nucleolytic reaction pathway (Galburt et al. 1999). The enzyme–substrate (ES) complex of I-PpoI has been trapped by substitution of a monovalent cation for an activating divalent cation in the active site and by substitution of alanine for the His 98 general base (H98A). Both of these substitutions inhibit DNA cleavage. The enzyme–product (EP) complex has been solved in the presence of magnesium.

3.3.1 Catalytic Metal Coordination

In the structure of the ES complex, the bound Mg^{2+} ion is coordinated in a six-fold geometry by the side-chain oxygen of Asn 119, the bridging 3′ oxygen and a single non-bridging oxygen atom of the scissile phosphate, and three well-ordered water molecules. An ideal octahedral coordination geometry is observed in the cleaved product complex. In contrast, both structures of the trapped substrate complex demonstrate that the bond angle from the non-bridging phosphate oxygen, through the bound cation to the 3′ bridging oxygen, is significantly strained.

The single bound metal ion in I-PpoI appears to serve three distinct roles in catalysis. Direct interaction of the bound metal ion with the scissile phosphate indicates that Mg^{2+} stabilizes the phosphoanion intermediate and the 3′ hydroxylate leaving group. Second, a water molecule in the inner coordination sphere of the metal is appropriately positioned to donate a proton to the 3′ hydroxyl leaving group. The metal ion decreases the pK_a of this water molecule and accelerates proton transfer. Finally, the bound metal forms a geometrically strained octahedral complex with surrounding protein, DNA and solvent atoms that is relaxed after DNA bond cleavage.

3.3.2 *Alignment and Activation of the Hydrolytic Water*

In both ES complexes, an ordered water molecule is observed positioned for in-line attack on the scissile phosphate (Fig. 4A, B). Thus, the appearance and position of the bound solvent molecule are not simply a result of the H98A mutation, but are consistent features of the uncleaved enzyme–DNA complex. The δN of His 98 is directly hydrogen-bonded to the water molecule. This histidine appears to act as a general base by activating the water molecule and may also participate in stabilization of the phosphoanion transition state.

Because the observed nucleophilic water molecule is not associated with a bound cation or any other electrophilic group, its pK_a is likely to be higher than the metal-bound water nucleophiles that are postulated for enzymes

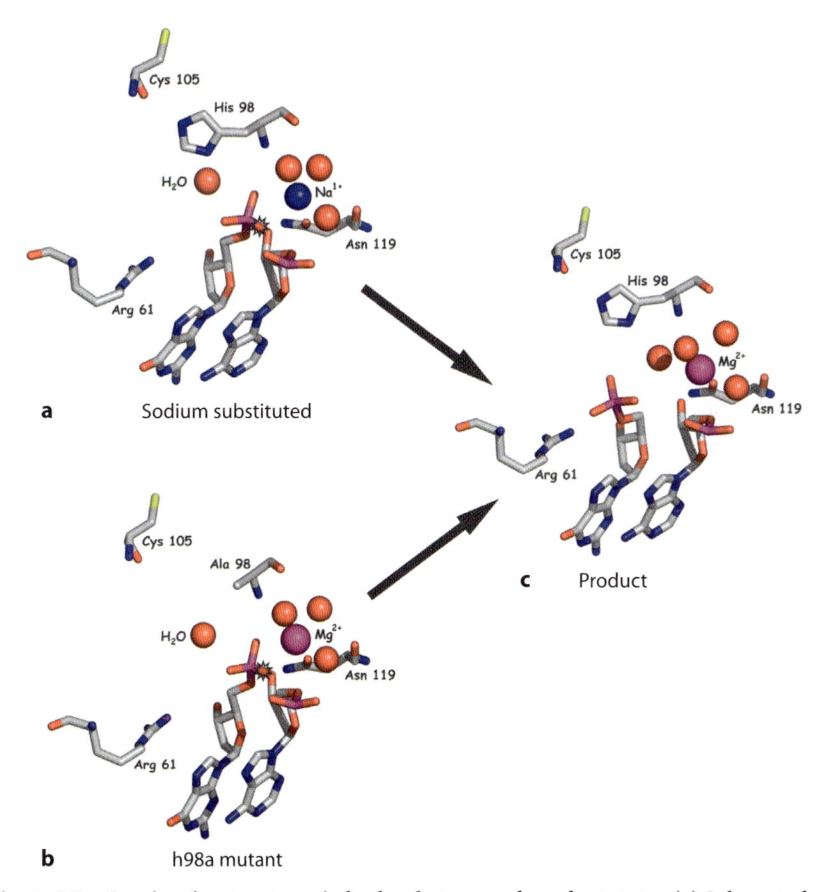

Fig. 4. I-PpoI active site structures in both substrate and product states. (**a**) Substrate form trapped by substituting sodium for magnesium. (**b**) Substrate form trapped by H98A mutation. (**c**) Product form

such as BamHI or EcoRV. However, the pK_a of an uncharged histidine residue is only about 6, so it would seem likely that such a side chain must be rendered a stronger base through an interaction with a hydrogen bond acceptor in order to effectively deprotonate a water molecule. In both substrate complexes the backbone carboxylate oxygen of Cys 105 is 2.8 Å from the His 98 εN, and is positioned to form a linear hydrogen bond. The structure of the Serratia nuclease displays a similar active site architecture (Miller and Krause 1996). Here, the δN of His 89, the putative general base, is similarly stabilized by Asn 106. However, there are currently no reported pH-dependence studies of the chemical step of the I-PpoI reaction, nor has the importance of this interaction been experimentally tested by mutation of Asn 106 in *Serratia* nuclease or by measurement of the pK_a of His 98 in I-PpoI.

3.3.3 Conformational Changes and Transition State Stabilization

A series of conformational changes are observed in the active site as a result of DNA bond cleavage (Fig. 4C). The free 5′-phosphate moves by over 2.5 Å from its position in the substrate complex and forms a 2.8-Å electrostatic bond with a guanido nitrogen of Arg 61, which moves by ~0.5 Å. The movement of the 5′-phosphate disrupts the interaction between its non-bridging oxygen and the bound metal ion. A fourth well-resolved water molecule is added to the inner metal coordination sphere, which assumes a more ideal octahedral geometry. The metal ion does not move significantly upon cleavage, and maintains interactions with Asn 119 and the 3′ oxygen leaving group of the cleaved phosphodiester bond.

These structures indicate that the phosphoanion transition state is stabilized through contacts with the bound metal ion and the imidazole ring of His 98. This contact exists in the ES complex as a polar interaction with the hydrolytic water molecule and is maintained in the free 5′-phosphate group of the EP complex. Arg 61 does not appear to play a role in transition state stabilization, because the distance from this side chain to the scissile phosphate before bond cleavage is too long, at 5.5 Å. Arg 61 does, however, appear to stabilize the final product complex, and thus may help to drive the reaction forward by inhibiting re-ligation on the enzyme.

All the important active site residues are conserved among the His-Cys box endonucleases, suggesting that even though their overall folded structure might differ, their active site architectures will be very similar. In addition to members of the His-Cys box family, there are examples of other nucleases with similar active site architectures. We have already mentioned the *Serratia* nuclease, but a recent structure of an HNH homing endonuclease family member reveals a similar active site geometry and chemistry.

4 HNH and His-Cys Box Homing Endonucleases

The HNH family of endonucleases is characterized by a defining motif that spans 25 residues of a well-conserved sequence. The homing endonucleases I-CmoeI, I-TevIII, I-HmuI, and I-HmuII (Eddy and Gold 1991; Goodrich-Blair and Shub 1996) are members of the HNH family, and I-CmoeI has been well characterized biochemically (Drouin et al. 2000; for a full description of these enzymes, see Keeble et al., this Volume). The non-specific DNase colicin E9 was the first HNH enzyme to have its structure solved (Kleanthous et al. 1999) and allowed biochemical and structural comparisons of the active sites of I-PpoI, *Serratia* nuclease, colicin E9, and HNH homing endonucleases. These studies led to the hypothesis that these enzymes are evolutionarily related and share similar catalytic mechanisms (Friedhoff et al. 1999; Kuhlmann et al. 1999). Recently, the first structure of an HNH homing endonuclease, I-HmuI, has been solved and confirms these relationships (Shen et al. 2004). The structure of I-HmuI is comprised of an N-terminal antiparallel DNA binding β-sheet (similar to those of the His-Cys box family), the ββα-Me catalytic motif (encoded by the HNH motif), two DNA-binding α-helices, and a C-terminal helix-turn-helix DNA-binding domain which has been previously observed in the structure the GIY-YIG homing endonuclease I-TevI (van Roey et al. 2001). The ββα-Me motif is composed of two antiparallel β-strands followed by an α-helix and is stabilized by the binding of a metal ion between the first strand and the helix (Kuhlmann et al. 1999).

Although the two families of homing endonucleases share both an active site motif and a DNA-binding domain, their use of these apparently modular structural domains are different. The His-Cys box enzymes have evolved into obligate homodimers whereby two of the β-sheet DNA-binding domains are juxtaposed along a two-fold symmetry axis. This requires pseudopalindromic homing sites, whereas the single copy of this DNA-binding motif in the functional unit of the HNH enzyme allows for more degenerate sequence recognition by removing the palindromic constraint. Furthermore, although HNH endonucleases, His-Cys box endonucleases, and the colicin enzymes all contain the ββα-Me active site motif, their structures indicate that they are used somewhat differently with respect to the chemical identity of metal-binding residues. Specifically, I-HmuI uses both an asparagine and aspartic acid as metal ligands, I-PpoI uses only an asparagine as part of an octahedral magnesium coordination, and the structure of colicin E9 shows a pair of histidine residues binding a metal ion.

The evidence described above supports the hypothesis that HNH and His-Cys box enzymes contain similar DNA-binding motifs and almost identical active sites and therefore are likely to have a common evolutionary ancestor. In fact, the two families are now considered to be divergent members of a single super family of homing endonucleases (see Keeble et al. this Vol).

Acknowledgements. The authors wish to thank P. Haugen for identifying additional His-Cys box endonucleases, and B. Shen and B.L. Stoddard for sharing unpublished results. E.A.G. is supported by a Jane Coffin Childs postdoctoral fellowship.

References

Argast GM (1998) I-PpoI and I-CreI homing site sequence degeneracy determined by random mutagenesis and sequential in vitro enrichment. J Mol Biol 280:345–353

Bhattacharya D, Friedl T, Helms G (2002) Vertical evolution and intragenic spread of lichen-fungal group I introns. J Mol Evol 55:74–84

Brehm SL, Cech SL, Cech TR (1983) Fate of an intervening sequence ribonucleic acid: excision and cyclization of the Tetrahymena ribosomal ribonucleic acid intervening sequence in vivo. Biochemistry 22:2390–2397

Cho Y, Qiu YL, Kuhlman P, Palmer JD (1998) Explosive invasion of plant mitochondria by a group I intron. Proc Natl Acad Sci USA 95:14244–14249

De Jonckheere JF (1994) Evidence for the ancestral origin of group I introns in the SSUrDNA of *Naegleria* spp. J Eukaryot Microbiol 41:457–463

De Jonckheere JF, Brown S (1998) Three different group I introns in the nuclear large subunit ribosomal DNA of the amoeboflagellate *Naegleria*. Nucleic Acids Res 26:456–461

Decatur WA, Einvik C, Johansen S, Vogt VM (1995) Two group I ribozymes with different functions in a nuclear rDNA intron. EMBO J 14:4558–4568

Decatur WA, Johansen S, Vogt VM (2000) Expression of the *Naegleria* intron endonuclease is dependent on a functional group I self-cleaving ribozyme. RNA 6:616–627

Drouin M, Lucas P, Otis C, Lemieux C, Turmel M (2000) Biochemical characterization of I-CmoeI reveals that this H-N-H homing endonuclease shares functional similarities with H-N-H colicins. Nucleic Acids Res 28:4566–4572

Eddy SR, Gold L (1991) The phage T4 nrdB intron: a deletion mutant of a version found in the wild. Genes Dev 5:1032–1041

Einvik C, Decatur WA, Embley TM, Vogt VM, Johansen S (1997) Naegleria nucleolar introns contain two group I ribozymes with different functions in RNA splicing and processing. RNA 3:710–720

Elde M, Haugen P, Willassen NP, Johansen S (1999) I-NjaI, a nuclear intron-encoded homing endonuclease from *Naegleria*, generates a pentanucleotide 3′ cleavage-overhang within a 19 base-pair partially symmetric DNA recognition site. Eur J Biochem 259:281–288

Elde M, Willassen NP, Johansen S (2000) Functional characterization of isoschizomeric His-Cys box homing endonucleases from *Naegleria*. Eur J Biochem 267:7257–7266

Ellison EL, Vogt VM (1993) Interaction of the intron-encoded mobility endonuclease I-PpoI with its target site. Mol Cell Biol 13:7531–7539

Flick KE, Jurica MS, Monnat RJ Jr, Stoddard BL (1998) DNA binding and cleavage by the nuclear intron-encoded homing endonuclease I-PpoI. Nature 394:96–101

Foley S, Bruttin A, Brussow H (2000) Widespread distribution of a group I intron and its three deletion derivatives in the lysin gene of *Streptococcus thermophilus* bacteriophages. J Virol 74:611–618

Friedhoff P, Franke I, Meiss G, Wende W, Krause KL, Pingoud A (1999) A similar active site for non-specific and specific endonucleases. Nat Struct Biol 6:112–113

Galburt EA, Chevalier B, Tang W, Jurica MS, Flick KE, Monnat RJ Jr, Stoddard BL (1999) A novel endonuclease mechanism directly visualized for I-PpoI. Nat Struct Biol 6:1096–1099

Galburt EA, Chadsey MS, Jurica MS, Chevalier BS, Erho D, Tang W, Monnat RJ Jr, Stoddard BL (2000) Conformational changes and cleavage by the homing endonuclease I-PpoI: a critical role for a leucine residue in the active site. J Mol Biol 300:877–887

Goddard MR, Burt A (1999) Recurrent invasion and extinction of a selfish gene. Proc Natl Acad Sci USA 96:13880–13885

Goodrich-Blair H, Shub DA (1996) Beyond homing: competition between intron endonucleases confers a selective advantage on flanking genetic markers. Cell 84:211–221

Haugen P, Huss VA, Nielsen H, Johansen S (1999) Complex group-I introns in nuclear SSU rDNA of red and green algae: evidence of homing-endonuclease pseudogenes in the *Bangiophyceae*. Curr Genet 36:345–353

Haugen P, de Jonckheere JF, Johansen S (2002) Characterization of the self-splicing products of two complex *Naegleria* LSU rDNA group I introns containing homing endonuclease genes. Eur J Biochem 269:1641–1649

Haugen P, Reeb V, Lutzoni F, Bhattacharya D (2004) The evolution of homing endonuclease genes and group I introns in nuclear rDNA. Mol Biol Evol 21:129–140

Johansen S, Haugen P (1999) A complex group I intron in *Nectria galligena* rDNA. Microbiology 145:516–517

Johansen S, Vogt VM (1994) An intron in the nuclear ribosomal DNA of *Didymium iridis* codes for a group I ribozyme and a novel ribozyme that cooperate in self-splicing. Cell 76:725–734

Johansen S, Embley TM, Willassen NP (1993) A family of nuclear homing endonucleases. Nucleic Acids Res 21:4405

Johansen S, Elde M, Vader A, Haugen P, Haugli K, Haugli F (1997) In vivo mobility of a group I twintron in nuclear ribosomal DNA of the myxomycete *Didymium iridis*. Mol Microbiol 24:737–745

Jurica MS, Stoddard BL (1999) Homing endonucleases: structure, function and evolution. Cell Mol Life Sci 55:1304–1326

Kleanthous C, Kuhlmann UC, Pommer AJ, Ferguson N, Radford SE, Moore GR, James R, Hemmings AM (1999) Structural and mechanistic basis of immunity toward endonuclease colicins. Nat Struct Biol 6:243–252

Kuhlmann UC, Moore GR, James R, Kleanthous C, Hemmings AM (1999) Structural parsimony in endonuclease active sites: should the number of homing endonuclease families be redefined? FEBS Lett 463:1–2

Lambowitz AM, Belfort M (1993) Introns as mobile genetic elements. Annu Rev Biochem 62:587–622

Le Hir H, Nott A, Moore MJ (2003) How introns influence and enhance eukaryotic gene expression. Trends Biochem Sci 28:215–220

Lin J, Vogt VM (1998) I-PpoI, the endonuclease encoded by the group I intron PpLSU3, is expressed from an RNA polymerase I transcript. Mol Cell Biol 18:5809–5817

Lin J, Vogt VM (2000) Functional alpha-fragment of beta-galactosidase can be expressed from the mobile group I intron PpLSU3 embedded in yeast pre-ribosomal RNA derived from the chromosomal rDNA locus. Nucleic Acids Res 28:1428–1438

Loizos N, Tillier ER, Belfort M (1994) Evolution of mobile group I introns: recognition of intron sequences by an intron-encoded endonuclease. Proc Natl Acad Sci USA 91:11983–11987

Lowery R, Hung L, Knoche K, Bandziulis R (1992) Properties of I-PpoI: a rare-cutting intron-encoded endonuclease. Promega Notes 38:8–12

Maniatis T, Reed R (2002) An extensive network of coupling among gene expression machines. Nature 416:499–506
Miller MD, Krause KL (1996) Identification of the Serratia endonuclease dimer: structural basis and implications for catalysis. Protein Sci 5:24–33
Mota EM, Collins RA (1988) Independent evolution of structural and coding regions in a Neurospora mitochondrial intron. Nature 332:654–656
Muller KM, Cannone JJ, Gutell RR, Sheath RG (2001) A structural and phylogenetic analysis of the group IC1 introns in the order *Bangiales* (Rhodophyta). Mol Biol Evol 18:1654–1667
Muscarella DE, Ellison EL, Ruoff BM, Vogt VM (1990) Characterization of I-Ppo, an intron-encoded endonuclease that mediates homing of a group I intron in the ribosomal DNA of *Physarum polycephalum*. Mol Cell Biol 10:3386–3396
Nielsen H, Fiskaa T, Birgisdottir AB, Haugen P, Einvik C, Johansen S (2003) The ability to form full-length intron RNA circles is a general property of nuclear group I introns. RNA 9:1464–1475
Nozaki H, Takahara M, Nakazawa A, Kita Y, Yamada T, Takano H, Kawano S, Kato M (2002) Evolution of rbcL group IA introns and intron open reading frames within the colonial *Volvocales (Chlorophyceae)*. Mol Phylogenet Evol 23:326–338
Ogawa S, Naito K, Angata K, Morio T, Urushihara H, Tanaka Y (1997) A site-specific DNA endonuclease specified by one of two ORFs encoded by a group I intron in *Dictyostelium discoideum* mitochondrial DNA. Gene 191:115–121
Pellenz S, Harington A, Dujon B, Wolf K, Schafer B (2002) Characterization of the I-Spom I endonuclease from fission yeast: insights into the evolution of a group I intron-encoded homing endonuclease. J Mol Evol 55:302–313
Perotto S, Nepote-Fus P, Saletta L, Bandi C, Young JP (2000) A diverse population of introns in the nuclear ribosomal genes of ericoid mycorrhizal fungi includes elements with sequence similarity to endonuclease-coding genes. Mol Biol Evol 17:44–59
Proudfoot NJ, Furger A, Dye MJ (2002) Integrating mRNA processing with transcription. Cell 108:501–512
Shen BW, Landthaler M, Shub DA, Stoddard BL (2004) DNA binding and cleavage by the HNH homing endonuclease I-HmuI. J Mol Biol 342:43–56
Tanabe Y, Yokota A, Sugiyama J (2002) Group I introns from *Zygomycota*: evolutionary implications for the fungal IC1 intron subgroup. J Mol Evol 54:692–702
Vader A (1998) Nuclear group I introns of the myxomycetes: organization, expression and evolution. University of Tromsù, Norway
Vader A, Nielsen H, Johansen S (1999) In vivo expression of the nucleolar group I intron-encoded I-dirI homing endonuclease involves the removal of a spliceosomal intron. EMBO J 18:1003–1013
Van Roey P, Waddling CA, Fox KM, Belfort M, Derbyshire V (2001) Intertwined structure of the DNA-binding domain of intron endonuclease I-TevI with its substrate. EMBO J 20:3631–3637
Wittmayer PK, Raines RT (1996) Substrate binding and turnover by the highly specific I-PpoI endonuclease. Biochemistry 35:1076–1083
Wittmayer PK, McKenzie JL, Raines RT (1998) Degenerate DNA recognition by I-PpoI endonuclease. Gene 206:11–21
Yokoyama E, Yamagishi K, Hara A (2002) Group-I intron containing a putative homing endonuclease gene in the small subunit ribosomal DNA of Beauveria bassiana IFO 31676. Mol Biol Evol 19:2022–2025
Zaug AJ, Grabowski PJ, Cech TR (1983) Autocatalytic cyclization of an excised intervening sequence RNA is a cleavage-ligation reaction. Nature 301:578–583

Group I Introns and Their Maturases: Uninvited, but Welcome Guests

Mark G. Caprara, Richard B. Waring

Abstract

Homing endonucleases are a class of invasive genetic elements that use several elegant solutions to ensure their survival in natural populations. Like all successful mobile entities, homing endonucleases must either reduce the deleterious effects of insertion within essential genes of host genomes or be lost. Many homing endonuclease genes have solved this problem by colonizing self-splicing group I introns. This association makes homing endonuclease genes phenotypically "silent" since they are spliced out and thus absent from the mature mRNA of the invaded gene. Through this union, homing endonucleases and introns have co-evolved into a "hybrid" mobile element providing the introns with a mechanism to propagate themselves in a population. Remarkably, in some cases within fungal mitochondrial genomes, homing endonucleases have adapted to facilitate splicing of their encoding introns and contribute to the host's regulation of the invaded gene. This novel adaptation, termed maturase activity, has likely served to ensure their fixture in mitochondrial and, perhaps, other genomes. In this chapter, we will review what is known concerning the mechanism of group I intron-encoded protein-assisted splicing. In addition, we will summarize new studies of both mobility and maturase functions that have resulted in a better understanding of how a single polypeptide carries out diverse and unrelated activities. Principles derived

M.G. Caprara (e-mail: mgc3@po.cwru.edu)
Center for RNA Molecular Biology, Case Western Reserve University School of Medicine, Cleveland, Ohio 44106, USA

R.B. Waring (e-mail: waring@temple.edu)
Department of Biology, Temple University, Philadelphia, Pennsylvania 19122, USA

Nucleic Acids and Molecular Biology, Vol. 16
Marlene Belfort et al. (Eds.)
Homing Endonucleases and Inteins

from maturase systems are likely to apply to numerous other multi-functional proteins that participate in diverse metabolic pathways.

1 Discovery of Self-Splicing Group I Introns

One of the first genomic projects, begun in the late 1970s, included efforts by several groups to deduce the sequences of the mitochondrial (mt) genomes of fungi. These projects were undertaken as it became clear that the expression of mt genes was controlled by both nuclear and other mt factors. To understand the regulatory mechanisms, the mt DNA of respiratory mutants in *S. cerevisiae* (which are viable under anaerobic conditions) as well as mt suppressor mutant strains were sequenced. One of the most remarkable observations that came from these projects was that mt genes were not contiguous: many open reading frames (ORFs) were interrupted by "inserts". Such a "mosaic" pattern of genes was concurrently discovered in viral and eukaryotic genes. With these discoveries, it became clear that, in such instances, mRNAs must be processed (spliced) to remove the intragenetic sequences (introns) and join together the coding sequences (exons). Importantly, the analysis of many respiratory deficient strains in *S. cerevisiae* revealed that mt mRNAs were not fully processed leading to truncated, inactive mt proteins.

Sequence analysis of the fungal mt introns showed that they fell into one of two structural classes, groups I and II (for a discussion of group II introns, see Lambowitz et al., this Vol.). Other group I introns were subsequently found in chloroplasts, cyanellar and nuclear genomes in eukaryotes, as well as bacteria and bacteriophages. Group I introns form a characteristic secondary structure, as shown in Fig. 1a. Biochemical analyses showed that group I introns were self-splicing, requiring only divalent cations and a guanosine cofactor for activity (reviewed in Cech 1990). The group I splicing mechanism consists of two sequential transesterification reactions. In the first step, the intron core binds an exogenous guanosine that attacks the 5′ splice site (SS) resulting in covalent attachment of the guanosine to the first intron nucleotide. In the second step, the 5′ exon attacks the 3′ SS, resulting in exon ligation and intron release (Cech 1990). The 5′ and 3′ SS are defined by pairing to an intron structure called the internal guide sequence, forming the P1 and P10 helices, respectively (Fig. 1a, b).

The group I intron catalytic core is made up of two extended helices called domains; the P4-P6 domain is formed by coaxial stacking of P5, P4, P6, and P6a helices while the P3-P9 domain is formed by stacking of P8, P3, P7, and P9 (Michel and Westhof 1990; Golden et al. 1998; Adams et al. 2004; Fig. 1a). Precise packing of the two domains via tertiary interactions forms the intron catalytic center that is capable of binding guanosine and the SS containing P1 and P10 helices. Interconnecting peripheral structures surround the intron core and stabilize the active conformation (Lehnert et al. 1996; Pan and Woodson 1999; Engelhardt et al. 2000; Fig. 1a). Folding of group I introns into an active conformation is often very slow due to long-lived, kinetically trapped intermediates (Woodson 2000). In vivo, folding can be accelerated by the binding of protein cofactors (Schroeder et al. 2002).

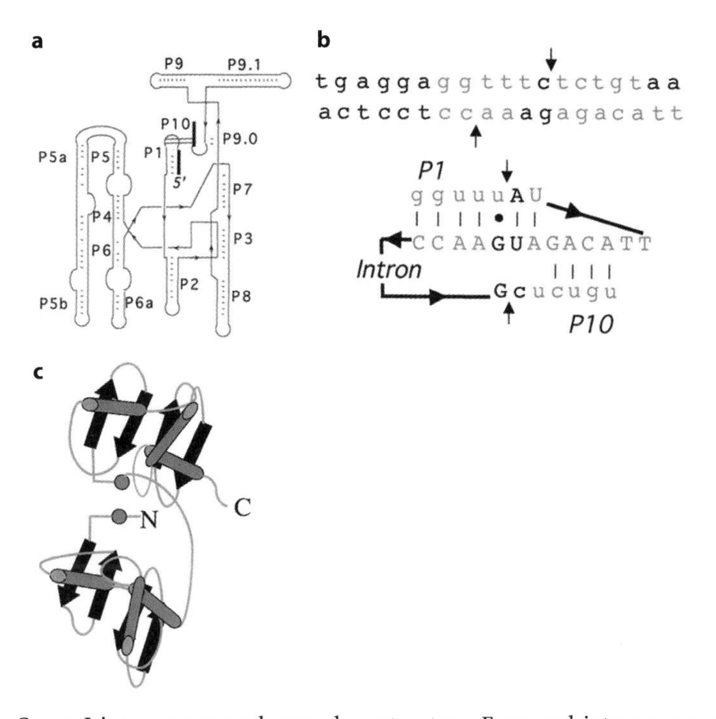

Fig. 1. **a** Group I intron conserved secondary structure. Exon and intron sequences are represented by *thick* and *thin lines*, respectively. In this diagram, the P2, P5a, b and P9.1b helices represent optional peripheral folding elements outside of the catalytic core. **b** Comparison of DNA target site sequences and the pre-RNA secondary structure around the 5′ and 3′ splice sites. For the DNA substrate, *arrows* indicate endonuclease cleavage sites used to initiate a homing event. For the RNA sequence, *upper-* and *lowercase letters* indicate intron and exon sequences, respectively. *Arrows* point to the splice sites. The sequences represent the I-AniI endonuclease/maturase substrates. *Gray bases* indicate the identical residues between the DNA site and P1/P10 pseudoknot. **c** Topology of I-AniI. The DNA-binding surface is composed of antiparallel β-sheets shown as *black arrows*. α-Helices are shown as gray cylinders and the LAGLIDADG helices are shown as *gray circles*. This “top” view emphasizes the non-DNA-binding surface of I-AniI that plays a major role in maturase activity. *C* and *N*, C- and N-termini, respectively

2 Genes Within Introns

In retrospect, since many group I introns are self-splicing, one might expect to find that the mutations that lead to processing defects would disrupt essential RNA structural elements required for folding and/or catalysis. However, genetic and sequencing analysis of three group I introns in the *S. cerevisiae* cytochrome b (cob) gene revealed a number of surprising features. First, the introns contain long coding regions in frame with the upstream 5′ exon ORFs. Second, many of the splicing impaired mutants carry non-sense codons in the intron ORF, suggesting that the full-length fusion protein or ribosome read through is required for splicing. Thirdly, disruption of upstream splicing events prevents expression of downstream intron ORFs, and thus, splicing of those introns. Finally, it was shown that mutants are rescued by providing the wild-type intron-encoding protein *in trans* (Lazowska et al. 1980, 1989; Groudinsky et al. 1981; Anziano et al. 1982; Weiss-Brummer et al. 1982; Delahodde et al. 1989). These observations provided a convincing case that intron-encoded proteins are required for group I intron splicing in vivo and were appropriately named "maturases".

3 How Many Maturases?

In total, genetic studies in *S. cerevisiae* identified three dedicated maturases in the cob gene designated bI2, bI3, and bI4. In each case, the maturase facilitates splicing of its cognate intron while the bI4 maturase also promotes splicing of the homologous aI4α intron in the cytochrome oxidase subunit I (COXI) gene (reviewed in Lambowitz et al. 1999). Interestingly, the aI4α intron also encodes a latent maturase protein that is activated by a single mutation in its ORF (*mim*2-1) or by a dominant suppressor mutant (*NAM*2-1) of an additional splicing cofactor, the nuclear encoded mt leucyl tRNA synthetase (Leu RS, see below; Dujardin et al. 1982; Lambouesse et al. 1987; Herbert et al. 1988). Outside of *S. cerevisiae*, four other maturases have also been discovered. The bI2 intron maturase from *S. capensis* differs by only four amino acids from the corresponding maturase from *S. cerevisiae* and, not surprisingly, the *S. capensis* maturase also functions in splicing (Szczepanek and Lazowska 1996). The first intron in the mt cytochrome c oxidase (cox1I1) from *S. pombe* encodes a maturase (Merlos-Lange et al. 1987). Finally, the protein (I-AniI) encoded by the single intron (*A.n.* COB) in the cob gene of *Aspergillus nidulans* can facilitate the splicing of the A.n. COB intron in vitro, as can the related maturase from *Venturia inaequalis* (Ho et al. 1997; Kwon and Waring, unpubl.; Table 1). Although other maturases most certainly exist, studies with this handful of intron-splicing systems have revealed significant insights with regard to the function of these unusual proteins.

Table 1. Maturases confirmed by experimental data

Species	Gene	Intron[a]	Maturase reference	Endonuclease	Endonuclease reference
S. cerevisiae	cyt. b	bI2	Lazowska et al. (1980)	No, unless two mutations made	Szczepanek and Lazowska (1996)
S. cerevisiae	cyt. b	bI3	Lazowska et al. (1989)	No	
S. cerevisiae	cyt. b	bI4	Anziano et al. (1982); de la Salle et al. (1982); Weiss-Brummer et al. (1982)	No, but stimulates recombination in E. coli	Goguel et al. (1989)
V. inaequalis	cyt. b	VibI5	Kwon and Waring (unpubl.)	?	
S. pombe	cyt. oxidase subunit I	COX1I1	Merlos-Lange et al. (1987)	Yes, I-*Spo*I	Pellenz et al. (2002)
S. capensis	cyt. b	bI2	Szczepanek and Lazowska (1996)	Yes, I-*Sca*I	Lazowska et al. (1992)
A. nidulans	cyt. b	An COB	Ho et al. (1997)	Yes, I-*Ani*I	Ho et al. (1997)
S. cerevisiae	cyt. oxidase subunit I	aI4α with *mim*2-1 mutation	Dujardin et al. (1982)	Yes, I-*Sce*II	Delahodde et al. (1989); Wenzlau et al. (1989)

[a] Strains can vary in the total number of introns and whether the indicated intron is present.

4 The Mechanism of Maturase-Assisted Splicing

Maturases are relatively small proteins (~30 kDa) that interact with much larger RNAs [e.g., ~380 kDa for the average size of maturase encoding introns, ~1200 nucleotides (nt)]. How might such a small protein recognize a large

RNA and facilitate splicing? Since maturases are intron-specific cofactors (see above), they must recognize idiosyncratic features of their cognate introns but what governs maturase specificity remains a mystery.

The yeast maturases are expressed as fusion proteins with the upstream exon and are proteolytically processed to remove the upstream peptide (Fig. 2a; Anziano et al. 1982; Weiss-Brummer et al. 1982; Lazowska et al. 1989), possibly with the aid of the mt ATP-dependent PIM1 protease (van Dyck et al. 1998). This processing may be obligatory as various N-terminal truncated forms of a bI4 fusion maturase expressed in the cytoplasm and imported into the mitochondria differ in their ability to complement mt bI4 maturase mutants. In particular, a protein with only 45 residues upstream of the mapped proteolytic cleavage site is not processed at the site and functions poorly as a maturase. In contrast, a maturase that mimics a processed protein with the N-terminal 45 residues deleted has efficient maturase activity (Banroques et al. 1986, 1987). Presumably, proteolytic processing is required to remove inhibitory, non-maturase protein sequence upstream of the intron-encoded residues.

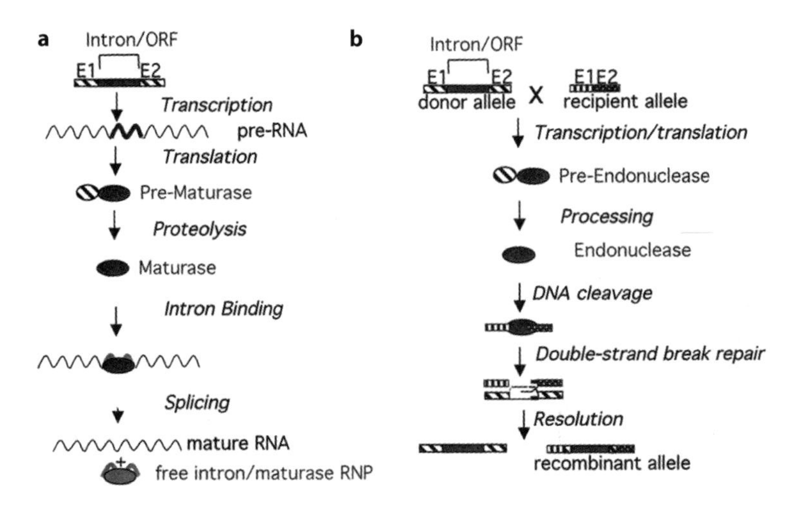

Fig. 2. Group I intron maturase-dependent splicing and mobility. **a** Schematic representation of the splicing pathway. A maturase is translated as a fusion protein with the upstream exon and is proteolytically processed to form the active maturase that binds specifically to the intron and promotes splicing. **b** Schematic representation of the homing pathway. In some cases where the endonuclease is expressed as a fusion protein, it is processed to remove upstream exon-derived residues (see Guo et al. 1995). The intron-containing (donor) allele encodes an endonuclease that recognizes a sequence in the region of the exon junction in the recipient allele and makes a double-strand break. Double-strand break repair results in the donor intron being copied into the recipient allele

Genetic studies with the bI4 intron showed that expression of the maturase is not sufficient for splicing. Mutations that disrupt either the P7 or P3 helix in the catalytic core of the bI4 intron abolish maturase-dependent splicing (de la Salle et al. 1982). Although it is not known if the splicing defect arises from impaired maturase binding or misfolding of the intron in the RNA/protein complex, these data suggest that maturase-dependent splicing requires the integrity of the intron's catalytic core.

Biochemical studies of I-AniI using recombinant protein purified from *E. coli* have increased our understanding of maturase/intron RNA complex assembly and activation of intron splicing. The protein binds tightly (K_d ~50 pM) and specifically to the COB intron-containing pre-mRNA in a 1:1 stoichiometry and is required for splicing at physiological concentrations of Mg^{2+} (Ho and Waring 1999; Solem et al. 2002). Kinetic analyses of binding and splicing are consistent with I-AniI first binding in a weak intermediate complex with the intron RNA that is subsequently resolved into a splicing-competent complex (Solem et al. 2002). Deletion analysis has shown that I-AniI requires most of the *A.n.* COB catalytic core and peripheral domains for binding, and yet ribonuclease T1 probing has shown that intron tertiary structure is virtually absent without bound protein (Ho and Waring 1999; Geese and Waring 2001; Solem et al. 2002). Taken together, these observations support a model in which I-AniI pre-associates with an unfolded COB intron via a "labile" interaction that facilitates correct folding of the intron's catalytic core. Specific binding of I-AniI to its intron interaction site "locks-in" the catalytic conformation of the intron (Solem et al. 2002).

This mode of interaction differs from other group I intron splicing cofactors that have been described previously. Two nuclear-encoded proteins have been extensively studied in vitro; the *S. cerevisiae* CBP2 protein that specifically promotes splicing of bI5 in the cob gene and the *Neurospora* mt tyrosyl tRNA synthetase (CYT-18 protein) that promotes splicing of a variety of group I introns (for a review, see Lambowitz et al. 1999). CBP2 promotes bI5 splicing by a tertiary capture mechanism. Here, the intron catalytic core is unstable but transiently folds, forming part of CBP2's interaction site whereupon the protein binds rapidly to "capture" the catalytically active intron (Weeks and Cech 1995, 1996). CYT-18-assisted splicing uses a scaffolding-like mechanism. Here, the secondary structure of the P4-P6 domain is largely formed but the rest of the intron structure is not stable. CYT-18 binds to the P4-P6 region and this RNA/protein complex provides a scaffold for assembly of the P3-P9 domain that contains additional protein contact sites (Saldanha et al. 1996; Caprara et al. 1996a,b). Thus, although all three proteins "stabilize" the catalytically active intron structure, the pathways for protein-facilitated RNA folding are quite distinct.

5 Maturases Cooperate with Other Proteins for Splicing

Maturases are not necessarily sufficient for promoting group I intron splicing, as additional proteins are also required. The *S. cerevisiae* bI3 intron also requires the nuclear-encoded MRS1 protein (Kreike et al. 1986; Bousquet et al. 1990). MRS1 is homologous to the mt CceI protein that resolves four-way DNA junctions. In vitro kinetic analyses reveal that the bI3 maturase binds the intron as a monomer while two MRS1 dimers bind cooperatively to separate sites within the intron. Binding of the maturase to its site likely occurs first and this accelerates the binding of the MRS1 dimers. Only after MRS1 binding does the intron catalyze the splicing reaction (Bassi et al. 2002; Bassi and Weeks 2003).

The bI4 and aI4α introns require the bI4 maturase as well as the nuclear-encoded leucyl tRNA synthetase, Leu RS. In mutant strains with a disrupted bI4 maturase, a single point mutation in Leu RS activates the latent maturase activity of aI4α protein (see above). Interestingly, the location of the mutation resides in the CP1 domain of the Leu RS which functions in the editing of misactivated amino acids and binding of the tRNA acceptor helix (Li et al. 1996; Rho et al. 2002). Remarkably, the expression of the CP1 peptide is sufficient for rescuing splicing-inactivated Leu RS protein suggesting that this domain is directly involved in splicing and may recognize an intron structure analogous to the tRNA acceptor helix (Rho et al. 2002). A combination of yeast two- and three-hybrid analyses suggest that the bI4 maturase and Leu RS do not interact and bind to independent sites on the bI4 intron (Rho and Martinis 2000; Rho et al. 2002). In light of these observations, it is still unknown how the *NAM2*-1 mutation activates the aI4α maturase. Nevertheless, it is apparent that, even in these more complex systems, maturases function by inducing conformational changes leading to the binding of other factors and/or RNA catalysis.

6 Maturases or DNA Endonucleases?

All group I intron maturases contain two copies of a conserved amino acid motif designated LAGLIDADG. Over 200 other proteins also contain one or two copies of this motif and most of these proteins are site-specific DNA endonucleases that function in genetic mobility (a process called "homing": Dujon 1989; Chevalier and Stoddard 2001). Genes coding for these proteins are found in self-splicing polypeptide inteins, and group I and II introns as well as free standing ORFs (Chevalier and Stoddard 2001; Toor and Zimmerly 2002). Group I introns that code for LAGLIDADG endonucleases are mobile; they spread through populations by a gene conversion process that copies them to cognate, intron-less alleles. Insertion of a group I intron/endonuclease gene into a specific site is initiated by a double-stranded cut in the recipient allele catalyzed by the homing endonuclease. Cleavage stimulates the double-strand break repair response that uses the donor allele as a template for DNA synthesis and ultimately results in a copy of the intron inserted into the recipient allele. Members of the LAGLIDADG endonuclease family recognize long target sites (≥18 bp) and make base-specific contacts in the major groove (Chevalier et al., this Vol.).

Many observations suggest that maturase function arose from homing endonucleases that had invaded pre-existing group I introns. First, most homing endonucleases are not required for intron splicing. Second, four of the eight confirmed RNA maturases are also homing endonucleases (Table 1). In addition, the *S. cerevisiae* bI2 maturase can gain homing endonuclease activity by mutation of just two non-adjacent amino acid residues (Szczepanek and Lazowska 1996). Third, the finding that identical or closely related introns contain different homing endonuclease genes supports the hypothesis that the proteins are ancestral elements that colonized group I introns. Given these observations, it seems likely that RNA maturase activity is a secondary adaptation of LAGLIDADG homing endonucleases (for further discussion, see Lambowitz and Belfort 1993).

7 How Intertwined Are the DNA and RNA Activities in Maturases?

The fact that bifunctional LAGLIDADG DNA endonucleases/RNA maturases exist raises the question of whether or not such proteins use a single binding site to interact with their nucleic acid ligands. Indeed, the similarity between respective DNA-binding sites and the RNA structure around the 5′ and 3′ SS (P1/P10 pseudoknot; Fig. 2b) has fueled speculation that such proteins may use analogous interactions to perform both functions (reviewed in Lambowitz et al. 1999). However, in vitro experiments with the *A.n.* COB I-AniI maturase showed that disruption of the COB pre-mRNA's P1/P10 by deletion of P1 or the 3′ exon has only modest effects on I-AniI binding (Solem et al. 2002). These observations suggest that protein interactions in and around this region, if they exist, contribute little to binding affinity and that maturases recognize their introns in a more complex manner than had originally been speculated.

Structural, genetic and biochemical evidence have been brought to bear on the question of how bifunctional endonuclease/maturases interact with their ligands. A cocrystal of I-AniI endonuclease/maturase complexed with its DNA substrate was recently solved (Bolduc et al. 2003; Fig. 1c). Overall, the structure of I-AniI is similar to those of other dedicated LAGLIDADG endonucleases that have also been crystallized (Chevalier and Stoddard 2001). The two LAGLIDADG motifs have α-helical conformation and pack against one another to form a central axis. Perpendicular to each helix is a curved, four-stranded, antiparallel β-sheet. The two β-sheets, one each from the N- and C-terminal domains, form an extended groove that binds to the DNA substrate making multiple specific base and non-specific phosphate contacts along 19 nucleotides (Fig. 1c). Remarkably, I-AniI shows no alterations or additions to the basic homing endonuclease framework despite its "extra" function.

Several lines of evidence suggest that maturases bind their cognate intron RNA using a distinct surface from their DNA-binding site. First, mutations in yeast bifunctional maturases have been recovered that differentially affect maturase and endonuclease functions. The homing endonuclease I-SceII, encoded by aI4α intron, functions as a latent maturase but still retains its homing activity (see above). Mutations in the third or ninth position of the first LAGLIDADG motif abolish endonuclease activity, while mutations in the same positions of the second motif only affect maturase activity (Henke et al. 1995). In addition, mutations in the acidic residue at the eighth position of the first and second LAGLIDADG motifs specifically affect DNA endonuclease activity in the endonuclease/maturases I-ScaI and I-AniI (Szczepanek et al. 2000; Chatterjee et al. 2003). Second, if maturases use the endonuclease extended groove to bind intron RNA, competition experiments between DNA substrate and intron RNA would show that binding of one substrate negates the binding of the other. Indeed, in the case of I-AniI, the *A.n.* COB intron RNA is an effective competitor for DNA cleavage and binding by I-AniI (Chatterjee et al. 2003; Geese et al. 2003). Surprisingly, however, the presence of the DNA substrate showed no observable effect on splicing of the pre-mRNA (Chatterjee et al. 2003; Geese et al. 2003). In addition, the presence of the DNA target site did not affect the binding of the full-length COB intron RNA or a mutant RNA that binds the protein ~30-fold less tightly than the competitor DNA (Chatterjee et al. 2003). Taken together, these observations suggest that maturases have functionally distinct RNA-binding sites. Presumably, I-AniI binding to the COB intron causes a conformational change in the protein that negatively affects DNA substrate binding (Chatterjee et al. 2003).

Clues to the location of the intron RNA-binding site have come from mutational analysis of the I-AniI maturase. Alignment of I-AniI with two of its closest maturase relatives (encoded by the *S. cerevisiae* bI3 and *V. inaequalis* bI5 introns) revealed conserved basic residues that fall on one side of I-AniI not in contact with the DNA (Bolduc et al. 2003). A single Arg to Glu substitution on this surface of the C-terminus of I-AniI reduced intron binding and splicing ~10-fold while having no effects on DNA substrate hydrolysis or affinity (Bolduc et al. 2003). Other maturase-specific mutations in the C-terminus have also been identified (Kwon and Waring, in prep.). Significantly, a C-terminal fragment of I-AniI that begins at the second LAGLIDADG motif does not cleave substrate DNA, as expected, but does facilitate splicing of COB intron almost as efficiently as the full-length protein. The corresponding N-terminal fragment, beginning at the first and ending at the second motif, has no splicing activity. Other experiments show that the C-terminal protein interacts with the COB intron in a manner directly analogous to the full-length protein (Downing et al. 2005). These observations are consistent with "domain" swapping experiments with the S. cerevisiae bI4 maturase and wild-type aI4α endonuclease that share extensive amino acid conservation (Delahodde et al. 1989). The analysis of hybrid proteins in vivo showed that while an N-terminal aI4α/C-terminal bI4 protein has robust RNA maturase activity, an N-terminal bI4/C-terminal aI4α protein has only weak maturase activity (Goguel et al. 1992). Taken together, these data show that splicing function is derived, for the most part, from the C-terminus of maturases on a surface opposite the DNA-binding site. These observations support a model in which homing endonucleases have developed maturase function by utilizing a previously "non-functional" protein surface.

8 Evolution of Maturase Activity

What might be the driving force behind a mobile DNA endonuclease evolving a splicing activity? Colonization of self-splicing group I introns by homing endonucleases provides a safe haven for these elements, as intron splicing removes the otherwise disruptive ORF from RNA transcripts. In addition, as the endonuclease and intron coevolved into a "hybrid" mobile element, the intron was equipped with a mechanism to propagate itself in a population. Phylogenetic studies suggest that homing intron/ORFs follow a "life cycle" that includes three steps: (1) invasion of an intron-less allele, (2) degeneration of the endonuclease ORF, and (3) intron loss (Fig. 3; Cho et al. 1998; Goddard and Burt 1999). Intron/ORFs are normally maintained in a population only through promiscuous invasion. However, these elements could be fixed if they offer a selective advantage to the host. In one scenario, splicing regulation could be a means by which the host controls the invaded gene's expression (Fig. 3). As detailed above, the splicing of many group I introns is regulated by nuclear-encoded cofactors, suggesting that intron integration has presented its hosts with an opportunity to regulate mt gene expression. While establishing a vital role in the host's metabolism would preclude intron loss it would not ensure retention of the freeloading ORF, which is naturally subject

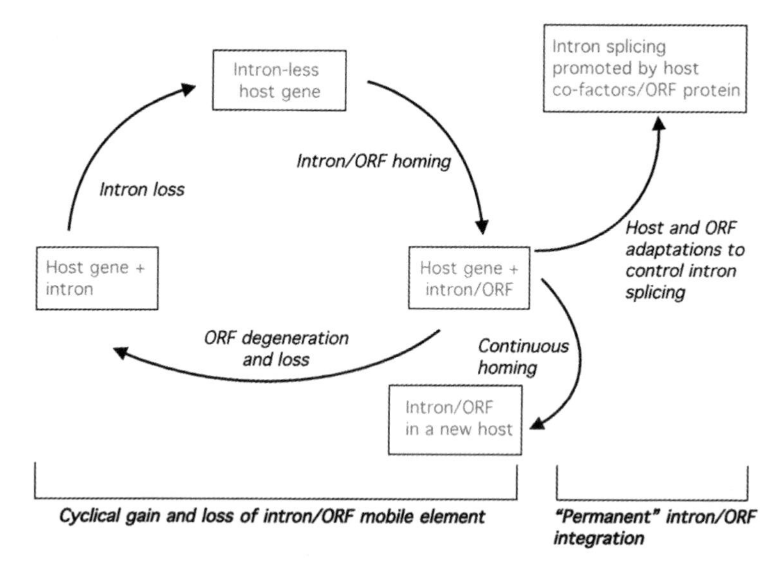

Fig. 3. Model for intron homing endonuclease ORF gain, loss and escape from the cycle by "permanent" fixation. The presence of the intron can confer an advantage if the host develops a means for controlling splicing. The endonuclease can maintain its presence if it develops an indispensable splicing cofactor function

to degeneration within the intron (Fig. 3). However, by establishing an essential role in intron splicing along with host cofactors, the intron-encoded protein may have assured its residency in the host. Therefore, selection for maturase function may be a means by which the selfish ORF protects itself from natural attrition. By gaining a functionally distinct site for RNA maturation, it can retain its native homing function, thus providing two powerful avenues to be sustained in a population.

9 Concluding Remarks

The adaptation of homing endonucleases to function in group I intron splicing is reminiscent of other proteins that bind both DNA and RNA, such as the homeodomain containing Bicoid protein, T4 DNA polymerase, and p53 tumor suppressor protein (Jeffery 1999; Cassiday and Maher 2002). Many of these so-called moonlighting proteins include factors involved in DNA replication or transcription that also bind mRNAs to repress translation. The determinants for both RNA and DNA binding are not known in almost all cases; however, there is evidence that ligand-binding sites overlap in transcription factor IIIA from *Xenopus* (reviewed in Cassiday and Maher 2002). Although the hypothesis that the DNA-binding sites of many of these moonlighting proteins are involved in their RNA function is very popular, our experience with homing endonuclease/maturases suggest that we can expect some surprises in this regard. Notably, it will be of great interest to see how widespread the phenomenon is of utilizing a "benign" but susceptible protein surface to develop a second function.

What remains to be learned from group I introns and their maturases? Although maturase and endonuclease function appear separate, the challenge remains to understand how members of this protein class bind with such high specificity to their cognate introns. Furthermore, the RNA-folding pathway(s) mediated by binding that leads to intron catalysis is unknown. To complicate matters, it is not clear if all maturases use the same mechanism of splicing facilitation. In particular, introns encoding maturases belong to separate structural subgroups and the maturases themselves show great divergence (Michel and Westhof 1990; Dalgaard et al. 1997). Furthermore, identical mutations in individual maturases can have opposite effects (e.g., Henke et al. 1995; Szczepanek et al. 2000). It will be essential to answer questions of maturase function by comparing and contrasting the activities of individual maturases and to identify common "maturase-motifs" if they exist. The identification of new group I intron maturases via genetic or biochemical means will be useful as the presence of the LAGLIDADG motif is not itself predictive of its RNA splicing or endonuclease function. Furthermore, it is not currently known if members of the other three major classes of intron-encoded endonucleases have adapted to facilitate splicing (Lambowitz et al. 1999; Chevalier and Stoddard 2001).

Finally, homing endonucleases show great, natural flexibility by changing their target specificity as a means to ensure their survival. This property is being exploited with regard to engineering enzymes with novel sequence specificities to be used in gene therapy applications (Chevalier and Stoddard 2001; Gimble, this Vol.). It will be of interest to see if the same adaptability holds true for maturase RNA recognition strategies. If so, it may be possible to use maturases as scaffolds to engineer new sequence- or structure-specific RNA-binding proteins for therapeutic or gene expression studies (e.g. Campisi et al. 2001).

Acknowledgements. We thank Kristina Brady for help in manuscript preparation. Research in the authors' laboratories is supported by National Institutes of Health grant GM-62853 to M.G.C. and National Science Foundation grant MCB-013099 to R.B.W.

References

Adams PL, Stahley MR, Kosek AB, Wang J, Strobel SA (2004) Crystal structure of a self-splicing group I intron with both exons. Nature 430:45–50

AnzianoPQ, Hanson DK, Mahler HR, Perlman PS (1982) Functional domains in introns: *trans*-acting and *cis*-acting regions of intron 4 of the cob gene. Cell 30:925–932

Banroques J, Delahodde A, Jacq C (1986) A mitochondrial RNA maturase gene transferred to the yeast nucleus can control mitochondrial mRNA splicing. Cell 46:837–844

Banroques J, Perea J, Jacq C (1987) Efficient splicing of two yeast mitochondrial introns controlled by a nuclear-encoded maturase. EMBO J 6:1085–1091

Bassi GS, Weeks KM (2003) Kinetic and thermodynamic framework for assembly of the six-component bI3 group I intron ribonucleoprotein catalyst. Biochemistry 42:9980–9988

Bassi GS, de Oliveira DM, White MF, Weeks KM (2002) Recruitment of intron-encoded and co-opted proteins in splicing of the bI3 group I intron RNA. Proc Natl Acad Sci USA 99:128–133

Bousquet I, Dujardin G, Poyton RO, Slonimski PP (1990) Two group I mitochondrial introns in the cob-box and coxI genes require the same MRS1/PET157 nuclear gene product for splicing. Curr Genet 18:117–124

Bolduc JM, Spiegel PC, Chatterjee P, Brady KL, Downing ME, Caprara MG, Waring RB, Stoddard BL (2003) Structural and biochemical analyses of DNA and RNA binding by a bifunctional homing endonuclease and group I intron splicing factor. Genes Dev 17:2875–2888

Campisi DM, Calabro V, Frankel AD (2001) Structure-based design of a dimeric RNA-peptide complex. EMBO J 20:178–186

Cassiday LA, Maher LJ III (2002) Having it both ways: transcription factors that bind DNA and RNA. Nucleic Acids Res 30:4118–4126

Caprara MG, Mohr G, Lambowitz AM (1996a) A tyrosyl-tRNA synthetase protein induces tertiary folding of the group I intron catalytic core. J Mol Biol 257:512–531

Caprara MG, Lehnert V, Lambowitz AM, Westhof E (1996b) A tyrosyl-tRNA synthetase recognizes a conserved tRNA-like structural motif in the group I intron catalytic core. Cell 87:1135–1145

Cech TR (1990) Self-splicing of group I introns. Annu Rev Biochem 59:543–568

Chatterjee P, Brady KL, Solem A, Ho Y, Caprara MG (2003) Functionally distinct nucleic acid binding sites for a group I intron encoded RNA maturase/DNA homing endonuclease. J Mol Biol 329:239–251

Chevalier BS, Stoddard BL (2001) Homing endonucleases: structural and functional insight into the catalysts of intron/intein mobility. Nucleic Acids Res 29:3757–3774

Cho Y, Qiu YL, Kuhlman P, Palmer JD (1998) Explosive invasion of plant mitochondria by a group I intron. Proc Natl Acad Sci USA 95:14244–14249

Dalgaard JZ, Klar AJ, Moser MJ, Holley WR, Chatterjee A, Mian IS (1997) Statistical modeling and analysis of the LAGLIDADG family of site-specific endonucleases and identification of an intein that encodes a site-specific endonuclease of the HNH family. Nucleic Acids Res 25:4626–4638

Delahodde A, Goguel V, Becam AM, Creusot F, Perea J, Banroques J, Jacq C (1989) Site-specific DNA endonuclease and RNA maturase activities of two homologous intron-encoded proteins from yeast mitochondria. Cell 56:431–441

De la Salle H, Jacq C, Slonimski PP (1982) Critical sequences within mitochondrial introns: pleiotropic mRNA maturase and cis-dominant signals of the box intron controlling reductase and oxidase. Cell 28:721–732

Downing, M.E, Brady, K.L and Caprara M.G (2005). A C-terminal fragment of an intron-encoded maturase is sufficient for promoting group I intron splicing. RNA, 11:437-446.

Dujardin G, Jacq C, Slonimski PP (1982) Single base substitution in an intron of oxidase gene compensates splicing defects of the cytochrome b gene. Nature 298:628–632

Dujon B (1989) Group I introns as mobile genetic elements: facts and mechanistic speculations – a review. Gene 82:91–114

Engelhardt MA, Doherty EA, Knitt DS, Doudna JA, Herschlag D (2000) The P5abc peripheral element facilitates preorganization of the *Tetrahymena* group I ribozyme for catalysis. Biochemistry 39:2639–2651

Geese WJ, Waring RB (2001) A comprehensive characterization of a group IB intron and its encoded maturase reveals that protein-assisted splicing requires an almost intact intron RNA. J Mol Biol 308:609–622

Geese WJ, Kwon YK, Wen X, Waring RB (2003) In vitro analysis of the relationship between endonuclease and maturase activities in the bi-functional group I intron-encoded protein, I-AniI. Eur J Biochem 270:1543–1554

Goddard MR, Burt A (1999) Recurrent invasion and extinction of a selfish gene. Proc Natl Acad Sci USA 96:13880–13885

Goguel V, Bailone A, Devoret R, Jacq C (1989) The bI4 RNA mitochondrial maturase of *Saccharomyces cerevisiae* can stimulate intra-chromosomal recombination in *Escherichia coli*. Mol Gen Genet 216:70–74

Goguel V, Delahodde A Jacq C (1992) Connections between RNA splicing and DNA intron mobility in yeast mitochondria: RNA maturase and DNA endonuclease switching experiments. Mol Cell Biol 12:696–705

Golden BL, Gooding AR, Podell ER, Cech TR (1998) A preorganized active site in the crystal structure of the *Tetrahymena* ribozyme. Science 282:259–264

Groudinsky O, Dujardin G, Slonimski PP (1981) Long range control circuits within mitochondria and between nucleus and mitochondria. II. Genetic and biochemical analyses of suppressors which selectively alleviate the mitochondrial intron mutations. Mol Gen Genet 184:493–503

Guo WW, Moran JV, Hoffman PW, Henke RM, Butow RA, Perlman PS (1995) The mobile group I intron 3α of the yeast mitochondrial *COXI* gene encoded a 35-kDa processed protein that is an endonuclease but not a maturase. J Biol Chem 270:15563–15570

Henke RM, Butow RA, Perlman PS (1995) Maturase and endonuclease functions depend on separate conserved domains of the bifunctional protein encoded by the group I intron aI4α of yeast mitochondrial DNA. EMBO J 14:5094–5099

Herbert CJ, Labouesse M, Dujardin G, Slonimski PP (1988) The NAM2 proteins from *S. cerevisiae* and *S. douglasii* are mitochondrial leucyl-tRNA synthetases, and are involved in mRNA splicing. EMBO J 7:473–483

Ho Y, Waring RB (1999) The maturase encoded by a group I intron from *Aspergillus nidulans* stabilizes RNA tertiary structure and promotes rapid splicing. J Mol Biol 292:987–1001

Ho Y, Kim SJ, Waring RB (1997) A protein encoded by a group I intron in *Aspergillus nidulans* directly assists RNA splicing and is a DNA endonuclease. Proc Natl Acad Sci USA 94:8994–8999

Jeffery CJ (1999) Moonlighting proteins. Trends Biochem Sci 24:8–11

Kreike J, Schulze M, Pillar T, Korte A, Rodel G (1986) Cloning of a nuclear gene MRS1 involved in the excision of a single group I intron (bI3) from the mitochondrial *COB* transcript in *S. cerevisiae*. Curr Genet 11:185–191

Labouesse M, Herbert CJ, Dujardin G, Slonimski PP (1987) Three suppressor mutations which cure a mitochondrial RNA maturase deficiency occur at the same codon in the open reading frame of the nuclear *NAM2* gene. EMBO J 6:713–721
Lambowitz AM, Belfort M (1993) Introns as mobile genetic elements. Annul Rev Biochem 62:587–622
Lambowitz AM, Caprara MG, Zimmerly, S, Perlman PS (1999) Group I and group II ribozymes as RNPs: clues to the past and guides to the future. In: Gesteland RF, Atkins JF, Cech TR (eds) The RNA world II. Cold Spring Harbor Laboratory Press, New York, pp 451–485
Lazowska J, Claude J, Slonimski PP (1980) Sequence of introns and flanking exons in wild-type and box3 mutants of cytochrome b reveals an interlaced splicing protein coded by an intron. Cell 22:333–348
Lazowska J, Claisse M, Gargouri A, Kotylak Z, Spyridakis A, Slonimski PP (1989) Protein encoded by the third intron of cytochrome b gene in *Saccharomyces cerevisiae* is an mRNA maturase. Analysis of mitochondrial mutants, RNA transcripts proteins and evolutionary relationships. J Mol Biol 205:275–289
Lazowska J, Szczepanek T, Macadre C, Dokova M (1992) Two homologous mitochondrial introns from closely related *Saccharomyces* species differ by only a few amino acid replacements in their open reading frames: one is mobile, the other is not. CR Acad Sci III Sci Vie 315:37–41
Lehnert V, Jaeger L, Michel F, Westhof E (1996) New loop-loop tertiary interactions in self-splicing introns of subgroup IC and ID: a complete 3D model of the *Tetrahymena thermophila* ribozyme. Chem Biol 12:993–1009
Li G-Y, Becam A-M, Slonimski PP, Herbert CJ (1996) In vitro mutagenesis of the mitochondrial leucyl tRNA synthetase of *Saccharomyces cerevisiae* shows that the suppressor activity of the mutant proteins is related to the splicing function of the wild-type protein. Mol Gen Genet 252:667–675
Merlos-Lange AM, Kanbay F, Zimmer M, Wolf K (1987). DNA splicing of mitochondrial group I and II introns in *Schizosacchromyces pombe*. Mol Gen Genet 206:273–278
Michel F, Westhof E (1990) Modelling of the three-dimensional architecture of group I catalytic introns based on comparative sequence analysis. J Mol Biol 216:585–610
Pan J, Woodson SA (1999) The effect of long-range loop-loop interactions on folding of the *Tetrahymena* self-splicing RNA. J Mol Biol 294:955–965
Pellenz S, Harington A, Dujon B, Wolf K, Schaefer B (2002) Characterization of the I-SpomI endonuclease from fission yeast: insights into the evolution of a group I intron-encoded homing endonuclease. J Mol Evol 55:302–313
Rho SB, Martinis SA (2000) The bI4 group I intron binds directly to both its protein splicing partners, a tRNA synthetase and maturase, to facilitate RNA splicing activity. RNA 6:1882–1894
Rho SB, Lincecum TL Jr, Martinis SA (2002) An inserted region of leucyl-tRNA synthetase plays a critical role in group I intron splicing. EMBO J 21:6874–6881
Saldanha R, Ellington A, Lambowitz AM (1996) Analysis of the CYT-18 protein binding site at the junction of stacked helices in a group I intron RNA by quantitative binding assays and in vitro selection. J Mol Biol 261:23–42
Schafer B, Wilde B, Massardo DR, Manna F, Giudice LD, Wolf K (1994) A mitochondrial group I intron in fission yeast encodes a maturase and is mobile in crosses. Curr Genet 25:336–341
Schroeder R, Grossberger R, Pichler A, Waldsich C (2002) RNA folding in vivo. Curr Opin Struct Biol 12:296–300
Solem A, Chatterjee P, Caprara MG (2002) A novel mechanism for protein-assisted group I intron splicing. RNA 8:412–425

SzczepanekT, Lazowska J (1996) Replacement of two non-adjacent amino acids in the *S. cerevisiae* bi2 intron-encoded RNA maturase is sufficient to gain a homing-endonuclease activity. EMBO J 15:3758–3767

Szczepanek T, Jamoussi K, Lazowska J (2000). Critical base substitutions that affect the splicing and/or homing activities of the group I intron bi2 of yeast mitochondria. Mol Gen Genet 264:137–144

Toor N, Zimmerly S (2002) Identification of a family of group II introns encoding LAGLI-DADG ORFs typical of group I introns. RNA 8:1373–1377

Van Dyck L, Neupert W, Langer T (1998) The ATP-dependent PIM1 protease is required for the expression of intron-containing genes in mitochondria. Genes Dev 12:1515–1524

Weeks KM, Cech TR (1995) Protein facilitation of group I intron splicing by assembly of the catalytic core and the 5′ splice site domain. Cell 82:221–230

Weeks KM, Cech TR (1996) Assembly of a ribonucleoprotein catalyst by tertiary structure capture. Science 271:345–348

Weiss-Brummer B, Rodel G, Schweyen RJ, Kaudewitz F (1982). Expression of the split gene cob in yeast: evidence for a precursor of a „maturase“ protein translated from intron 4 and preceding exons. Cell 29:527–536

Wenzlau JM, Saldanha RJ, Butow RA, Perlman PS (1989) A latent intron-encoded maturase is also an endonuclease needed for intron mobility. Cell 56:421–430

Woodson SA (2000) Recent insights on RNA folding mechanisms from catalytic RNA. Cell Mol Life Sci 57:796–808

Group II Intron Homing Endonucleases: Ribonucleoprotein Complexes with Programmable Target Specificity

Alan M. Lambowitz, Georg Mohr, Steven Zimmerly

1 Introduction

Group II intron homing endonucleases are ribonucleoproteins (RNPs) consisting of a catalytically active intron RNA and an intron-encoded protein (IEP), with reverse transcriptase (RT) and/or DNA endonuclease (En) activity. The RNP is formed when the IEP binds to the intron in unspliced RNA and promotes its splicing by stabilizing the catalytically active RNA structure. Afterwards, the IEP remains tightly bound to the excised intron RNA to constitute the homing endonuclease. The homing endonuclease promotes intron mobility by a remarkable mechanism in which the intron RNA reverse splices directly into a target DNA and is then reverse-transcribed by the IEP. Importantly, the target site for intron insertion is determined mainly by base pairing between short sequence elements in the intron RNA and target DNA, making it straightforward to change the target specificity of the homing endonuclease simply by modifying the intron RNA. This feature combined with their very high specificity and insertion frequencies have made it possible to develop mobile group II introns into gene targeting vectors, called "targetrons", with programmable target specificity.

Here, we focus on the structure, function, and interaction of the IEP and intron RNA in the homing endonuclease. We discuss in detail how group II

A.M. Lambowitz (e-mail: lambowitz@mail.utexas.edu), G. Mohr
Institute for Cellular and Molecular Biology, Department of Chemistry and Biochemistry, and Section of Molecular Genetics and Microbiology, School of Biological Sciences, University of Texas at Austin, Austin, Texas 78712 USA

S. Zimmerly
University of Calgary, Department of Biological Sciences, 2500 University Drive NW, Calgary, AB, T2N 1N4, Canada

Nucleic Acids and Molecular Biology, Vol. 16
Marlene Belfort et al. (Eds.)
Homing Endonucleases and Inteins

intron homing endonucleases recognize DNA target sites and how they can be used for targeted gene disruption and site-specific DNA insertion in bacteria, or potentially, in eukaryotes.

2 Group II Intron Structure

Group II introns have a widespread phylogenetic distribution, being found in bacteria and in mitochondrial (mt) and chloroplast (cp) genomes of lower eukaryotes and higher plants. They are relatively common in Gram-negative and Gram-positive bacteria, but rare in archaea. Mobile group II introns that form homing endonucleases range in size from 2–4 kb, and consist of a core catalytic RNA (ribozyme) of 500–1000 nt and an open-reading frame (ORF) of 1.5–2 kb. Group II intron RNAs have a conserved secondary structure consisting of six domains (DI–DVI), with three major subclasses (IIA, IIB, and IIC) distinguished by specific features (Michel et al. 1989; Michel and Ferat 1995; Toro 2003). This structure is illustrated in Fig. 1a for the *Lactococcus lactis* Ll.LtrB intron, an important model system, which belongs to sublcass IIA. DI is the largest domain, and DV is the most highly conserved in sequence. DVI contains the branch-point nucleotide residue, usually a bulged A (Michel et al. 1989). DIV encodes the IEP in a large "loop" and also contains a high-affinity binding site for the IEP in subdomain DIVa (Wank et al. 1999).

In most mobile group II introns, the IEP is a multifunctional protein with RT activity (described in detail below). A few fungal mtDNA group II introns differ in encoding unrelated LAGLIDADG-type DNA homing endonucleases (Toor and Zimmerly 2002), and many mt and cp group II introns do not contain ORFs and are not independently mobile. In Ll.LtrB and other mobile bacterial group II introns, the IEP has its own Shine-Dalgarno (SD) sequence and initiation codon, typically located in or near intron subdomain DIVa (Fig. 1a; Shearman et al. 1996; Singh et al. 2002). By contrast, in yeast mtDNA *coxI*-I1 and -I2 and in many other mt group II introns, the ORF extends upstream from DIV and is translated in frame with the upstream exon, yielding a fusion protein that is processed proteolytically to generate the active IEP (see Fig. 2; Michel and Ferat 1995).

3 Catalytic Properties of the Intron RNA

Group II intron RNAs are ribozymes capable of self-splicing in vitro via two sequential transesterification reactions that result in the formation of an intron lariat (Fig. 1b). Chemically, the splicing reactions are the same as those of spliceosomal introns, which are believed to be evolutionary descendants

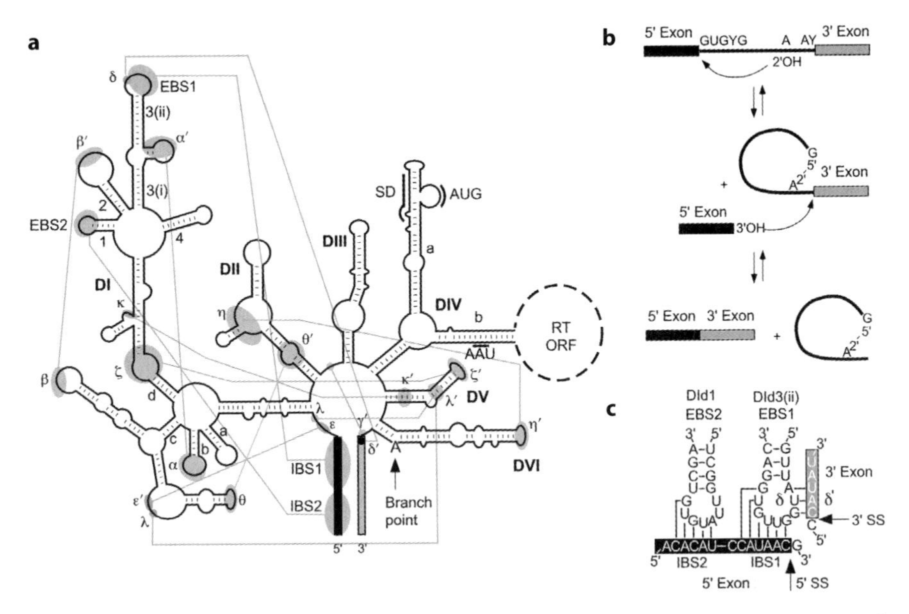

Fig. 1. Group II intron RNA secondary structure and splicing mechanism. **a** Conserved RNA secondary structure. The structure consists of six double-helical domains (*DI–DVI*) radiating from a central wheel; subdomains are indicated by *lowercase letters* (e.g., DIVa). The ORF is encoded within DIV (*dotted loop*), and DIVa is a high-affinity binding site for the IEP. The locations of the Shine-Dalgarno (SD) sequence, AUG initiation codon, and UAA termination codon are indicated. *Greek letters* and *gray shading* indicate sequences involved in tertiary interactions (*light gray lines*). *EBS* and *IBS* denote exon- and intron-binding sites, respectively. The structure shown is for the *L. lactis* Ll.LtrB intron (Mills et al. 1996; Shearman et al. 1996). **b** Splicing mechanism. In the first transesterification, nucleophilic attack at the 5′-splice site by the 2′ OH of a bulged A-residue in DVI results in cleavage of the 5′-splice site coupled to formation of lariat intermediate. In the second transesterification, nucleophilic attack at the 3′-splice site by the 3′ OH of the cleaved 5′ exon results in exon ligation and release of the intron lariat. **c** EBS/IBS and δ–δ′ interactions between the Ll.LtrB intron and flanking 5′- and 3′-exon sequences in unspliced precursor RNA. The 5′ and 3′ exons are indicated by *black* and *gray shading*, respectively. SS, splice site

of group II introns (Michel and Ferat 1995). To catalyze splicing, group II intron RNAs, like protein enzymes, fold into a specific three-dimensional structure that forms an active site (Qin and Pyle 1998). For group II introns, the active site is thought to include Mg^{2+} ions bound at specific positions in DV, in combination with parts of DI (Sigel et al. 2000; Gordon and Piccirili 2001). The active site aligns and activates specific chemical bonds, including the 2′ OH of the branch-point A in DVI and the phosphodiester bonds at the 5′- and 3′-splice sites, which must be broken and joined during splicing. The folding

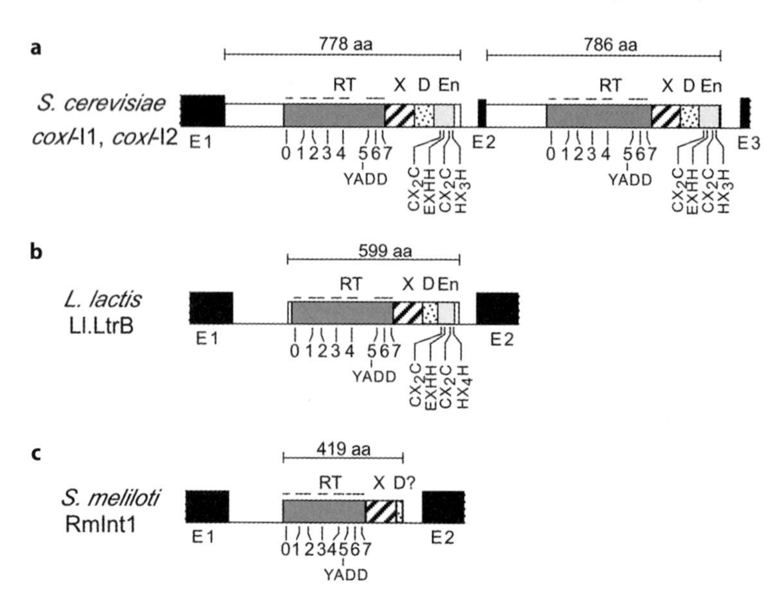

Fig. 2. Group II intron-encoded proteins. Protein coding regions are shown as *rectangles*, with *different shading*, indicating conserved regions or domains. Exons (*E*) are *black boxes*. Protein domains are: *RT*, with conserved sequence blocks RT-0 to -7 indicated below and delineated above; *X*, associated with maturase activity; *D*, DNA-binding; and *En*, DNA endonuclease. **a** *S. cerevisiae* mtDNA introns *coxI*-I1 and -I2. **b** *L. lactis* Ll.LtrB intron. **c** *S. meliloti* RmInt1 intron, which encodes a protein lacking the En domain. *D?* in RmInt1 indicates that the function of this region has not been established

of the intron RNA to form the active site involves both intra- and interdomain tertiary interactions, of which over a dozen have been identified (Fig. 1a; IBS1–3/EBS1–3, α-α' to λ–λ'). NMR and X-ray crystal structures have been determined for DV and DV+DVI, respectively (Zhang and Doudna 2002; Sigel et al. 2004), and structural models have been proposed for the active site (Costa et al. 2000; Swisher et al. 2001).

An important feature of the group II intron splicing reaction is that 5′- and 3′-exon sequences flanking the splice sites are bound at the active site by base-pairing interactions with specific sequence elements in DI (Fig. 1a,c; Michel and Ferat 1995; Qin and Pyle 1998; Costa et al. 2000). The sequence elements EBS1 and EBS2 (exon-binding sites 1 and 2), located in two different stem-loops in DI, base pair with 5′-exon sequences IBS1 and IBS2 (intron-binding sites 1 and 2), directly upstream of the 5′-splice site. The 3′ exon is recognized by another short pairing, which differs between group IIA and IIB introns. In group IIA introns, the sequence δ, adjacent to EBS1, base pairs with the first few nucleotide residues of the 3′ exon (δ'; Fig. 1c), while in group IIB and probably IIC introns, a different sequence element, EBS3, base pairs with the 3′ exon (IBS3; not shown). Because the transesterification reactions catalyzed by the intron RNA are reversible, the same base-pairing interactions between the intron and the 5′ and 3′ exons are required not only for RNA splicing, but also for reverse splicing into RNA, or into DNA target sites during intron mobility. This feature enables the facile engineering of homing endonucleases with different target specificities (see below).

Finally, although group II introns are intrinsically catalytic, most if not all require proteins to help fold the RNA into the active structure for efficient splicing (reviewed in Lambowitz et al. 1999; Lehmann and Schmidt 2003). In the case of ORF-containing group II introns, the key splicing factor is the IEP, which binds specifically to its own intron RNA and stabilizes the catalytically active structure ("maturase" activity; Carignani et al. 1983; Saldanha et al. 1999; Matsuura et al. 2001; Noah and Lambowitz 2003). The dependence of mobile group II introns on a single major splicing factor encoded within the intron enables them to function in different organisms. The *L. lactis* Ll.LtrB intron, for example, is spliced efficiently in both its natural host and a number of other Gram-positive or Gram-negative bacteria (Mills et al. 1996; Shearman et al. 1996; Matsuura et al. 1997; Belhocine et al. 2004; Staddon et al. 2004).

4 Biochemical Activities of Group II Intron-Encoded Proteins

Group II IEPs, examples of which are shown in Fig. 2, contain several domains associated with different biochemical activities. The RT domain at the N-terminus contains conserved sequence blocks RT-1 to -7, characteristic of the fingers and palm of retroviral RTs, along with an additional upstream motif, RT-0 (Xiong and Eickbush 1990; Malik et al. 1999; Zimmerly et al. 2001). The latter is found only in non-LTR-retroelement RTs and may be part of an extended fingers subdomain used for specific binding of template RNA (Chen and Lambowitz 1997; Bibillo and Eickbush 2002). Domain X, which is a site of mutations affecting maturase activity, is located in the position corresponding to the thumb and part of the connection domain of retroviral RTs and likely corresponds to the RT thumb (Mohr et al. 1993; Blocker et al. 2005). The RT and X domains together bind the intron RNA both as a substrate for RNA splicing and as a template for reverse transcription (Kennell et al. 1993; Cui et al. 2004; Blocker et al. 2005). A functional RT active site is not required for RNA splicing, and the RT and maturase activities are readily separated by mutation (Moran et al. 1995; Zimmerly et al. 1995b; Cui et al. 2004).

The C-terminal D and En domains are not required for RNA splicing, but contribute to interaction with the DNA target site during intron mobility. Domain D is required for reverse splicing of the intron RNA into double-stranded DNA and is thought to recognize specific nucleotide residues in the DNA target, leading to local DNA unwinding (Guo et al. 1997; San Filippo and Lambowitz 2002). Studies with the Ll.LtrB IEP identified two functionally important regions, which were also identified in other IEPs; one of these regions contains a cluster of basic amino acid residues and the other a predicted α-helix (San Filippo and Lambowitz 2002).

The En domain catalyzes second-strand DNA cleavage to generate the primer for target DNA-primed reverse transcription (TPRT) of the intron RNA (Zimmerly et al. 1995a,b; Yang et al. 1996). It contains conserved sequence motifs characteristic of the HNH family of DNA endonucleases, typically interspersed with two pairs of conserved cysteine residues, similar to an arrangement found in the phage T4 endonuclease VII subfamily of HNH endonucleases (Gorbaleyna 1994; Shub et al. 1994; San Filippo and Lambowitz 2002). In the Ll.LtrB IEP, the HNH active site contains a single catalytically essential Mg^{2+} ion, while the conserved cysteine motifs appear to stabilize the higher-order structure of the domain, but, unlike endonuclease VII, do not contain a coordinated Zn^{2+} ion, at least in the purified protein (San Filippo and Lambowitz 2002). In both yeast *coxI*-I1 and Ll.LtrB, deletion of the En domain or mutations in critical active-site residues result in loss of RT activity, implying a functional interaction with the RT domain (San Filippo and Lambowitz 2002). Notably, many group II introns, particularly in bacteria, encode proteins lacking the En domain, and at least some of these are mobile, using alternate mechanisms for priming reverse transcription (see below).

Different lineages of mobile group II introns can be distinguished based on their RNA structures and phylogenetic analysis of their RT sequences (Toor et al. 2001; Lambowitz and Zimmerly 2004). Importantly, within these lineages, the IEPs appear to have coevolved with their intron RNAs, with little if any exchange of IEPs between different introns (Toor et al. 2001). As a result, group II IEPs generally splice and mobilize only the intron in which they are encoded and not other group II introns (Saldanha et al. 1999). This situation differs markedly from that for mobile group I introns, which have been invaded by several different types of homing endonucleases that are indiscriminate in their ability to mobilize different introns (Lambowitz and Belfort 1993; Belfort et al. 2002).

5 Mobility Mechanisms

All characterized mobile group II introns use a mobility mechanism in which the intron RNA in the homing endonuclease uses its ribozyme activity to reverse splice into a DNA target site and is then reverse transcribed by the IEP (reviewed in Lambowitz and Zimmerly 2004). Different variations of this mechanism are used both for "retrohoming" to a specific target site, and for "retrotransposition" to ectopic sites. For retrohoming, group II introns that encode proteins with an En domain use this domain after reverse splicing to cleave the opposite DNA strand a short distance downstream from the intron-insertion site (Fig. 3a). For the Ll.LtrB and yeast mtDNA introns, this cleavage occurs at either 3′-exon position +9 (Ll.LtrB; Matsuura et al. 1997) or +10 (yeast introns; Zimmerly et al. 1995a; Yang et al. 1996). The 3′ end of the cleaved strand is then used as a primer for reverse transcription of the inserted intron RNA. In the case of the lactococcal intron, the resulting intron cDNA is integrated by a DNA repair mechanism independent of RecA function (Mills et al. 1997; Cousineau et al. 1998), while, for the yeast mtDNA introns, DNA recombination mechanisms play a greater role in cDNA integration (Eskes et al. 1997, 2000). Notably, yeast mtDNA introns with mutations that inhibit RT activity can still home by using the En domain to cleave the

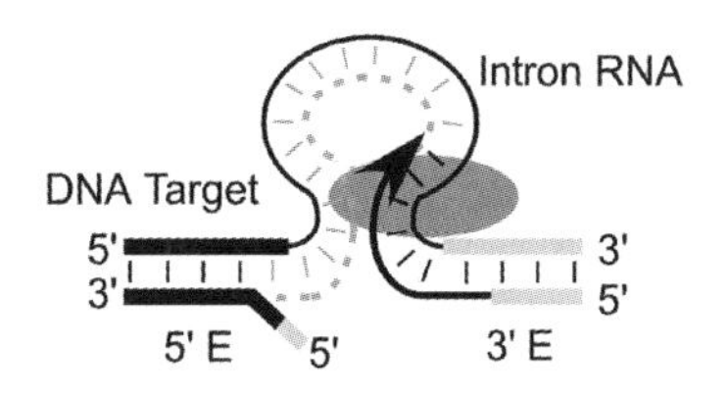

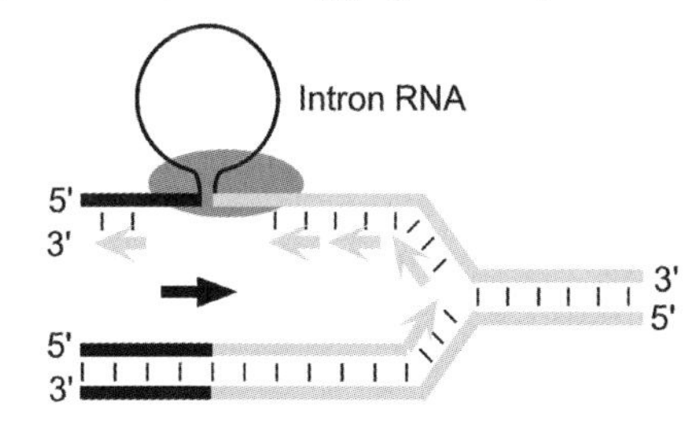

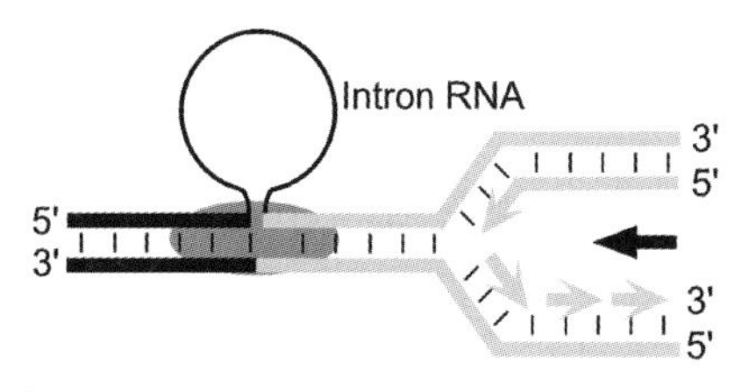

Fig. 3. Group II intron mobility mechanisms. **a** Retrohoming by group II introns that encode proteins containing a C-terminal En domain. The excised intron RNA in the homing endonuclease reverse splices into one DNA strand, while the IEP cleaves the opposite strand and uses the cleaved 3′ end as a primer for reverse transcription of the inserted intron RNA. **b, c** Retrohoming by group II introns that encode proteins lacking En activity. In **b**, the intron RNA reverse splices into transiently single-stranded DNA at a replication fork and uses a nascent lagging strand as the primer for reverse transcription. In **c**, the intron RNA reverse splices into double-stranded DNA prior to passage of a replication fork and uses a nascent leading strand as the primer. *Thick black and gray lines* indicate the 5′-exon (5′E) and 3′-exon (3′E) DNA, respectively. *Thin black lines* indicate intron RNA, and *gray dashes* indicate future cDNA. *Black arrows* indicate the direction of DNA replication, and *gray arrows* indicate newly synthesized DNA

DNA target site, thereby initiating double-strand break repair (DSBR) recombination, analogous to the mechanism of group I intron mobility (Eskes et al. 1997, 2000).

Mobile group II introns that encode proteins lacking the En domain also initiate mobility by reverse splicing into DNA sites, but are unable to carry out second-strand cleavage (Muñoz-Adelantado et al. 2003). Recent studies with the *Sinorhizobium meliloti* RmInt1 intron indicate that a major mechanism involves the use of a nascent lagging strand at a DNA replication fork as the primer for reverse transcription (Fig. 3b; Martínez-Abarca et al. 2004). A similar mechanism has been proposed for retrotransposition of the Ll.LtrB intron in *L. lactis* (Ichiyanagi et al. 2002, 2003). In both cases, the use of nascent lagging strand primers may be facilitated by reverse splicing into transiently single-stranded DNA at a replication fork. En^- mutants of the Ll.LtrB intron retrohome at decreased but still appreciable frequencies by a similar mechanism, involving reverse splicing into double-strand DNA with preferred use of nascent leading strands as primers (Fig. 3c; Zhong and Lambowitz 2003). This orientation bias is thought to reflect that, after reverse splicing into double-stranded DNA, the intron RNA is positioned to directly use a leading-strand primer, without requiring passage of the replication fork through the region containing the inserted RNP. Additionally, it has been suggested, but not demonstrated, that En^- introns might also retrohome by using non-specific opposite strand nicks to prime reverse transcription (e.g., see Schäfer et al. 2003). Group II intron mobility pathways are described in greater detail in Lambowitz and Zimmerly (2004).

6 Binding of the IEP to the Intron RNA

The properties of group II intron homing endonucleases are dictated by the interaction between the intron RNA and the IEP. This interaction starts when the IEP binds the intron in unspliced precursor RNA and promotes its splicing by stabilizing the catalytically active RNA structure. The Ll.LtrB IEP binds to the intron RNA as a dimer, and the binding of the cognate intron is very tight, with the apparent K_d measured as k_{off}/k_{on} in the pM range (Saldanha et al. 1999; Rambo and Doudna 2004). Importantly, the activities of the IEP and intron RNA are interdependent. The binding of the IEP stabilizes the active structure of the intron RNA and, conversely, the binding of the intron RNA stabilizes the active structure of the IEP, which in its free form is prone to rapid, irreversible denaturation (Matsuura et al. 1997; Saldanha et al. 1999; Zimmerly et al. 1999). In the absence of intron RNA, the IEP is not capable of binding specifically to DNA target sites and has no detectable DNA endonuclease activity (Matsuura et al. 1997).

Studies with *L. lactis* Ll.LtrB and yeast *coxI*-I2 have revealed a number of features of the RNA–protein interaction. In both introns, the IEP has a high-affinity binding site in intron subdomain DIVa, a small stem-loop structure at the beginning of DIV (see Fig. 1a), and makes weaker contacts with conserved catalytic core regions that stabilize the active RNA structure (Wank et al. 1999; Matsuura et al. 2001; Singh et al. 2002; Huang et al. 2003; Watanabe and Lambowitz 2004). The binding of the IEP to DIVa contributes to the intron specificity of maturases (Huang et al. 2003). In Ll.LtrB, DIVa contains the Shine-Dalgarno sequence and initiation codon for the IEP, and the initial binding of the IEP to DIVa downregulates translation (Singh et al. 2002). RNA footprinting experiments with Ll.LtrB identified potential secondary binding sites in DI, DII, and DVI (Matsuura et al. 2001). Additionally, DI, DIV, and DIVa separately form stable complexes with the IEP (Wank et al. 1999; Rambo and Doudna 2004). The IEP uses different regions of the RT and X domains to bind different regions of the intron RNA, with recent studies suggesting a model in which the N-terminus binds DIVa, while other parts of the RT and X domains bind catalytic core regions (Blocker et al. 2005; Cui et al. 2004).

Although DIVa is required for high-affinity binding of the IEP, significant residual splicing of both the yeast and lactococcal introns can occur in the absence of DIVa by direct binding of the IEP to the catalytic core (Wank et al. 1999; Matsuura et al. 2001). In vivo, ΔDIVa derivatives of the yeast and lactococcal introns splice at 70% and 6–10% wild-type efficiency, respectively (Huang et al. 2003; Cui et al. 2004). By contrast, the deletion of DIVa almost completely abolishes the mobility of both introns (D'Souza and Zhong 2002; Huang et al. 2003). The greater effect on mobility may reflect in part that the binding of the IEP to DIVa plays a key role in correctly positioning the RT to initiate reverse transcription near the 3′ end of the intron (Wank et al. 1999).

After splicing, the IEP presumably uses similar interactions with DIVa and catalytic core regions to remain associated with the excised intron lariat RNA in the homing endonuclease. Indeed, it is essential that the IEP continues to stabilize the active RNA structure so that the intron can reverse splice into the DNA target site. Since interactions with the intron RNA involve primarily the RT and X domains, the C-terminal D and En domains should be more or less free to engage in DNA interactions required for intron mobility. For mobility, the IEP must recognize the DNA target site using domain D and/or other regions, promote reverse splicing of the intron into one strand of the DNA, cleave the other strand using the En domain, and then reposition to place the cleaved 3′ DNA end at the RT active site to prime reverse transcription. The ability of the IEP to carry out sequential steps in RNA splicing and intron mobility may be facilitated by a mechanism in which the N-terminus of the protein remains tightly bound to DIVa, while other regions are relatively free to engage in different interactions.

7 DNA Target Site Recognition by Group II Intron Homing Endonucleases

Thus far, DNA target sites have been defined experimentally for the yeast introns *coxI*-I1 (Yang et al. 1998) and *coxI*-I2 (Guo et al. 1997), the *L. lactis* Ll.LtrB intron (Guo et al. 2000; Mohr et al. 2000; Perutka et al. 2004), and the *S. meliloti* RmInt1 intron (Jiménez-Zurdo et al. 2003; Fig. 4). All four introns use the same basic mechanism of DNA target site recognition involving both the recognition of specific nucleotide residues by the IEP and base pairing of the

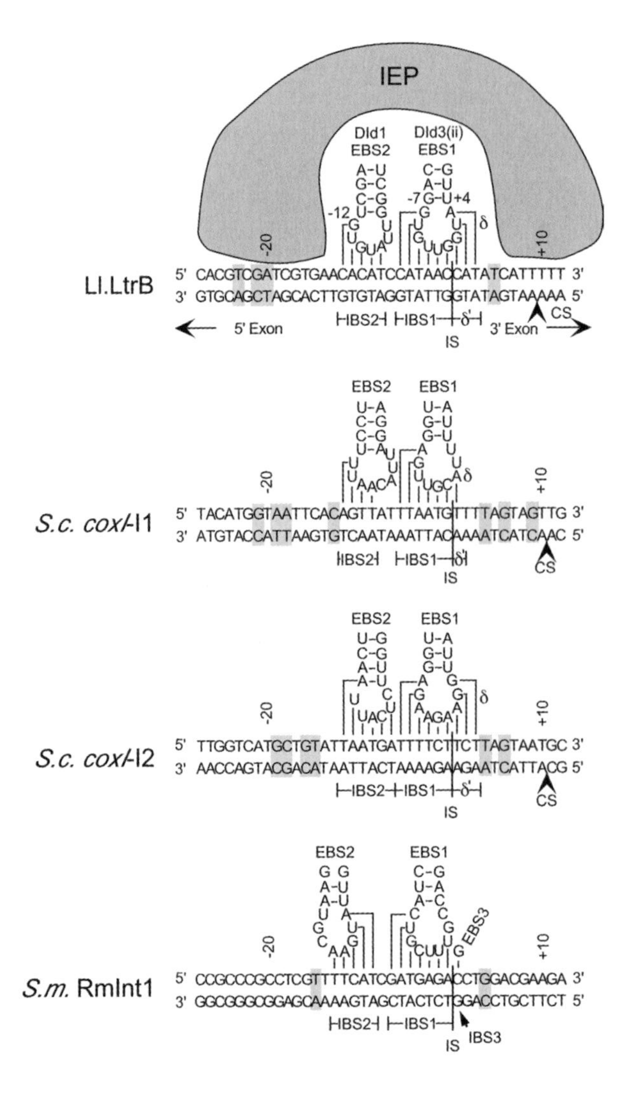

Fig. 4. Group II intron DNA target site interactions. Shown are target sites for the *L. lactis* Ll.LtrB intron (Mohr et al. 2000; Singh and Lambowitz 2001; Perutka et al. 2004), *S. cerevisiae coxI*-I1 (Yang et al. 1998), *S. cerevisiae coxI*-I2 (Guo et al. 1997), and the *S. meliloti* RmInt1 intron (Jiménez-Zurdo et al. 2003). Critical positions that affect intron mobility reactions are *shaded* (see references for details). The intron-insertion site (IS) is in the top strand at the exon junction (*vertical line*); the bottom-strand cleavage site (*CS*) is indicated by an *arrowhead.* Intron RNA sequences base pairing with the DNA target site are shown above. The IEP is shown only for the Ll.LtrB intron

intron RNA to the 5′-exon sequences IBS2 and IBS1, and the 3′-exon sequence δ′ (group IIA introns) or IBS3 (group IIB introns). For each of the base-pairing interactions, mutations in the DNA target site that inhibit reverse splicing in vitro or intron mobility in vivo could be rescued by compensatory mutations in the intron RNA. Notably, analysis of reverse splicing of a group II intron into RNA substrates showed that single-base mismatches affect the k_{cat} for reverse splicing as well as K_d and thus have much stronger inhibitory effects than are expected from decreased binding affinity alone (Xiang et al. 1998). This feature likely contributes to the very high target specificity for intron insertion.

For the yeast *coxI*-I1 and -I2 and Ll.LtrB introns, which contain C-terminal D and En domains, the IEP recognizes specific sequences both upstream and downstream of the IBS and δ sequences. For all three introns, mutations in key nucleotide residues recognized by the IEP in the distal 5′-exon region inhibit both reverse splicing and second-strand cleavage, while mutations at 3′-exon positions recognized by the IEP inhibit only second-strand cleavage. RmInt1, whose IEP lacks the En domain as well as all or part of domain D, uses the same basic mode of DNA target site recognition. However, recognition by the IEP appears more limited, with mutations at only two positions (−15 and +4) strongly inhibiting mobility (Jiménez-Zurdo et al. 2003). A caveat is that the deletion of 5′-exon sequences between positions −20 and −16 decreased the homing efficiency by >95%, suggesting that there could also be significant recognition determinants in this region (Jiménez-Zurdo et al. 2003). RmInt1 does not carry out second-strand cleavage, so the 3′-exon mutations must inhibit mobility either by affecting the initial DNA target site recognition or reverse splicing steps. The latter situation differs from that for the yeast and lactoccal introns where 3′-exon mutations inhibit only second-strand cleavage (see above).

Bacterial class C introns, which also lack the En domain, are found inserted downstream of palindromic rho-independent transcription terminators at sites that can potentially form only the EBS1/IBS1 pairing with the intron RNA (Granlund et al. 2001; Yeo et al. 2001; Dai and Zimmerly 2002). These introns may use a variation of the DNA target site recognition mechanism in which the IEP has adapted to recognize a specific higher-order DNA structure. Indeed, the formation of this palindromic structure may be favored in transiently single-stranded DNA regions at replication forks, thus targeting the introns to regions where they can use nascent lagging strands to prime reverse transcription (see above; Lambowitz and Zimmerly 2004).

Comparison of different target sites shows that even closely related group II introns, like yeast *coxI*-I1 and -I2, have different target specificities both in the region recognized by base pairing of the intron RNA and in the flanking regions recognized by the IEP. These findings suggest that the IEPs can readily evolve or adapt to recognize different target site sequences, enabling group II introns to establish themselves at new sites.

8 DNA Target Site Recognition by Ll.LtrB Intron RNPs

Thus far, the most detailed studies of DNA target site recognition have been carried out for the Ll.LtrB intron. Figure 5a shows a model of the target site interactions for the Ll.LtrB homing endonuclease, identified by DNA footprinting, modification-interference, and missing-base analysis (Singh and Lambowitz 2001). Figure 5b shows the information content of different positions in the target site and intron RNA (Zhong et al. 2003; Perutka et al. 2004). Critical bases in the target site were also identified by extensive mutagenesis and selection experiments (Guo et al. 2000; Mohr et al. 2000; Zhong and Lambowitz 2003).

Biochemical analysis showed that Ll.LtrB RNPs bind DNA non-specifically in a rapid diffusion-limited process and then search for DNA target sites, presumably via a facilitated diffusion mechanism like those used by site-specific DNA-binding proteins (Aizawa et al. 2003). Figure 5 shows that most of the interactions with the IEP are in the distal 5′-exon region of the DNA target site (positions –23 to –14), with one set of critical bases, T-23, G-21, and A-20, being recognized via major groove interactions (Singh and Lambowitz 2001). The recognition of these bases may be the initial step in DNA target site recognition. Ethylnitrosourea modification-interference experiments showed that the IEP also makes functionally important phosphate-backbone contacts along one face of the helix between positions –24 and –13. The IEP crosses the minor groove between –17 and –13, and there are preferences for specific bases in this region (positions –17, -15, and –14), but it is not clear if these reflect direct base recognition or indirect readout of DNA structure. Notably, mutations at critical nucleotide residues in the distal 5′-exon region (positions –23, –21, –20, and –15) strongly inhibit reverse splicing into double-stranded but not otherwise identical single-stranded DNA target sites, implying that their

→

Fig. 5. Model and information content analysis of the *L. lactis* Ll.LtrB intron DNA target site. **a** Target site model summarizing critical residues identified by DNA footprinting, modification-interference and missing-base analysis, reproduced from Singh and Lambowitz (2001). The Ll.LtrB target site is represented simply as a B-form helix; its actual conformation in the complex is unknown. Critical bases recognized by the IEP are colored *red*, and critical phosphates are colored *dark and light purple*, indicating strong and weak interference, respectively, by ethylnitrosourea modification. *Blue* indicates bases protected

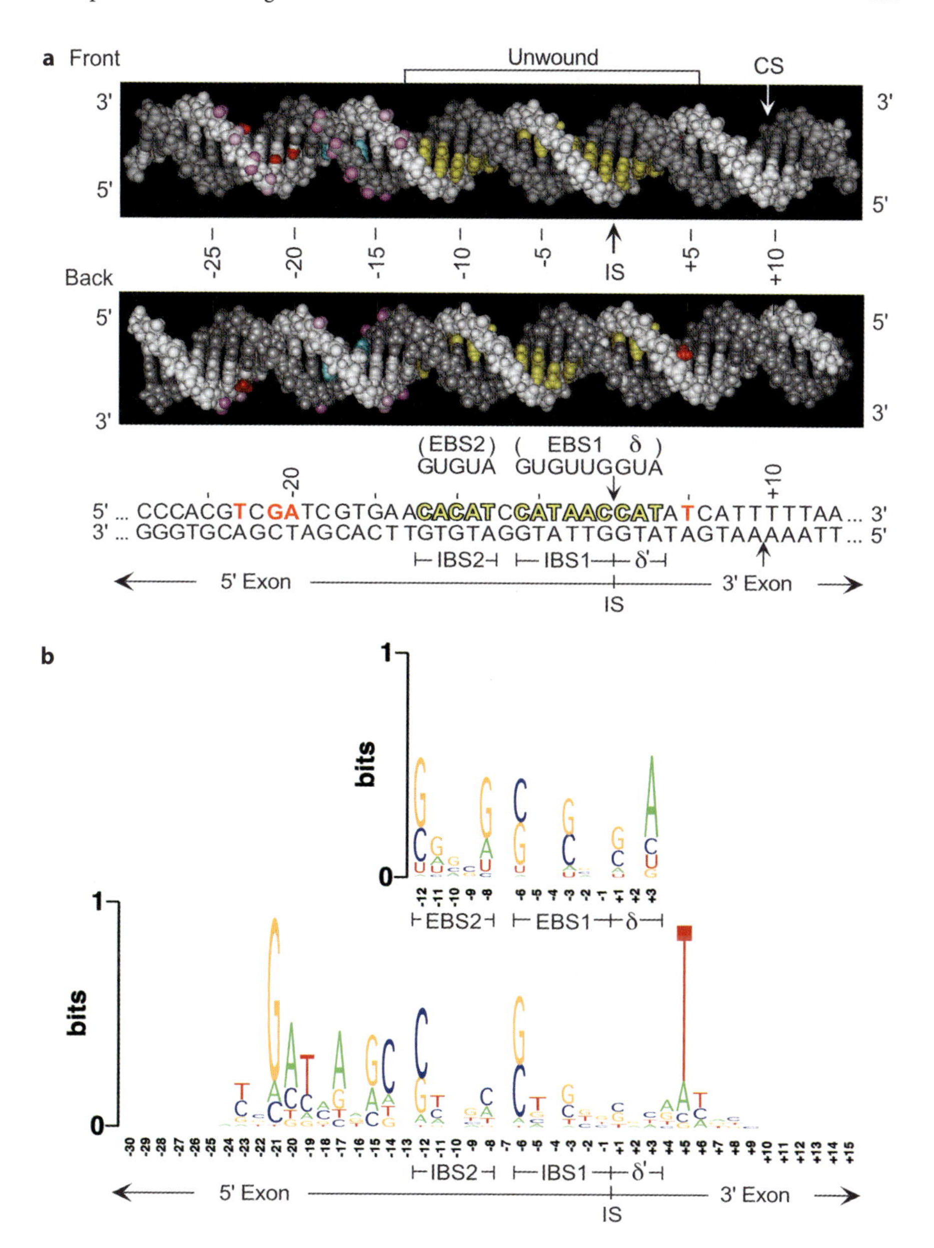

of the DNA target site is indicated by a *bracket* above. The intron-insertion site (*IS*) and bottom-strand cleavage site (*CS*) are indicated by *arrows*. **b** Information content analysis of DNA target site and intron RNA positions involved in target site recognition. The analysis is based on 88 random insertion sites in the *E. coli* genome, obtained using an Ll.LtrB intron with randomized EBS and δ sequences (Zhong et al. 2003). The logo shows the information content (bits) for each position represented by the size of the letter (http://weblogo.berkeley.edu/)

recognition by the IEP is required primarily for local DNA unwinding (Zhong and Lambowitz 2003).

The intron RNA's EBS2 and EBS1/δ sequences are located in two separate stem-loop structures (denoted DId1 and DId3(ii) in the Ll.LtrB intron; Figs. 1a,c, and 4). Base pairing of these sequences to the IBS and δ′ sequences in the target DNA may occur concomitantly with and help drive local DNA unwinding (Singh and Lambowitz 2001). Detailed analysis showed that the DNA positions recognized by base pairing are −12 to −8 (IBS2), −6 to −1 (IBS1), and +1 and +2 (δ′; Perutka et al. 2004). The complementary EBS2, EBS1 and δ positions in the intron RNA are numbered according to the DNA target site position with which they pair. Although nucleotide residues at intron RNA positions −13, −7 and +4 could also potentially base pair with the DNA target site, the U at RNA position −13 instead pairs with the opposite G in the intron RNA to form the top of the DId1 stem, and the G and U at RNA positions −7 and +4, respectively, pair with each other to form the top of DId3(ii) stem (see Fig. 4, top). Additionally, an A-residue is favored at the intron RNA +3 position regardless of whether it can base pair with the target site, suggesting that the contribution of a favorable loop structure can override any additional contribution from base pairing at this position. The information content analysis and mutagenesis experiments show that base pairing at some positions is more important than others and that there are preferences for specific base pairs at certain positions (Guo et al. 2000; Mohr et al. 2000; Zhong et al. 2003; Perutka et al. 2004). For example, positions −12 and −6 at the beginning of IBS2 and IBS1 show strong preferences for GC or CG base pairs, possibly to anchor the ends of these duplexes (see Fig. 5 and references cited above).

Bottom-strand cleavage occurs after reverse splicing and requires the same IEP and base-pairing interactions that are required for reverse splicing, as well as additional interactions between the IEP and the 3′ exon. Experiments using RNPs containing lariat RNA whose 3′ OH had been blocked by periodate oxidation showed that reverse splicing *per se* is not required for second-strand cleavage (Aizawa et al. 2003). For the Ll.LtrB intron, the most critical IEP interaction with the 3′ exon is at T+5 (Fig. 5). Mutation of this base has no effect on reverse splicing but strongly inhibits second-strand cleavage (Mohr et al. 2000). The bases flanking T+5 appear to make smaller contributions (Fig. 5b). Although interactions between the IEP and the 3′ exon are not required for DNA unwinding, the locally unwound region extends into the 3′ exon and includes T+5, perhaps facilitating its recognition for bottom-strand cleavage (Singh and Lambowitz 2001).

The En domain of the IEP also interacts at the bottom-strand cleavage site between positions +9 and +10, but there are no strong sequence requirements in this region, and single phosphate-backbone modifications are not inhibitory in ethylnitrosourea modification-interference experiments (Singh and Lambowitz 2001). Thus, positioning of the En active site for second-strand cleavage appears to be determined mainly by the recognition of T+5 and, to a lesser extent, neighboring bases, with IEP interactions around the cleavage site being relatively weak and non-specific. This also appears to be the case for the *Bombyx mori* non-LTR-retrotransposon R2Bm, which uses an analogous TPRT mechanism in which a separate En domain of the RT cleaves the DNA target site to generate the primer for reverse transcription (Christensen and Eickbush 2004). As in that case, weak, non-specific binding at the cleavage site may facilitate transfer of the cleaved 3′ end to the RT active site for initiation of TPRT.

9 Group II Introns as Gene-Targeting Vectors

The novel DNA insertion mechanism used by group II introns has enabled the development of a new technology for manipulating DNA. Group II introns adapted for gene targeting have been dubbed "targetrons". The introns have a number of characteristics that make them well suited for gene targeting. First, they have very high insertion frequencies, which can approach 100%. Second, they have long (~30 bp) target sequences, which result in high specificity. Third, the introns can be programmed to insert into new target sequences simply by modifying the RNA sequences that pair with the DNA target. Fourth, foreign sequences can be inserted within the loop of domain IV and then delivered site-specifically with the intron. Finally, because the introns are largely self-contained and require only commonly available DNA recombination or repair enzymes to complete cDNA integration, they have a wide host-range and can potentially function in any organism.

In bacteria, targetrons are expressed from a donor plasmid, such as pACD3, which is used for targetron expression in *E. coli* (Fig. 6; Karberg et al. 2001; Zhong et al. 2003). pACD3 employs an inducible T7*lac* promoter to express an Ll.LtrB-ΔORF intron and short flanking exons, with the IEP expressed from a position just downstream of the 3′ exon. The IEP expressed from this position promotes efficient splicing of the intron RNA, forming the RNP homing endonuclease that promotes mobility, but when the intron inserts into a new location, it does not carry the ORF, and thus is unable to splice. The use of the ΔORF intron also greatly increases the insertion frequency by making the intron RNA less susceptible to nuclease digestion (Guo et al. 2000). Analogous vectors have been developed for the Ll.LtrB targetron in other bacteria with appropriate modifications of the promoter and Shine-Dalgarno sequence used for intron RNA and IEP expression (e.g. Frazier et al. 2003), and the same approaches should be readily adaptable for other group II introns.

The intron is targeted to specific sites with the help of a computer algorithm that scans the target sequence for the best matches to the positions recognized by the IEP and then designs primers to modify the intron's EBS and δ sequences to insert into those sites (Perutka et al. 2004). The positions recognized by the IEP are sufficiently few and flexible that the program readily identifies multiple rank-ordered target sites in any gene. The IBS sequences in the 5′ exon of the precursor RNA must also be made complementary to the retargeted EBS sequences for efficient RNA splicing. The required modifications are introduced into the donor plasmid via polymerase chain reaction (PCR; Zhong et al. 2003). In *E. coli*, targetrons commonly insert in the desired chro-

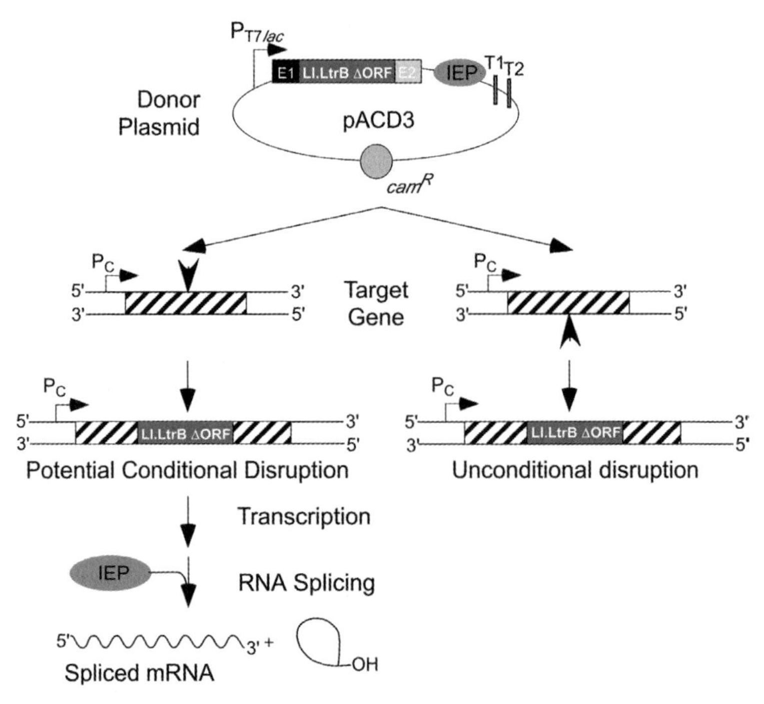

Fig. 6. Use of targetrons for gene disruption. The targetron donor plasmid (*pACD3*) uses a T7*lac* promoter to express a ΔORF-derivative of the Ll.LtrB intron and short flanking exons, with the *IEP* expressed from a position just downstream of the 3′ exon. Introns targeted to sites in the top (sense) or bottom (antisense) strands insert in different orientations relative to target gene transcription. A ΔORF intron inserted in the sense orientation can potentially yield a conditional disruption by linking its splicing to the expression of the IEP from a separate construct (Karberg et al. 2001; Frazier et al. 2003). An intron inserted in the antisense orientation cannot be spliced, yielding an unconditional disruption. P_C Promoter of targeted chromosomal gene

mosomal target sites at frequencies >1% without selection, readily detectable by colony PCR screening (Perutka et al. 2004).

Targetrons containing conventional or retrotransposition-activated selectable markers (RAM) inserted in DIV can be used to genetically select intron-integration events. RAM markers are patterned after previously described retrotransposition-indicator gene (RIG) markers (Ichiyanagi et al. 2002). The prototype RAM marker used for gene targeting was a small trimethoprim-resistance (*Tp*^{R}) gene inserted in group II intron DIV in the reverse orientation, but containing an efficiently self-splicing group I intron (the phage T4 *td* intron) inserted in the forward orientation. The group I intron is excised during retrotransposition, activating the *Tp*^{R} gene (Zhong et al. 2003). Typically, nearly 100% of the *Tp*^{R} colonies have correctly targeted introns, with few if any having non-specific insertions. A *kan*^{R}-RAM marker has also been used for gene targeting (unpubl. data), and, in principle, the same approach could be used for any other selectable or screenable marker.

Gene Disruption by Insertional Mutagenesis. The Ll.LtrB targetron has been used for targeted gene disruption in *E. coli* and other Gram-negative and Gram-positive bacteria (Karberg et al. 2001; Frazier et al. 2003). Because the targetron contains multiple stop codons in all reading frames, suitably placed disruptions totally ablate gene function. The intron RNA can be targeted to insert in either strand, resulting in different orientations relative to target gene transcription. An intron that inserts in the antisense orientation cannot be spliced and gives an unconditional disruption, whereas an intron that inserts in the sense orientation can potentially yield a conditional disruption by linking its splicing to the expression of the IEP from a separate construct containing an inducible promoter (Fig. 6; Karberg et al. 2001; Frazier et al. 2003). In practice, the splicing efficiency of the retargeted intron is generally less than that of the wild-type intron, likely due to a combination of suboptimal exon contexts at new target sites and decreased splicing efficiency when the IEP is expressed *in trans*. Consequently, conditional disruptions can be obtained in some cases but not others. This situation may be improved by optimizing the IEP for expression *in trans*.

By incorporating a RAM marker, an Ll.LtrB intron with randomized target site recognition sequences was used to obtain disruptions at sites distributed throughout the *E. coli* genome, analogous to global transposon mutagenesis (Zhong et al. 2003). Despite clustering of insertions near the chromosome replication origin, the resulting library was sufficiently complex to contain most viable *E. coli* gene disruptions. Recent studies on Ll.LtrB retrotransposition suggest that altering the growth conditions may lead to a more uniform distribution of insertion sites (Coros et al. 2005). In addition, group II introns that are inserted into any gene can be "fished" from the library by PCR and inserted into a donor plasmid to obtain single disruptions in the same or different host strains, an important advantage over libraries generated with conventional transposons (Zhong et al. 2003; J. Yao, J. Zhong, A.M. Lambowitz, in prep.). The incorporation of a RAM marker also makes it possible to automate whole genome library construction by designing introns targeted to each gene and using robots to select antibiotic-resistant colonies. The advantages are that each gene can be targeted individually, with multiple introns if necessary, and that high insertion frequencies in the absence of a selectable marker facilitate the construction of strains having multiple disruptions or other desirable combinations of traits.

Site-Specific DNA Insertion. In addition to disrupting genes, group II introns can be used to integrate cargo genes inserted in DIV at desired chromosomal locations (e.g., Frazier et al. 2003). The insertion of extra DNA in DIV may decrease the integration frequency by making the intron more susceptible to degradation by host nucleases (see Guo et al. 2000). In most cases, however, the frequency remains sufficiently high to detect the desired integration. In the example shown in Fig. 7, this approach was used to engineer lactobacteria by inserting a commercially important phage resistance gene (*abiD*) into a specific genomic locus, thereby conferring regulatable phage resistance. The desired insertion was identified at a frequency of 0.5–2% by colony PCR screening. The ability to obtain desired insertions without selection for antibiotic resistance is particularly important in food-grade organisms like lactobacteria.

Targeted Double-Strand Breaks. Experiments with rare-cutting DNA endonucleases have shown that the introduction of double-strand breaks in bacterial, plant, or animal chromosomal DNA greatly stimulates homologous recombination with a cotransformed DNA fragment, enabling the introduction of desired mutations by gene replacement (Jasin 1996; Donoho et al. 1998). However, despite efforts at protein engineering (Chevalier et al. 2002; Epinat et al. 2003), the power of this approach remains limited by the fixed specificity of protein endonucleases. Group II introns can be used similarly, with the advantage that the double-strand break can be targeted to any desired chromosome region. The proof-of-principle experiment shown in Fig. 8 used an Ll.LtrB intron with a point mutation abolishing the RT activity of the IEP to prevent cDNA synthesis at the cleaved target site. The RT-deficient intron was targeted to introduce a double-strand break (reverse splicing plus second-strand cleavage) in the *E. coli* chromosomal *thyA* gene, which was then repaired by homologous recombination with a cotransformed plasmid containing a segment of the *thyA* gene with stop codons at desired locations (Karberg et al. 2001). In the absence of cDNA synthesis, the reverse-spliced intron RNA was presuma-

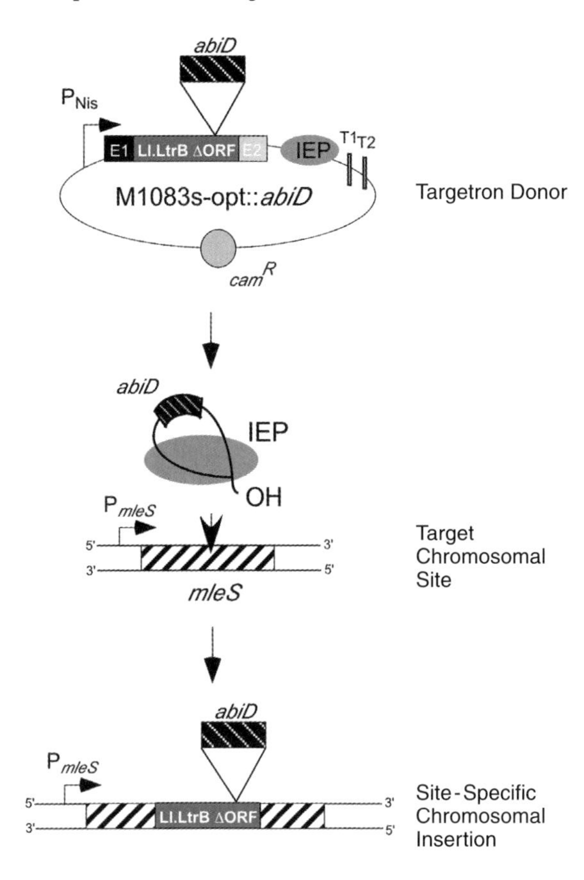

Fig. 7. Use of targetrons for site-specific insertion of a cargo gene at a desired genomic location. In the example shown, the *abiD* gene, which confers phage resistance, was inserted at a site within the regulatable *mleS* gene in the *L. lactis* chromosome (Frazier et al. 2003). An Ll.LtrB-ΔORF intron targeted to the *mleS* gene with the *abiD* gene inserted in intron DIV in place of the IEP was expressed from an *L. lactis* donor plasmid using a nisin-inducible promoter (P_{Nis}). The intron is flanked by short exon sequences (*E1* and *E2*). The *IEP*, which splices and mobilizes the intron, is cloned downstream of *E2*. Splicing results in the formation of an RNP containing the IEP and intron lariat RNA with the *abiD* gene in DIV, which is then inserted into the chromosomal DNA target site by retrohoming

bly degraded by cellular enzymes, leaving a double-strand break that could be acted on by cellular enzymes. The introduction of the double-strand break by targeted Ll.LtrB introns stimulated homologous recombination frequencies in *E. coli* by one to two orders of magnitude (Karberg et al. 2001).

Use of Targeted Group II Introns in Eukaryotes. If appropriate methods can be developed for eukaryotes, mobile group II introns would have potentially wide medical, agricultural, and commercial applications, including the introduction of genetically stable gene disruptions for functional genomics, and the site-specific introduction or repair of genes for genetic engineering and gene therapy. In initial work establishing the potential of the approach, group II introns were targeted to insert into the HIV-1 provirus and the human gene encoding CCR5, an important target site in anti-HIV-1 therapy (Guo et al. 2000). The retargeted introns were shown to insert at high frequencies in-

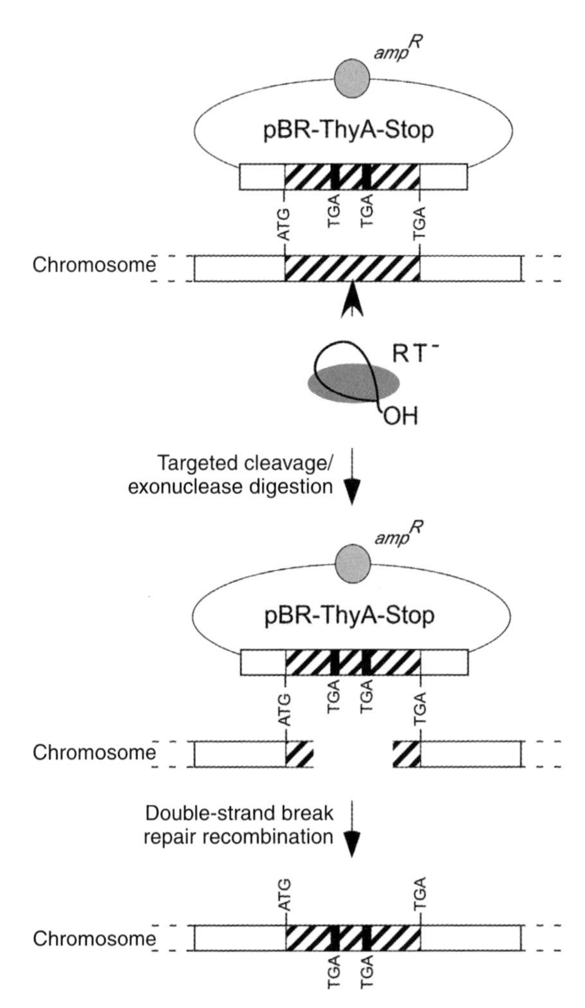

Fig. 8. Use of targetrons to introduce targeted double-strand breaks that stimulate homologous recombination with a cotransformed DNA fragment. In the example shown, an Ll.LtrB intron targeted to the *E. coli thyA* gene was used to introduce stop codons into the *thyA* gene in the *E. coli* chromosome (Karberg et al. 2001). The intron carrying an RT$^-$ mutation to prevent cDNA synthesis at the target site was expressed from a donor plasmid (not shown) to produce an RNP that makes a double-strand break at the chromosomal *thyA* target site (*arrow*). The double-strand break then stimulates homologous recombination with a cotransformed plasmid (*pBRR-ThyA-Stop*) carrying a segment of the *thyA* gene with stop codons at desired sites (*black boxes*)

to their desired target sites in an *E. coli* plasmid assay, and the intron RNPs retained activity in human cells inserting into a plasmid target site in a liposome-mediated transfection assay. Current work focuses on developing more efficient methods for introducing or expressing RNPs in mammalian cells, in order to obtain a sufficiently high concentration of RNPs in the nucleus for efficient chromosome integration.

Acknowledgements. Work in the author's laboratories was supported by NIH grants GM37949 and GM3791 to A.M.L. and CIHR, NSERC, and AHFMR grants to S.Z.

References

Aizawa Y, Xiang Q, Lambowitz AM, Pyle AM (2003) The pathway for DNA recognition and RNA integration by a group II intron retrotransposon. Mol Cell 11:795–805

Belfort M, Derbyshire V, Parker MM, Cousineau B, Lambowitz AM (2002) Mobile introns: pathways and proteins. In: Craig NL, Craigie R, Gellert M, Lambowitz AM (eds) Mobile DNA II. ASM Press, Washington, DC, pp 761–783

Belhocine K, Plante I, Cousineau B (2004) Conjugation mediates transfer of the Ll.LtrB group II intron between different bacterial species. Mol Microbiol 51:1459–1469

Bibillo A, Eickbush TH (2002) The reverse transcriptase of the R2 non-LTR retrotransposon: continuous synthesis of cDNA on non-continuous RNA templates. J Mol Biol 316:459–473

Blocker FJH, Mohr G, Conlan LH, Qi L, Belfort M, Lambowitz AM (2005) Domain structure and three-dimensional model of a group II intron-encoded reverse transcriptase. RNA 11:14–28

Carignani G, Groudinsky O, Frezza D, Schiavon, E, Bergantino E, Slonimski PP (1983) An mRNA maturase is encoded by the first intron of the mitochondrial gene for the subunit I of cytochrome oxidase in S. cerevisiae. Cell 35:733–742

Chen B, Lambowitz AM (1997) De novo and DNA primer-mediated initiation of cDNA synthesis by the Mauriceville retroplasmid reverse transcriptase involve recognition of a 3′ CCA sequence. J Mol Biol 271:311–332

Chevalier BS, Korteme T, Chadsey MS, Baker D, Monnat RJ Jr, Stoddard B (2002) Design, activity, and structure of a highly specific artificial endonuclease. Mol Cell 10:895–905

Christensen S, Eickbush TH (2004) Footprint of the retrotransposon R2Bm protein on its target site before and after cleavage. J Mol Biol 336:1035–1045

Coros CJ, Landthaler M, Piazza CL, Beauregard A, Esposito D, Perutka J, Lambowitz AM, Belfort M (2005) Retrotransposition strategies of the *Lactococcus lactis* LI.LtrB group II intron are dictated by host identity and cellular environment. Mol Microbiol 56:509–529

Costa M, Michel F, Westhof E (2000) A three-dimensional perspective on exon binding by a group II self-splicing intron. EMBO J 19:5007–5018

Cousineau B, Smith D, Lawrence-Cavanagh S, Mueller JE, Yang J, Mills D, Manias D, Dunny G, Lambowitz AM, Belfort M (1998) Retrohoming of a bacterial group II intron: mobility via complete reverse splicing, independent of homologous DNA recombination. Cell 94:451–462

Cui X, Matsuura M, Wang Q, Ma H, Lambowitz AM (2004) A group II intron-encoded maturase functions preferentially in *cis* and requires both the reverse transcriptase and X domains to promote RNA splicing. J Mol Biol 340:211–231

D'Souza LM, Zhong J (2002) Mutations in the *Lactococcus lactis* Ll.LtrB group II intron that retain mobility *in vivo*. BMC Mol Biol 3:17–25

Dai L, Zimmerly S (2002) Compilation and analysis of group II intron insertions in bacterial genomes: evidence for retroelement behavior. Nucleic Acids Res 30:1091–1102

Donoho G, Jasin M, Berg P (1998) Analysis of gene targeting and intrachromosomal homologous recombination stimulated by genomic double-strand breaks in mouse embryonic stem cells. Mol Cell Biol 18:4070–4078

Epinat JC, Arnould S, Chames P, Rochaix P, Desfontaines D, Puzin C, Patin A, Zanghellini A, Pâques F, Lacroix E (2003) A novel engineered meganuclease induces homologous recombination in yeast and mammalian cells. Nucleic Acids Res 31:2952–2962

Eskes R, Yang J, Lambowitz AM, Perlman PS (1997) Mobility of yeast mitochondrial group II introns: engineering a new site specificity and retrohoming via full reverse splicing. Cell 88:865–874

Eskes R, Liu L, Ma H, Chao MY, Dickson L, Lambowitz AM, Perlman PS (2000) Multiple homing pathways used by yeast mitochondrial group II introns. Mol Cell Biol 20:8432–8446

Frazier CL, San Filippo J, Lambowitz AM, Mills DA (2003) Genetic manipulation of *Lactococcus lactis* by using targeted group II introns: generation of stable insertions without selection. Appl Environ Microbiol 69:1121–1128

Gorbalenya AE (1994) Self-splicing group I and group II introns encode homologous (putative) DNA endonucleases of a new family. Protein Sci 3:1117–1120

Gordon PM, Piccirilli JA (2001) Metal ion coordination by the AGC triad in domain 5 contributes to group II intron catalysis. Nature Struct Biol 8:893–898

Granlund M, Michel F, Norgren M (2001) Mutually exclusive distribution of IS*1548* and GBSi1, an active group II intron identified in human isolates of group B streptococci. J Bacteriol 183:2560–2569

Guo H, Zimmerly S, Perlman PS, Lambowitz AM (1997) Group II intron endonucleases use both RNA and protein subunits for recognition of specific sequences in double-stranded DNA. EMBO J 16:6835–6848

Guo H, Karberg M, Long M, Jones JP 3rd, Sullenger B, Lambowitz AM (2000) Group II introns designed to insert into therapeutically relevant DNA target sites in human cells. Science 289:452–457

Huang HR, Chao MY, Armstrong B, Wang Y, Lambowitz AM, Perlman PS (2003) The DIVa maturase binding site in the yeast group II intron aI2 is essential for intron homing but not for in vivo splicing. Mol Cell Biol 23:8809–8819

Ichiyanagi K, Beauregard A, Lawrence S, Smith D, Cousineau B, Belfort M (2002) Retrotransposition of the Ll.LtrB group II intron proceeds predominantly via reverse splicing into DNA targets. Mol Microbiol 46:1259–1272

Ichiyanagi K, Beauregard A, Belfort M (2003) A bacterial group II intron favors retrotransposition into plasmid targets. Proc Natl Acad Sci USA 100:15742–15747

Jasin M (1996) Genetic manipulation of genomes with rare-cutting endonucleases. Trends Genet 12:224–228

Jiménez-Zurdo JI, García-Rodríguez FM, Barrientos-Durán A, Toro N (2003) DNA target site requirements for homing *in vivo* of a bacterial group II intron encoding a protein lacking the DNA endonuclease domain. J Mol Biol 326:413–423

Karberg M, Guo H, Zhong J, Coon R, Perutka J, Lambowitz AM (2001) Group II introns as controllable gene targeting vectors for genetic manipulation of bacteria. Nature Biotechnol 19:1162–1167

Kennell JC, Moran JV, Perlman PS, Butow RA, Lambowitz AM (1993) Reverse transcriptase activity associated with maturase-encoding group II introns in yeast mitochondria. Cell 73:133–146

Lambowitz AM, Belfort M (1993) Introns as mobile genetic elements. Annu Rev Biochem 62:587–622

Lambowitz AM, Zimmerly S (2004) Mobile group II introns. Annu Rev Genet 38:1–35

Lambowitz AM, Caprara MG, Zimmerly S, Perlman PS (1999) Group I and group II ribozymes as RNPs: clues to the past and guides to the future. In: Gesteland R, Cech TR, Atkins J (eds) The RNA world, 2nd edn. Cold Spring Harbor Lab Press, Plainview, NY, pp 451–485

Lehmann K, Schmidt U (2003) Group II introns: structure and catalytic versatility of large natural ribozymes. Crit Rev Biochem Mol Biol 38:249–303

Malik HS, Burke WD, Eickbush TH (1999) The age and evolution of non-LTR retrotransposable elements. Mol Biol Evol 16:793–805

Martínez-Abarca F, Barrientos-Durán A, Fernández-López M, Toro N (2004) The RmInt1 group II intron has two different retrohoming pathways for mobility using predomi-

nantly the nascent lagging strand at DNA replication forks for priming. Nucleic Acids Res 32:2880–2888
Matsuura M, Saldanha R, Ma H, Wank H, Yang J, Mohr G, Cavanagh S, Dunny GM, Belfort M, Lambowitz AM (1997) A bacterial group II intron encoding reverse transcriptase, maturase, and DNA endonuclease activities: biochemical demonstration of maturase activity and insertion of new genetic information within the intron. Genes Dev 11:2910–2924
Matsuura M, Noah JW, Lambowitz AM (2001) Mechanism of maturase-promoted group II intron splicing. EMBO J 20:7259–7270
Michel F, Ferat JL (1995) Structure and activities of group II introns. Annu Rev Biochem 64:435–461
Michel F, Umesono K, Ozeki H (1989) Comparative and functional anatomy of group II catalytic introns – a review. Gene 82:5–30
Mills DA, McKay LL, Dunny GM (1996) Splicing of a group II intron involved in the conjugative transfer of pRS01 in lactococci. J Bacteriol 178:3531–3538
Mills DA, Manias DA, McKay LL, Dunny GM (1997) Homing of a group II intron from *Lactococcus lactis* subsp. *lactis* ML3. J Bacteriol 179:6107–6111
Mohr G, Perlman PS, Lambowitz AM (1993) Evolutionary relationships among group II intron-encoded proteins and identification of a conserved domain that may be related to maturase function. Nucleic Acids Res 21:4991–4997
Mohr G, Smith D, Belfort M, Lambowitz AM (2000) Rules for DNA target-site recognition by a lactococcal group II intron enable retargeting of the intron to specific DNA sequences. Genes Dev 14:559–573
Moran JV, Zimmerly S, Eskes R, Kennell JC, Lambowitz AM, Butow RA, Perlman PS (1995) Mobile group II introns of yeast mitochondrial DNA are novel site-specific retroelements. Mol Cell Biol 15:2828–2838
Muñoz-Adelantado E, San Filippo J, Martínez-Abarca F, García-Rodríguez FM, Lambowitz AM, Toro N (2003) Mobility of the *Sinorhizobium meliloti* group II intron RmInt1 occurs by reverse splicing into DNA, but requires an unknown reverse transcriptase priming mechanism. J Mol Biol 327:931–943
Noah JW, Lambowitz AM (2003) Effects of maturase binding and Mg^{2+} concentration on group II intron RNA folding investigated by UV cross-linking. Biochemistry 42:12466–12480
Perutka J, Wang W, Goerlitz D, Lambowitz AM (2004) Use of computer-designed group II introns to disrupt *Escherichia coli* DExH/D-box protein and DNA helicase genes. J Mol Biol 336:421–439
Qin PZ, Pyle AM (1998) The architectural organization and mechanistic function of group II intron structural elements. Curr Opin Struct Biol 8:301–308
Rambo RP, Doudna JA (2004) Assembly of an active group II intron-maturase complex by protein dimerization. Biochemistry 43:6486–6497
Saldanha R, Chen B, Wank H, Matsuura M, Edwards J, Lambowitz AM (1999) RNA and protein catalysis in group II intron splicing and mobility reactions using purified components. Biochemistry 38:9069–9083
San Filippo J, Lambowitz AM (2002) Characterization of the C-terminal DNA-binding/DNA endonuclease region of a group II intron-encoded protein. J Mol Biol 324:933–951
Schäfer B, Gan L, Perlman PS (2003) Reverse transcriptase and reverse splicing activities encoded by the mobile group II intron *COB*I1 of fission yeast mitochondrial DNA. J Mol Biol 329:191–206
Shearman C, Godon JJ, Gasson M (1996) Splicing of a group II intron in a functional transfer gene of *Lactococcus lactis*. Mol Microbiol 21:45–53

Shub DA, Goodrich-Blair H, Eddy SR (1994) Amino acid sequence motif of group I intron endonucleases is conserved in open reading frames of group II introns. Trends Biochem Sci 19:402–404
Sigel RK, Vaidya A, Pyle AM (2000) Metal ion binding sites in a group II intron core. Nature Struct Biol 7:1111–1116
Sigel RK, Sashital DG, Abramovitz DL, Palmer AG 3rd, Butcher SE, Pyle AM (2004) Solution structure of domain 5 of a group II intron ribozyme reveals a new RNA motif. Nature Struct Mol Biol 11:187–192
Singh NN, Lambowitz AM (2001) Interaction of a group II intron ribonucleoprotein endonuclease with its DNA target site investigated by DNA footprinting and modification interference. J Mol Biol 309:361–386
Singh RN, Saldanha RJ, D'Souza LM, Lambowitz AM (2002) Binding of a group II intron-encoded reverse transcriptase/maturase to its high affinity intron RNA binding site involves sequence-specific recognition and autoregulates translation. J Mol Biol 318:287–303
Staddon JH, Bryan EM, Manias DA, Dunny GM (2004) Conserved target for group II intron insertion in relaxase genes of conjugative elements of gram-positive bacteria. J Bacteriol 186:2393–2401
Swisher J, Duarte CM, Su LJ, Pyle AM (2001) Visualizing the solvent-inaccessible core of a group II intron ribozyme. EMBO J 20:2051–2061
Toor N, Zimmerly S (2002) Identification of a family of group II introns encoding LAGLIDADG ORFs typical of group I introns. RNA 8:1373–1377
Toor N, Hausner G, Zimmerly S (2001) Coevolution of group II intron RNA structures with their intron-encoded reverse transcriptases. RNA 7:1142–1152
Toro N (2003) Bacteria and archaea group II introns: additional mobile genetic elements in the environment. Environ Microbiol 5:143–151
Wank H, San Filippo J, Singh RN, Matsuura M, Lambowitz AM (1999) A reverse transcriptase/maturase promotes splicing by binding at its own coding segment in a group II intron RNA. Mol Cell 4:239–250
Watanabe K, Lambowitz AM (2004) High-affinity binding site for a group II intron-encoded reverse transcriptase/maturase within a stem-loop structure in the intron RNA. RNA 10:1433–1443
Xiang Q, Qin PZ, Michels WJ, Freeland K, Pyle AM (1998) Sequence specificity of a group II intron ribozyme: multiple mechanisms for promoting unusually high discrimination against mismatched targets. Biochemistry 37:3839–3849
Xiong Y, Eickbush TH (1990) Origin and evolution of retroelements based upon their reverse transcriptase sequences. EMBO J 9:3353–3362
Yang J, Zimmerly S, Perlman PS, Lambowitz AM (1996) Efficient integration of an intron RNA into double-stranded DNA by reverse splicing. Nature 381:332–335
Yang J, Mohr G, Perlman PS, Lambowitz AM (1998) Group II intron mobility in yeast mitochondria: target DNA-primed reverse transcription activity of aI1 and reverse splicing into DNA transposition sites in vitro. J Mol Biol 282:505–523
Yeo CC, Yin S, Tan BH, Poh CL (2001) Isolation and characterization of group II introns from *Pseudomonas alcaligenes* and *Pseudomonas putida*. Plasmid 45:233–239
Zhang L, Doudna JA (2002) Structural insights into group II intron catalysis and branch-site selection. Science 295:2084–2088
Zhong J, Lambowitz AM (2003) Group II intron mobility using nascent strands at DNA replication forks to prime reverse transcription. EMBO J 22:4555–4565
Zhong J, Karberg M, Lambowitz AM (2003) Targeted and random bacterial gene disruption using a group II intron (targetron) vector containing a retrotransposition-activated selectable marker. Nucleic Acids Res 31:1656–1664

Zimmerly S, Guo H, Perlman PS, Lambowitz AM (1995a) Group II intron mobility occurs by target DNA-primed reverse transcription. Cell 82:545–554
Zimmerly S, Guo H, Eskes R, Yang J, Perlman PS, Lambowitz AM (1995b) A group II intron RNA is a catalytic component of a DNA endonuclease involved in intron mobility. Cell 83:529–538
Zimmerly S, Moran JV, Perlman PS, Lambowitz AM (1999) Group II intron reverse transcriptase in yeast mitochondria. Stabilization and regulation of reverse transcriptase activity by the intron RNA. J Mol Biol 289:473–490
Zimmerly S, Hausner G, Wu X (2001) Phylogenetic relationships among group II intron ORFs. Nucleic Acids Res 29:1238–1250

Free-Standing Homing Endonucleases of T-even Phage: Freeloaders or Functionaries?

David R. Edgell

1 Introduction

The ability of group I and II introns, and of inteins, to promote their own mobility to cognate genes lacking the intron or intein is a property that is now well documented in the scientific literature. This so-called homing of introns and inteins is mediated by homing endonucleases encoded within the genetic elements. Other chapters within this book have detailed the different classes of homing endonucleases, how homing endonuclease specifically recognize their target sequences, and the mechanisms of DNA cleavage. This chapter deals with the surprising observation that homing endonucleases are not always found encoded within introns or inteins, and are often found inserted between genes. Such free-standing homing endonucleases are particularly abundant in the well-studied *Escherichia coli* bacteriophage T4. In phage T4, the endonucleases themselves are mobile genetic elements, promoting their spread to related T-even phage genomes that lack the endonuclease gene by a double-strand break (DSB) repair pathway. This process is remarkably similar to endonuclease-mediated intron homing, and has been termed intronless homing. Free-standing endonucleases have been identified in most sequenced genomes, but it is unlikely that all practice intronless homing, as some have been co-opted by cellular genomes to function in pathways unrelated to endonuclease mobility. Here, I will focus on free-standing endonucleases of phage genomes, and highlight differences in DNA-binding and cleavage strategies of intron-encoded versus free-standing endonucleases as it relates to promoting mobility between genomes.

D.R. Edgell (e-mail: dedgell@uwo.ca)
Department of Biochemistry, University of Western Ontario,
London, ON N6A 5C1, Canada

Nucleic Acids and Molecular Biology, Vol. 16
Marlene Belfort et al. (Eds.)
Homing Endonucleases and Inteins

2 When Is a Free-Standing Endonuclease a Homing Endonuclease?

Free-standing endonucleases are usually identified by BLAST or other similarity-based search algorithms, exhibiting similarity to well-characterized intron-encoded homing endonucleases, most often of the GIY-YIG or HNH families. More precisely, it is the N-terminal catalytic domain of intron-encoded versions that shares extensive similarity with free-standing versions, whereas the C-terminal domains of free-standing versions (presumably the DNA-binding domains) show little similarity to intron-encoded counterparts. While this similarity provides a convenient and facile method of identifying potential free-standing endonucleases, caution must be exercised in over-interpreting the function of a free-standing endonuclease based on similarity to intron-encoded endonucleases alone.

For instance, standard BLAST searches with the catalytic domain of GIY-YIG or HNH endonucleases reveal similarity to many proteins that are not intron- or intein-encoded. Perhaps the best-characterized prokaryotic example of similarity between *bona fide* intron endonucleases and free-standing versions is UvrC, an endonuclease involved in nucleotide excision repair (Sancar 1996; Kowalski et al. 1999). The N-terminus of UvrC is indistinguishable in sequence from intron-encoded GIY-YIG endonucleases, but is fused to a domain unrelated to the C-terminal DNA-binding domains of intron endonucleases (Kowalski et al. 1999; Sitbon and Pietrokovski 2003). Furthermore, mutation of a homologous arginine residue in UvrC and intron-encoded endonucleases results in a loss of DNA-cleavage activity (Kowalski et al. 1999; Verhoeven et al. 2000; Truglio et al. 2005). One evolutionary scenario to explain the relatedness of UvrC, intron-encoded GIY-YIG endonucleases, and other proteins containing the GIY-YIG module is that of domain shuffling, whereby the GIY-YIG catalytic module has repeatedly been fused to distinct N- or C-terminal domains, with each fusion protein under selection for a distinct biochemical activity. Similar arguments can be made for intron-encoded HNH family endonucleases and free-standing endonucleases of bacterial and phage genomes that possess the HNH endonuclease motif (Gorbalenya 1994; Shub et al. 1994; Kutter et al. 1995). Such structural and mechanistic similarity is exemplified by comparisons of HNH intron endonucleases and colicins, non-specific nucleases encoded within bacterial genomes (Pommer et al. 2001; Maté and Kleanthous 2004; Shen et al. 2004). These and other examples lend support to the notion that endonuclease domains have been co-opted many times by cellular genomes to function in pathways unrelated to intron or endonuclease mobility.

While the definitive proof of endonuclease mobility must result from experimental studies, a preliminary criterion for judging whether or not a free-standing endonuclease functions as a mobile element is often the genomic location, and its distribution in related genomes. For example, T4-encoded SegG, a GIY-YIG family endonuclease, is inserted in a polymorphic region of T4 between genes *59* and *32*, each of which are highly conserved among T-even phage (Repoila et al. 1994; Liu et al. 2003). In phage T2 and other T-even phage, the *segG*-coding sequence is absent, replaced by an unrelated intergenic spacer that includes the gene *32* regulatory elements (Loayza et al. 1991). Such sporadic distribution of *segG* within T-even genomes might suggest that *segG* was once common to all T-even genomes, and has since been deleted from select genomes. Alternatively, the sporadic distribution of *segG* suggests that the endonuclease is actually a mobile genetic element, as its distribution parallels that of mobile group I introns of T-even phage (Eddy 1992; Sandegren and Sjoberg 2004), where the intron exists within a population of intronless alleles, a prerequisite for maintenance of a functional homing endonuclease (Quirk et al. 1989b; Goddard and Burt 1999).

As another cautionary note, the genomic location taken as the sole indicator for a homing function can often be a false predictor of endonuclease mobility. Endonuclease II (*denA*) of phage T4 possesses all the characteristics of a mobile free-standing endonuclease; its N-terminal region is similar to that of intron-encoded GIY-YIG endonucleases (Carlson and Kosturko 1998), and it is inserted between conserved protein-coding genes (Ray et al. 1972). However, the experimental evidence to date demonstrates that the function of endonuclease II is not to promote its own mobility, but to restrict cytosine-containing host DNA that is recycled into the phage nucleotide synthesis pathway (T-even phage DNA contains hydroxymethylcytosine and is resistant to endonuclease II cleavage; Carlson and Wiberg 1983). Thus, there is a danger in over-interpreting the potential function of a free-standing endonuclease based on similarity to homing endonucleases, and on its genomic location.

3 Intronless Homing and Marker Exclusion

The first clue that free-standing endonucleases of phage genomes are mobile genetic elements came from 30-year-old studies on the inheritance of genetic markers in crosses of the closely related phages T2 and T4 (Pees and de Groot 1970; Russell and Huskey 1974). In measuring the frequency of T2 genes in the progeny of mixed infection with phage T4, Russell and Huskey found that the corresponding T4 genetic markers predominated in the progeny (Fig. 1). This result was unexpected as T4 and T2 are closely related T-even phage, and the expectation was that T2 markers would be inherited in 50% of progeny. Russell and Huskey called this phenomenon partial marker exclusion, whereby T2 markers were, on average, present in only 20% of progeny. Interestingly, two specific regions of the T2 genome, centered on genes *32* and *56*, were

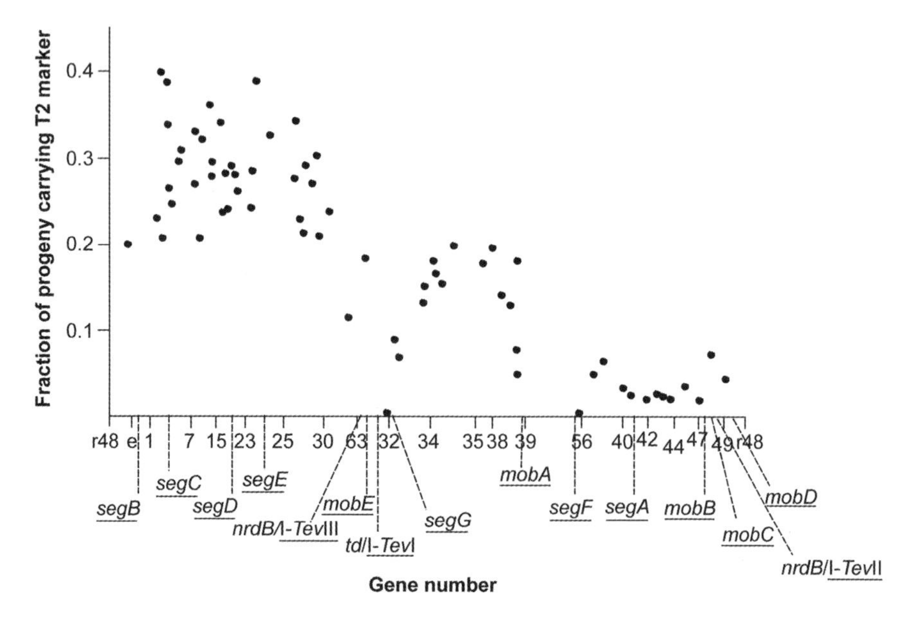

Fig. 1. Exclusion of T2 markers from the progeny of mixed infection with T4 (adapted from Russell and Huskey 1974). Shown is a linear representation of the T4 genetic map with the position of amber markers in essential genes indicated mainly by *numbers*, as well as the positions of T4-encoded endonucleases (*underscored*). For each genetic cross, the frequency of progeny carrying the T2 markers is indicated by a *dot* at the appropriate map position

almost completely excluded, present in <1% of progeny (Fig. 1). In the absence of DNA sequence data or the knowledge of site-specific DNA endonucleases, Russell and Huskey hypothesized that localized marker exclusion was likely the result of a T4 gene that specified a protein with excluding activity, and that would be capable of recognizing specific DNA sequences.

We now know that the regions of localized marker exclusion in genes *32* and *56* correspond to locations in the T4 genome that are near the insertion sites of the GIY-YIG free-standing endonucleases, *segG* and *segF*, respectively (Miller et al. 2003). Indeed, SegG and SegF make DSBs in T2 genes *32* and *56*, respectively (Belle et al. 2002; Liu et al. 2003). In the case of SegG-induced DSBs, analysis of the co-conversion tracts in progeny phage reveals that co-conversion is quite extensive 3′ to the SegG cleavage site, which is correlated with the very low level of inheritance of T2 markers (Liu et al. 2003). Mutation of *segG* or *segF* activity reduces the dominance of T4 over T2 markers in the gene *32* or *56* regions, consistent with a role for SegG and SegF in the marker exclusion phenomenon. Because of the similarity to endonuclease-medi-

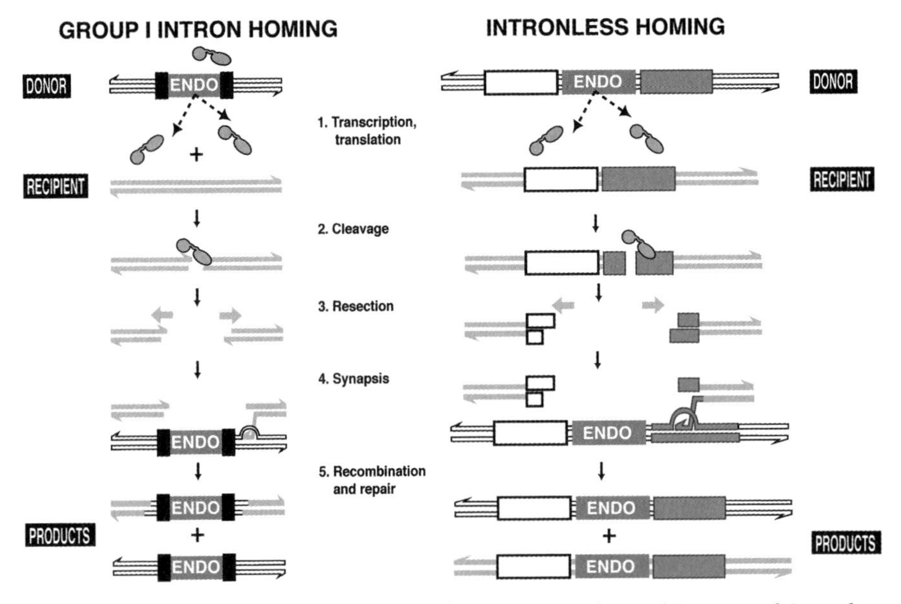

Fig. 2. Endonuclease-mediated homing pathways. Comparison of intron and intronless homing. Homing is a DNA-dependent pathway, initiated by the intron-encoded homing endonuclease (*ENDO, red rectangle*) encoded within a group I intron (indicated by *black rectangles*). The endonuclease (*grey dumbbell*) initiates the homing pathway by binding to and cleaving a sequence in recipient or intronless alleles (*grey rectangles*). For intronless homing, the endonuclease gene (*red rectangle*) is not intron- or intein-encoded but is instead free standing, inserted between genes (*open and dark grey rectangles*) conserved between different phage genomes. The endonuclease (*grey dumbbell*) binds and cleaves in a gene adjacent to the endonuclease insertion site

ated homing of group I introns, the spread of free-standing endonucleases to genomes that lacked the endonuclease gene was named intronless homing (Belle et al. 2002; Fig. 2).

Of the remaining free-standing endonucleases in T4, only *segE*, adjacent to *uvsW* (Miller et al. 2003), has also been shown to exclude genetic markers in crosses between T4 and RB30, a related T-even phage (Kadyrov et al. 1997). Interestingly, Russell and Huskey's data shows that the *uvsW* region of the T4 genome does not completely dominate over that of T2 in progeny phage, as the frequency of T2 markers in this region is ~20%, rather than <1% expected from localized marker exclusion (Russell and Huskey 1974). This may reflect the paucity of essential genes in this region of the T4 and T2 genomes for which Russell and Huskey could obtain amber mutants, and thus measure exclusion. The same is true of three other *seg* genes (*segB*, *segC* and *segD*) encoded in the same genomic region as *segE* (Miller et al. 2003), where none of the surrounding genetic markers dominate over the corresponding T2 markers in progeny phage (Russell and Huskey 1974), but none of these *seg* genes have been shown to encode active endonucleases.

4 The Recognition Sites of Free-Standing Endonucleases Are Distinct from the Endonuclease Insertion Site

Phage-encoded free-standing endonucleases that have been characterized to date all introduce a DSB, leaving a 3′ 2-nucleotide overhang (Belle et al. 2002; Liu et al. 2003), similar to GIY-YIG intron-encoded endonucleases (Bell-Pedersen et al. 1989; Loizos et al. 1996; Edgell and Shub 2001). While the biochemistry of DNA cleavage may be similar to intron-encoded versions, free-stand-

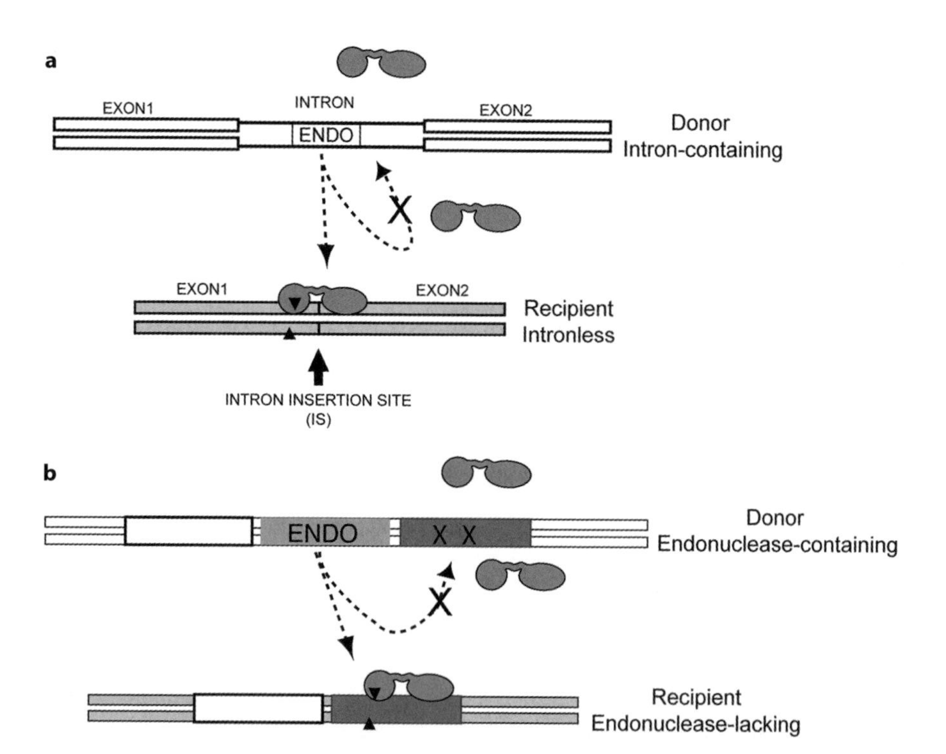

Fig. 3. Mechanisms to minimize cleavage of host genomes differ for **a** intron-encoded or **b** free-standing endonucleases. *Black triangles* represent endonuclease cleavage sites. *X* indicates a nucleotide substitution that may prevent cleavage by free-standing endonucleases

ing endonucleases differ from intron endonucleases in one key property, the separation of endonuclease cleavage and endonuclease gene insertion sites, which are often hundreds of base pairs distant (Fig. 3; Edgell 2002). This arrangement contrasts with that of intron-encoded endonucleases, where the endonuclease cleavage and intron insertion sites are separated by 2–25 base pairs (Chevalier and Stoddard 2001).

For intron- or intein-encoded endonucleases, the recognition site and cleavage sites are intimately linked to the insertion site of the genetic element. This is due to the fact that intron- or intein-endonuclease recognition sites encompass sequence up- and down-stream of the insertion site. Compilation of the insertion sites of self-splicing introns and inteins has revealed that many interrupt conserved protein-coding genes (Dalgaard et al. 1997; Edgell et al. 2000; Pietrokovski 2001; Gogarten et al. 2002). Furthermore, the intron or intein insertion site often lies within, or nearby, nucleotide sequence that corresponds to functionally critical residues of the host protein. Thus, the recognition site of an intron- or intein-encoded endonuclease will include the conserved nucleotide sequence(s) that surround the insertion site (Edgell et al. 2000, 2003). These nucleotide sequences are those that are likely to be conserved between related genomes, thus maximizing the potential of the endonuclease to spread to related genomes.

In theory, free-standing endonucleases could use the junction of the endonuclease insertion site as a recognition sequence, in analogy to the recognition sequences of intron- or intein-encoded endonucleases. Such sequences, however, are likely to be poor choices for a recognition site because of the potential for variability in intergenic regions of phage DNA. Mapping of the cleavage sites for SegE, SegF and SegG all demonstrated that they lie within genes adjacent to the Seg insertion site, and that the target genes are conserved between T-even phage genomes (Kadyrov et al. 1997; Belle et al. 2002; Liu et al. 2003). This strategy of selection of a recognition site within a conserved gene is similar to that employed by intron-encoded versions, as it maximizes the potential that the site will be present in related genomes.

5 How Do Free-Standing Endonucleases Prevent Cleavage of Their Host Genome?

The genomes of organisms harboring intron- or intein-encoded endonucleases are protected from cleavage by the endonuclease primarily because the endonuclease's recognition site is disrupted by the presence of the intron or intein, preventing the endonuclease from binding and cleaving intron- or intein-containing alleles (Fig. 3A). In addition, the length of a homing endonuclease's recognition sequence (14–40 bp) minimizes potential interactions with the host genome, as the recognition site is likely to present in a single copy per genome (Jurica and Stoddard 1999).

Free-standing endonucleases must minimize cleavage of host genomes by different mechanisms, however, because the recognition sequence of a free-standing endonuclease is not interrupted by the endonuclease gene (Fig. 3B). Thus, potential cleavage sites exist for the free-standing endonuclease in both the host (self) and target (non-self) genome. Experimental evidence to date suggests that SegE, SegG and SegF cleave their host T4 genome inefficiently as compared to T2 or other T-even genomes (Kadyrov et al. 1997; Belle et al. 2002; Liu et al. 2003). While the exact mechanism that protects T4 DNA from efficient cleavage by SegE, SegF or SegG is unknown, examination of T4 and T2 DNA sequences surrounding the endonuclease's cleavage sites suggests that nucleotide polymorphisms may provide a means to discriminate T4 from T2 DNA. In particular, the cleavage sites of SegG map to a polymorphic region of gene *32*, and it was shown that discrimination between substrates is at the level of DNA cleavage rather than DNA binding (Liu et al. 2003). Furthermore, the resistant site would be inherited in progeny phage of an intronless homing event along with the endonuclease gene, thus rendering progeny phage immune to cleavage by the newly inherited endonuclease.

6 The Separation of Cleavage and Insertion Sites Influences Endonuclease Mobility

Endonuclease-mediated homing of introns or inteins is dependent on the host DSB-repair machinery (Fig. 2), which uses the intron-containing (or donor) allele as a template to repair the break in the intronless (or recipient) allele (reviewed in Belfort et al. 2002). The distance between the endonuclease's insertion and cleavage sites determines the extent of exonucleolytic processing of the DSB that is required in the intronless allele (Huang et al. 1999). Exonucleolytic processing will proceed until sufficient homology occurs between the donor and recipient alleles to promote strand invasion and subsequent repair, which in phage T4 can be as little as 30 base pairs of sequence identity (Bautz and Bautz 1967; Singer et al. 1982; Parker et al. 1999). For T4 intron-encoded endonucleases, the amount of exonucleolytic degradation required to initiate repair is limited to the distance that separates the cleavage sites and the intron insertion site, 23 base pairs in the case of the *td* intron and its encoded endonuclease, I-TevI (Huang et al. 1999). This requirement for limited processing of DSBs is reflected in the high frequency of inheritance of the phage T4 introns (Quirk et al. 1989a; Liu et al. 2003; Sandegren and Sjoberg 2004).

The distance separating the cleavage sites and the insertion site of the free-standing endonuclease gene is much greater than that for intron-encoded endonucleases. This distance, which is often hundreds of base pairs, influences the extent of exonucleolytic degradation required in intronless homing. The extent of exonucleolytic degradation is also influenced by the degree of nucleotide sequence identity in the genomic region between the endonuclease gene insertion site and the cleavage sites. In regions of high nucleotide identity, strand invasion and subsequent repair can be initiated close to the endonuclease's cleavage sites. For SegG-mediated events, repair of the DSB in T2 gene *32* can be initiated close to the SegG cleavage sites, as they lie within a region of gene *32* that shares high nucleotide identity between T2 and T4. Such repair events are evidenced by the ~90% inheritance of the *segG* gene in progeny phage, whereas sequence immediately surrounding the cleavage sites is inherited in 100% of progeny phage. This contrasts with intronless homing of T4-encoded SegF, whose cleavage sites lie within a region of patchy sequence identity between T2 and T4 genes *56* (Belle et al. 2002). Insufficient homology exists between T2 and T4 gene *56* to initiate strand invasion, meaning that exonucleolytic degradation must extend past gene *56* on both sides of the DSB to the neighboring and well-conserved *dam* and *soc* genes.

7 The Sporadic Distribution of Free-Standing Homing Endonucleases

Experiments have shown that free-standing endonucleases can be inherited at a very high frequency in progeny phage (Kadyrov et al. 1997; Liu et al. 2003), consistent with their role as selfish genetic elements promoting spread to related genomes by a DSB-repair mechanism. Furthermore, free-standing homing endonucleases are more abundant in multi-copy genomes, such as bacteriophage, viral and organellar genomes than, for instance, bacterial genomes. The propensity of homing endonucleases to reside in multi-copy genomes may reflect the greater number of genome equivalents available to repair DSBs as compared to lower copy number genomes. This is particularly evident when considering the case of bacteriophage T4, as it is infested with an astounding number of homing endonucleases, 15 in total (16, if endonuclease II/*denA* is included; Miller et al. 2003). This equates to approximately 15 kb of endonuclease coding sequence and represents, on average, an endonuclease gene every 11 kb of T4 DNA, or ~8% of the T4 genome. Three of the endonucleases are intron-encoded, I-TevI, -II or -III. The remaining are free-standing endonucleases of the GIY-YIG or HNH families, termed Seg (similarity to endonucleases of group I intron; Sharma et al. 1992) and Mob (*mobility*; Kutter et al. 1995) genes, respectively.

As noted previously, T4 appears to be an oddity among T-even phage with respect to number of endonucleases (Eddy 1992; Edgell et al. 2000; Sandegren and Sjoberg 2004). Although very few systematic studies have been undertaken to assess the distribution of endonucleases in T-even phage genomes (Eddy 1992; Sandegren and Sjoberg 2004), it appears that free-standing and intron-encoded endonucleases are relatively rare in T-even phage. While many attempts have been made to rationalize the natural distribution of endonucleases in T-even phage with seemingly contradictory laboratory demonstrations of efficient homing (Edgell et al. 2000), the sporadic distribution of free-standing endonucleases is consistent with the notion that they are selfish genetic elements that promote their own spread to related genomes when the opportunity arises. As has been shown by Goddard and Burt (1999), endonuclease-mediated homing of a group I intron in yeast mitochondria follows a cyclical process of invasion, degeneration, and loss, followed by reinvasion. It is possible that the current distribution of endonucleases in T4 and other T-even phage represents an early step in the invasion stage of Goddard and Burt's model.

Regardless, it is clear that all 15 T4 endonucleases have successfully integrated themselves into the genome by various means, and that T4 appears to be a haven for endonucleases as compared to other phage genomes. An interesting theory as to why endonucleases persist in the T4 genome was raised by Shub and colleagues (Belle et al. 2002). They postulated that T4 possesses extremely efficient repair machinery that would be necessary to accommodate spurious DSBs generated throughout the T4 genome by each of the 15 endonucleases binding at non-cognate sites. Shub and coworkers suggested that one explanation for the paucity of endonucleases in other T-even genomes is that those phages might possess inefficient repair machinery as compared to T4, and can thus only tolerate a limited number of endonucleases.

8 The "Function" of Free-Standing Endonucleases

Selfish genetic elements are not commonly thought of as providing either a selective benefit or burden to a host genome, but are considered to be phenotypically neutral with respect to host fitness (Doolittle and Sapienza 1980). This assumption is supported by the experimental observation that none of the characterized homing endonucleases of phage T4 are essential; phage carrying mutations in the endonuclease genes are viable and do not exhibit reduced burst size or altered plaque morphology as compared to wild-type phage, a result likely to hold true for the remaining uncharacterized endonucleases in T4. In the absence of an apparent selective advantage for the retention of a mobile element gene in the host genome, mobile endonucleases will suffer the fate of extinction unless they can successfully spread between genomes (Goddard and Burt 1999). Thus, it is perhaps not surprising that endonucleases thrive in genomic populations where recombination is rampant, such as T-even phage, because recombination provides the biochemical means for spread of the elements. Alternatively, once established in a genome, endonucleases can avoid deletion from that genome by integrating themselves into some aspect of host physiology. The positioning of the promoters for the essential gene *32* within the *segG*-coding sequence (Loayza et al. 1991; Liu et al. 2003), making deletion of *segG* from the T4 genome a lethal event, may represent such an adaptation.

One, however, is struck by the observation that phage T4 possesses a total of 15 homing endonucleases, 12 of which are free standing (Miller et al. 2003). Retention of approximately 15 kb of non-essential endonuclease coding sequence in the size-constrained phage genome represents a significant commitment to replicate that DNA by phage T4, leading one to question whether the endonucleases do in fact function in some manner that confers an advantage to T4 over its relatives. Because intron-encoded or free-standing endonucleases have not yet been shown to provide an essential function for T4, the question of whether or not homing endonucleases provide a benefit to their host must be based on a consideration of the consequences of the endonuclease's biochemical activity. If the combined actions of the phage T4-encoded endonucleases resulted in complete exclusion of T2 genetic markers from the progeny of mixed infection, then the answer would be yes, because this would provide a clear-cut example of phage exclusion and selective advantage to T4. T2 markers are not, however, completely excluded from progeny, and contribute significantly to the genome of progeny phage. It is possible that the selective advantages conferred upon T4 by its encoded endonucleases are more subtle than have yet been experimentally measured, but it is unclear what advantage(s) the endonucleases might confer, and how to design experiments to test this.

9 The Diversity of Endonuclease Function

This chapter has dealt solely with free-standing endonucleases that function in intronless homing. There is much to learn about the mechanisms of how free-standing endonucleases promote their mobility between genomes, specifically with regard to mechanisms of cleavage specificity, and how free-standing endonucleases have integrated themselves into the transcriptional and translational program of the host genome. This chapter has not touched on the many proteins encoded within bacterial, phage, organellar or eukaryotic genomes that possess domains resembling the catalytic domains of free-standing or intron-encoded endonucleases. Study of these proteins would not only provide insight into the functionality and adaptability of catalytic domains of homing endonucleases, but also illustrate how the biochemical activities of site-specific DNA endonucleases have been harnessed by host genomes over evolutionary time to function in cellular pathways unrelated to mobility.

References

Bautz FA, Bautz EK (1967) Transformation in phage T4: minimal recognition length between donor and recipient DNA. Genetics 57:887–895

Belfort M, Derbyshire V, Cousineau B, Lambowitz A (2002) Mobile introns: pathways and proteins. In: Craig N, Craigie R, Gellert M, Lambowitz A (eds) Mobile DNA II. ASM Press, Washington, DC, pp 761–783

Bell-Pedersen D, Quirk SM, Aubrey M, Belfort M (1989) A site-specific endonuclease and co-conversion of flanking exons associated with the mobile td intron of phage T4. Gene 82:119–126

Belle A, Landthaler M, Shub DA (2002) Intronless homing: site-specific endonuclease SegF of bacteriophage T4 mediates localized marker exclusion analogous to homing endonucleases of group I introns. Genes Dev 16:351–362

Carlson K, Kosturko L. D (1998) Endonuclease II of coliphage T4: a recombinase disguised as a restriction endonuclease? Mol Microbiol 27:671–676

Carlson K, Wiberg JS (1983) In vivo cleavage of cytosine-containing bacteriophage T4 DNA to genetically distinct, discretely sized fragments. J Virol 48:18–30

Chevalier BS, Stoddard BL (2001) Homing endonucleases: structural and functional insight into the catalysts of intron/intein mobility. Nucleic Acids Res 29:3757–3774

Dalgaard JZ, Moser MJ, Hughey R, Mian IS (1997) Statistical modeling, phylogenetic analysis and structure prediction of a protein splicing domain common to inteins and hedgehog proteins. J Comput Biol 4:193–214

Doolittle WF, Sapienza C (1980) Selfish genes, the phenotype paradigm and genome evolution. Nature 284:601–603

Eddy SR (1992) Introns in the T-even bacteriophages. University of Colorado, Boulder

Edgell DR (2002) Selfish DNA: new abode for homing endonucleases. Curr Biol 12: R276–R278

Edgell DR, Shub DA (2001) Related homing endonucleases I-BmoI and I-TevI use different strategies to cleave homologous recognition sites. Proc Natl Acad Sci USA 98:7898–7903

Edgell DR, Belfort M, Shub DA (2000) Barriers to intron promiscuity in bacteria. J Bacteriol 182:5281–5289

Edgell DR, Stanger MJ, Belfort M (2003) Importance of a single base pair for discrimination between intron-containing and intronless alleles by endonuclease I-BmoI. Curr Biol 13:973–978

Goddard MR, Burt A (1999) Recurrent invasion and extinction of a selfish gene. Proc Natl Acad Sci USA 96:13880–13885

Gogarten JP, Senejani AG, Zhaxybayeva O, Olendzenski L, Hilario E (2002) Inteins: structure, function, and evolution. Annu Rev Microbiol 56:263–287

Gorbalenya AE (1994) Self-splicing group I and group II introns encode homologous (putative) DNA endonucleases of a new family. Protein Sci 3:1117–1120

Huang YJ, Parker MM, Belfort M (1999) Role of exonucleolytic degradation in group I intron homing in phage T4. Genetics 153:1501–1512

Jurica MS, Stoddard BL (1999) Homing endonucleases: structure, function and evolution. Cell Mol Life Sci 55:1304–1326

Kadyrov FA, Shlyapnikov MG, Kryukov VM (1997) A phage T4 site specific endonuclease, SegE, is responsible for a non-reciprocal genetic exchange between T-even-related phages. FEBS Lett 415:75–80

Kowalski JC, Belfort M, Stapleton MA, Holpert M, Dansereau JT, Pietrokovski S, Baxter SM, Derbyshire V (1999) Configuration of the catalytic GIY-YIG domain of intron endonuclease I-TevI: coincidence of computational and molecular findings. Nucleic Acids Res 27:2115–2125

Kutter E, Gachechiladze K, Poglazov A, Marusich E, Shneider M, Aronsson P, Napuli A, Porter D, Mesyanzhinov V (1995) Evolution of T4-related phages. Virus Genes 11:285–97

Liu Q, Belle A, Shub DA, Belfort M, Edgell DR (2003) SegG endonuclease promotes marker exclusion and mediates co-conversion from a distant cleavage site. J Mol Biol 334:13–23

Loayza D, Carpousis AJ, Krisch HM (1991) Gene 32 transcription and mRNA processing in T4-related bacteriophages. Mol Microbiol 5:715–725

Loizos N, Silva GH, Belfort M (1996) Intron-encoded endonuclease I-TevII binds across the minor groove and induces two distinct conformational changes in its DNA substrate. J Mol Biol 255:412–424

Maté MJ, Kleanthous C (2004) Structure-based analysis of the metal-dependent mechanism of H-N-H endonucleases. J Biol Chem 279:34763–34769

Miller ES, Kutter E, Mosig G, Arisaka F, Kunisawa T, Ruger W (2003) Bacteriophage T4 genome. Microbiol Mol Biol Rev 67:86–156

Parker MM, Belisle M, Belfort M (1999) Intron homing with limited exon homology. Illegitimate double-strand-break repair in intron acquisition by phage T4. Genetics 153:1513–1523

Pees E, de Groot B (1970) Partial exclusion of genes of bacteriophage T2 with T4-glucosylated DNA in crosses with bacteriophage T4. Genetica 41:541–550

Pietrokovski S (2001) Intein spread and extinction in evolution. Trends Genet 17:465–472

Pommer AJ, Cal S, Keeble AH, Walker D, Evans SJ, Kuhlmann UC, Cooper A, Connolly BA, Hemmings AM, Moore GR, James R, Kleanthous C (2001) Mechanism and cleavage specificity of the H-N-H endonuclease colicin E9. J Mol Biol 314:735–749

Quirk SM, Bell-Pedersen D, Belfort M (1989a) Intron mobility in the T-even phages: high frequency inheritance of group I introns promoted by intron open reading frames. Cell 56:455–465

Quirk SM, Bell-Pedersen D, Tomaschewski J, Ruger W, Belfort M (1989b) The inconsistent distribution of introns in the T-even phages indicates recent genetic exchanges. Nucleic Acids Res 17:301–315

Ray U, Bartenstein L, Drake JW (1972) Inactivation of bacteriophage T4 by ethyl methanesulfonate: influence of host and viral genotypes. J Virol 9:440–447

Repoila F, Tetart F, Bouet JY, Krisch HM (1994) Genomic polymorphism in the T-even bacteriophages. EMBO J 13:4181–4192

Russell RL, Huskey RJ (1974) Partial exclusion between T-even bacteriophages: an incipient genetic isolation mechanism. Genetics 78:989–1014

Sancar A (1996) DNA excision repair. Annu Rev Biochem 65:43–81

Sandegren L, Sjoberg BM (2004) Distribution, sequence homology, and homing of group I introns among T-even-like bacteriophages: evidence for recent transfer of old introns. J Biol Chem 279:22218–22227

Sharma M, Ellis RL, Hinton DM (1992) Identification of a family of bacteriophage T4 genes encoding proteins similar to those present in group I introns of fungi and phage. Proc Natl Acad Sci USA 89:6658–6662
Shen BW, Landthaler M, Shub DA, Stoddard BL (2004) DNA binding and cleavage by the HNH homing endonuclease I-HmuI. J Mol Biol 342:43–56
Shub DA, Goodrich-Blair H, Eddy SR (1994) Amino acid sequence motif of group I intron endonucleases is conserved in open reading frames of group II introns. Trends Biochem Sci 19:402–404
Singer BS, Gold L, Gauss P, Doherty DH (1982) Determination of the amount of homology required for recombination in bacteriophage T4. Cell 31:25–33
Sitbon E, Pietrokovski S (2003) New types of conserved sequence domains in DNA-binding regions of homing endonucleases. Trends Biochem Sci 28:473–477
Truglio JJ, Rhau B, Croteau DL, Wang L, Skorvaga M, Karakas E, Della Vecchia MJ, Wang H, VanHouten B, Kisker C (2005) Structural insights into the first incision reaction during nucleotide excision repair. EMBO J 24:885-894.
Verhoeven EE, van Kesteren M, Moolenaar GF, Visse R, Goosen N (2000) Catalytic sites for 3′ and 5′ incision of *Escherichia coli* nucleotide excision repair are both located in UvrC. J Biol Chem 275:5120–5123

Function and Evolution of HO and VDE Endonucleases in Fungi

James E. Haber, Kenneth H. Wolfe

1 Introduction

The site-specific HO and VDE endonucleases are unusual members of a family of so-called group I LAGLIDADG homing endonucleases that are generally implicated in the homing of intron and intein sequences. The great majority of these endonucleases are found in mitochondria and plastids of eukaryotes, but in budding yeast two members of this family apparently "escaped" into the nucleus. In each case, the endonuclease creates a site-specific double-strand break (DSB) in a target and promotes mobility of DNA sequences by homologous recombination requiring the Rad52 and Rad51 group of recombination proteins.

The HO endonuclease has been the subject of a great deal of interest, both because of its remarkable evolution and because of its great utility in the detailed analysis of DSB-mediated recombination. Unlike most homing endonucleases, the HO endonuclease does not promote its own amplification, but rather catalyzes the switching/replacement of yeast's mating-type (*MAT*) genes.

In addition, budding yeasts harbor the VDE endonuclease, which is found as an intein in the *VMA1* gene. VDE promotes its propagation from a *VMA1* gene containing the VDE intein into a *VMA1* gene lacking such a sequence. VDE is remarkable also in sharing a close sequence and presumably evolutionary relationship with HO.

J.E. Haber (e-mail: haber@brandeis.edu)
Department of Biology and Rosenstiel Center, Brandeis University, Waltham, Massachusetts 02454–9110, USA

K.H. Wolfe (e-mail: khwolfe@tcd.ie)
Department of Genetics, Smurfit Institute, University of Dublin, Trinity College, Dublin 2, Ireland

Nucleic Acids and Molecular Biology, Vol. 16
Marlene Belfort et al. (Eds.)
Homing Endonucleases and Inteins

2 Mating-Type Genes in *S. cerevisiae*

Many fungi use some form of homothallic mating-type switching to produce diploids from haploid meiotic segregants. In budding yeast *MAT***a**/*MAT*α diploids are created by conjugation of haploid *MAT***a** and *MAT*α cells. The *MAT***a**

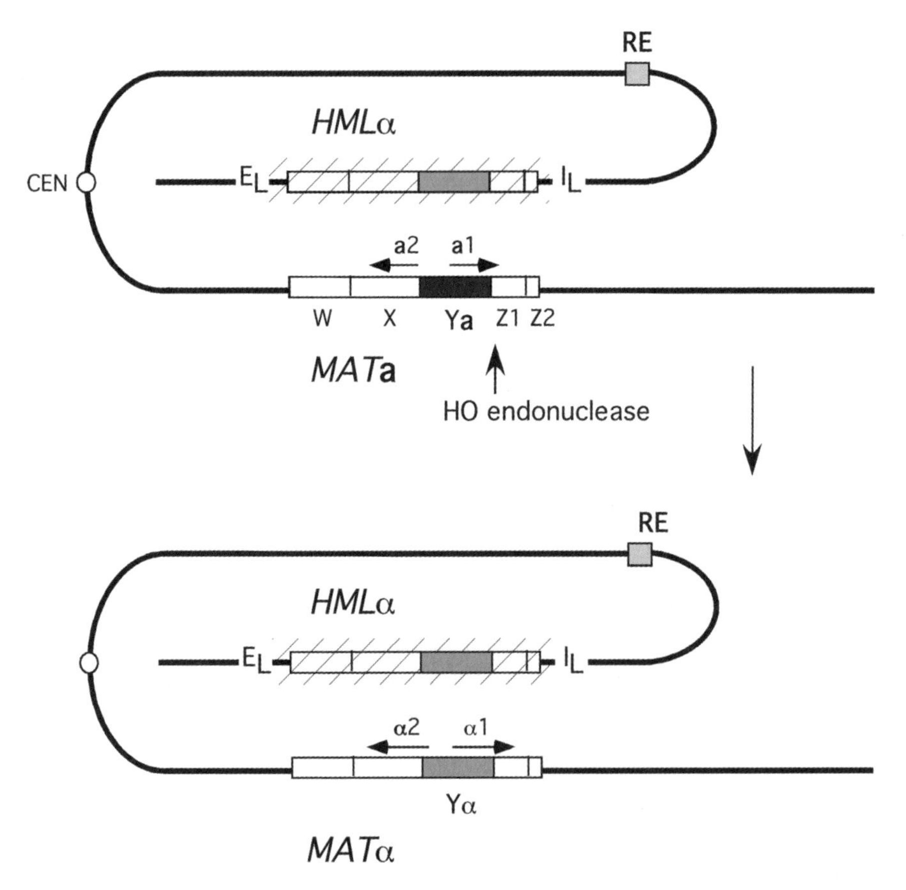

Fig. 1. Arrangement of mating-type genes in *S. cerevisiae*. The *MAT* locus is approximately 2400 bp. Mating-type-specific Y**a** and Yα sequences are about 650 and 750 bp, respectively, and are surrounded by 700-bp W and X regions, a 230-bp Z1 region and a 90-bp Z2 region. The orientation of *MAT***a** and *MAT*α transcripts are shown. The donor sequences *HML*α and *HMR***a** are unexpressed and maintained in a silent, heterochromatic state by action of the Sir2 histone deacetylase and accessory Sir1, Sir3 and Sir4 proteins, which are first recruited to E and I silencer sites surrounding the donors. *MAT* switching is initiated by HO endonuclease cleavage in Z1, cutting 7 and 3 bp on the top and bottom strands from the Y/Z1 border, creating 4-bp 3′ overhanging ends. The donor loci cannot be cut. Donor preference is governed by the recombination enhancer (RE), which controls recombination efficiency along the entire left arm of chromosome III

and *MAT*α alleles are located in the middle of chromosome III (Fig. 1) and differ from each other in a 643-bp (Y**a**) or 748-bp (Yα) region that contains different DNA sequences encoding regulators of mating type (reviewed by Haber 2002). *MAT*α encodes two genes, *MAT*α1 and *MAT*α2. *MAT*α1 encodes a positive transcriptional regulator of α-specific genes (e.g. the pheromone α-factor, and an **a**-pheromone receptor, Ste3). *MAT*α2 encodes a repressor with two modes of action. In *MAT*α cells, *MAT*α2 pairs with the general transcriptional regulator, Mcm1, to repress **a**-specific genes (e.g. those encoding the pheromone, a-factor, and the α-factor receptor, Ste2). In *MAT***a**/*MAT*α diploid cells, *MAT*α2 and *MAT***a**1 proteins form a repressor that turns off haploid-specific genes, including the pheromone signal transduction pathway, the nonhomologous end-joining (NHEJ) pathway of DNA repair and a repressor of meiosis.

Most *S. cerevisiae* strains carry two additional copies of mating-type genes, *HML*α and *HMR***a**, at distant locations on the same chromosome as the *MAT* locus (Fig. 1). In some strains the specific mating-type sequences at these loci are reversed (i.e. *HML***a** or *HMR*α). These genes are, however, not expressed, as both *HML* and *HMR* are surrounded by silencer sequences (designated E and I) that recruit the specialized histone deacetylase, Sir2, and other silencing factors (Sir1, Sir3 and Sir4) to create a heterochromatic domain that is not transcribed (Rusche et al. 2003). These silent loci are the key to the ability of HO endonuclease to promote the switching of the *MAT* locus.

3 HO-Induced *MAT* Switching in *S. cerevisiae*

Homothallic haploid cells, carrying an active HO endonuclease gene, can switch to the opposite mating-type allele. When the *MAT* locus is cleaved by HO endonuclease, the DSB is repaired by homologous recombination (gene conversion) using one of two unexpressed donor loci, *HML*α or *HMR***a**, which are located 200 and 100 kb 5′ and 3′, respectively, on chromosome III. *HML* and *HMR* differ from each other in that *HML* shares more extensive (regions W and Z2) homology with *MAT* than does *HMR*. *MAT* switching displays donor preference, in which *MAT***a** selectively recombines with *HML* whereas *MAT*α preferentially recombines with *HMR*.

Expression of HO endonuclease is strongly regulated, so that it is turned on only in the late G1 stage of the cell cycle and only in cells that have undergone at least one cell division cycle. After two cell divisions, two *MAT*α cells are juxtaposed to two *MAT***a** cells and efficient conjugation produces two *MAT***a**/MATα zygotes. Coexpression of *MAT***a** and *MAT*α turns off the HO gene itself. Thus, once cells switch, the process is complete until meiosis creates haploid segregants.

Much of the research studying HO-mediated *MAT* switching has relied on freeing the HO gene from its transcriptional controls. Jensen and Herskowitz (1984) fused the HO gene to a galactose-regulated promoter, so that all cells in the population can be induced to switch in a synchronous fashion. The use of the *GAL::HO* gene has proven to be invaluable in studying not only *MAT* switching, but in characterizing other forms of DSB repair (NHEJ and break-induced replication) as well as in the detailed study of the DSB-induced DNA damage checkpoint. The applications of HO as a regulated source of defined DSBs to study these different processes have been well reviewed (Moore and Haber 1996a,b; Toczyski et al. 1997; Lee et al. 1998; Nickoloff and Hoekstra 1998; Haber 2000; Melo and Toczyski 2002; Aylon and Kupiec 2004). Here, we focus only on *MAT* switching itself.

4 Mechanism of *MAT* Switching

The ability to induce synchronous HO cleavage at *MAT* made it possible to follow the process of *MAT* switching in "real-time" by isolating DNA samples at intervals after the creation of the DSB. A combination of Southern blot and polymerase chain reaction (PCR) analysis to analyze DNA intermediates (White and Haber 1990) as well as chromatin immunoprecipitation (ChIP) to identify recombination proteins that participate in the gene conversion process (Sugawara et al. 2003; Wolner et al. 2003; Wang and Haber 2004) have identified a number of steps in the process (here shown as *MAT***a** switching to *MAT*α in Fig. 2): (1) the resection of the DSB ends by 5′ to 3′ exonucleases to create single-strand DNA (ssDNA) ends; (2) assembly of a Rad51 filament on ssDNA to facilitate a search for homologous donor sequences to repair the DSB (i.e. *HML* or *HMR*); (3) strand invasion and the synapsis between the donor and *MAT* sequences; (4) use of the 3′ end of the invading ssDNA as a primer to promote new DNA synthesis, a process that involves PCNA and either DNA polymerases δ or ε; (5) displacement of the newly synthesized strand to pair with the second end of the DSB and the completion of the replacement of the original Y**a** sequence with Yα. During the process, the donor is left unchanged and all the newly synthesized DNA is found at the *MAT* locus (G. Ira and J.E. Haber, unpubl.).

5 Donor Preference Associated with *MAT* Switching

There is one other remarkable aspect of *MAT* switching: the mating-type regulated choice of *HML* or *HMR* as a donor in repairing the HO-induced DSB. A *MAT***a** haploid recombines with *HML*α about 90% of the time, whereas a

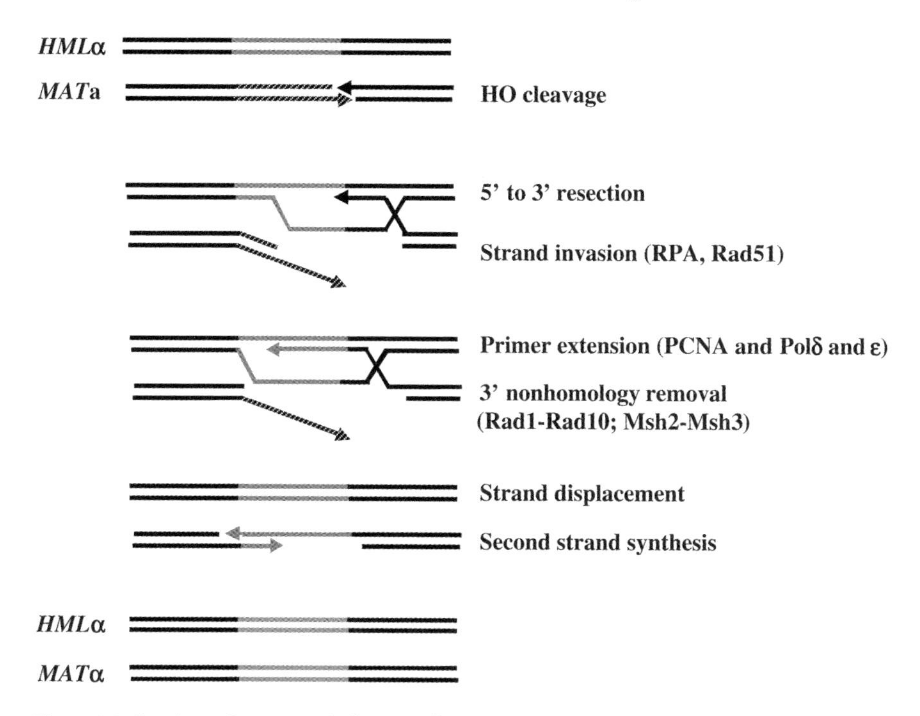

Fig. 2. Mechanism of *MAT***a** switching. Following the creation of an HO-induced DSB, the 5′ ends are resected to create 3′-ended single-stranded tails that are bound first by the single-stranded binding protein complex RPA and then by the Rad51 recombinase that promotes strand invasion of the Z region of the *HML*α donor. The 3′ end of the invading DNA is used as a primer to initiate new DNA synthesis. The newly synthesized strand is apparently displaced and pairs with resected DNA on the other side of the DSB. *MAT* switching is completed, almost always without an accompanying crossover, by copying the second strand

*MAT*α haploid selectively chooses *HMR***a**. This ensures that, most of the time, the process of switching will produce an equal number of juxtaposed cells of opposite mating type after two cell divisions. "Donor preference" is not determined by the differences between the silencer regions surrounding *HML* and *HMR*, nor by the Y**a** or Yα sequences themselves; in a strain in which *HML*α is deleted and replaced by a cloned *HMR*α locus, *MAT***a** cells continue to select the left-arm donor (Weiler and Broach 1992). Control of donor preference lies with the recombination enhancer (RE), a small, cis-acting sequence located 17 kb away from *HML* (Wu and Haber 1996; Wu et al. 1998). In *MAT***a** cells, RE is active and facilitates the use of *HML* over *HMR*. In *MAT*α cells, RE is repressed, and the left-arm donor becomes in some way inaccessible; consequently, the right-arm donor is used preferentially. RE acts over the entire 115 kb of the left arm of chromosome III (Wu and Haber 1996; Sun et al. 2002). A donor placed anywhere along this arm can be used selectively in *MAT***a** cells. When RE is deleted, the left-arm donor becomes inaccessible in *MAT***a** cells. Most of the activity of RE is contained in a segment of about 750 bp. Further analysis, including comparisons between the REs of three *Saccharomyces* species (Wu et al. 1998) and the creation of synthetic RE sequences from multimers of conserved subregions (Wu and Haber 1996; Sun et al. 2002), showed that RE consists of several highly conserved domains, including two regions with ten or more iterations of TTT(A/G) and a highly conserved MAT "Greek alpha" 2-Mcm1 "operator region" which shares strong sequence identity with sequences that control expression of "**a**-specific genes".

Repression of RE in *MAT*α cells occurs by binding of the α2-Mcm1 repressor that also turns off **a**-specific genes to the operator site and leads to the establishment of highly positioned nucleosomes across the 2.5-kb region containing RE (there are no open reading frames in this region; there is no change in the chromatin structure of any of the genes along the chromosome arm nor changes in the silencing of *HML*). In *MAT***a**, the positioned nucleosomes are absent and several DNaseI-hypersensitive sites indicative of protein binding are found (Wu et al. 1998). Activation of RE depends on Mcm1 (which is also an activator of expression of **a**-specific genes). Mcm1 binding facilitates the binding of the Fkh1 transcription regulator (Sun et al. 2002). The complex RE sequences can be replaced by as few as four 22-bp copies of one of the Fkh1-binding sites located in several conserved subregions of the RE. Small effects on *MATa*'s use of *HML* have been seen when the cohesin assembly protein Chl1 is deleted (Weiler et al. 1995).

How RE acts remains unknown, but experiments recruiting the LacI-GFP fusion protein to LacO arrays inserted near *HML*, *MAT* or *HMR* have provided some clues. Live-cell imaging of GFP-tagged chromosomes suggests that the left arm of chromosome III is more confined in *MAT*α (or RE-deleted) cells than in *MAT*a (Bressan et al. 2004). The molecular basis of this tethering remains under investigation.

6 Evolutionary Origins of the *HO* Gene and Other Components of the *MAT* Switching System

The evolutionary origin of *HO* is enigmatic because the gene has a very limited phylogenetic distribution. Apart from *S. cerevisiae* and its close relatives such as *S. bayanus* (the *Saccharomyces sensu stricto* group of species), the *HO* gene is present only in *Candida glabrata, Saccharomyces castellii,* and *Zygosaccharomyces rouxii* (Butler et al. 2004). In these species, mating-type switching probably occurs by HO-catalyzed switching between an active *MAT* locus and silent *HM* cassettes, similar to the *S. cerevisiae* paradigm, although a direct demonstration of such switching has yet to be shown. A second group of yeasts, exemplified by *Kluyveromyces lactis*, have *HM* cassettes but do not have an ortholog of *HO*. *K. lactis* can switch mating types at a very low frequency (Herman and Roman 1966; Zonneveld and Steensma 2003), similar to HO-negative *ho* strains of *S. cerevisiae*. Other species with cassettes but lacking *HO* include *K. waltii* and *A. gossypii*. Switching in these species probably proceeds by homologous recombination between the cassettes and the *MAT* locus, without a specific endonuclease to initiate the recombination process. A third group of more distantly related species, such as *Yarrowia lipolytica*, do not have silent cassettes or an *HO* gene and do not switch mating type; they are heterothallic. The *HO* gene seems to have been gained by yeasts shortly before the whole genome duplication (WGD) that occurred in the shared ancestor of several yeast species (Wolfe and Shields 1997; Dietrich et al. 2004; Kellis et al. 2004), because all known species that are descended from the WGD event have *HO* (Butler et al. 2004). The origin of *HO* can be inferred to pre-date the WGD because *HO* is present in *Z. rouxii*, which is a representative of the lineage of yeasts that is sister to the WGD lineage (Kurtzman and Robnett 2003). Gene order at the *HO* locus in different yeast species is well conserved once the WGD event is taken into account (Butler et al. 2004), and does not provide much in the way of clues to *HO*'s origins. The gene simply seems to materialize in the interval between the genes *RIO2* and *SSB*. This interval seems to have had an unusually dynamic history but the significance of this observation is unclear. In several species the interval contains a very rapidly evolving gene (*YNL208W*) with prion-like amino acid composition, and in *Z. rouxii* a gene for an endoribonuclease of the T_2 family (distantly related to *S. cerevisiae RNY1*; MacIntosh et al. 2001) is present.

7 Linkage of an *HM* Cassette with the Recombination Enhancer (RE)

In *C. glabrata* one *HM* cassette is near a telomere of the same chromosome as the *MAT* locus (chromosome 2) but the other is on chromosome 5, also near a telomere (Dujon et al. 2004). The chromosome 2 interval between *HM* and *MAT* spans the region (*KAR4–SPB1*) where the RE is located in *S. cerevisiae*, but the RE sequence motifs themselves are apparently not conserved outside the *Saccharomyces sensu stricto* species (Zhou et al. 2001). The most conserved element within the RE sequence – a binding site for the α2-Mcm1 repressor – is notably absent in the more distantly related (*sensu lato*) species. The stretch of DNA spanning *MAT*, the putative RE, and an *HM* cassette is one of the largest unrearranged genomic regions shared by *K. lactis* and

K. waltii, and is also essentially unrearranged in *A. gossypii*, which suggests that the chromosomal structure has been preserved by natural selection. Much of the gene order in this interval is also shared with *S. cerevisiae* and *C. glabrata*, but there have been some additional rearrangements in those species as well as the effect of WGD.

It should be noted that in an efficiently switching species such as *S. cerevisiae*, RE actually does not seem very important, as random use of *HML*α and *HMR***a** would yield predominantly **a**/α diploids in a few generations. But, at least in *S. cerevisiae*, the absence of RE does not give random switching; instead both *MAT***a** and *MAT*α cells use *HMR* 90% of the time (Wu and Haber 1996). Whether this preferential "default" use of one donor is yet another aspect of the evolution of chromosome III is not yet clear. If a donor on the same chromosome arm is much more likely to be used in repairing a DSB and if HO cleavage is less efficient in other species, RE would continue to be an important feature of homothallic switching.

8 The HO Endonuclease Site in *MAT*α1

The origin of HO is intimately associated with the structure of the *MAT* locus. In *S. cerevisiae*, the **a** and α idiomorphs (versions) of the *MAT* locus differ only by their Y**a** or Yα regions. Y**a** contains the complete coding region of the **a**-specific *MAT***a**1 gene, but Yα contains only the 5′ portions of the α-specific genes *MAT*α1 and *MAT*α2 which are transcribed divergently (Fig. 1). This arrangement ensures that each of the three *MAT* genes is only expressed in the correct cell type, but it also means that the 3′ ends of the *MAT*α1 and *MAT*α2 genes are located in DNA common to both idiomorphs. Thus, the junctions between unique (Y) and flanking (X or Z1) DNA occur within the coding regions of the *MAT*α1 and *MAT*α2 genes (Fig. 1). The HO cleavage site abuts the Y/Z1 junction. Cleavage of *MAT* by HO has not been experimentally demonstrated in species other than *S. cerevisiae*, but, in every species that has an *HO* gene, the Y/Z1 junction occurs at a conserved point in *MAT*α1 which corresponds to the HO cleavage site in *S. cerevisiae*. This arrangement strongly suggests that mating-type switching in those species occurs by a mechanism analogous to the *S. cerevisiae* mechanism. In contrast, in species that do not have an *HO* gene, such as *K. lactis* and *S. kluyveri*, the Y/Z1 junction is downstream of the stop codon of *MAT*α1 and is not at a conserved position (Butler et al. 2004). Thus, the co-option of HO into the switching process seems to have "locked" the Y/Z1 junction to a particular site in *MAT*α1, whereas previously its location could drift.

Recent sequence data from *Saccharomyces sensu stricto* species have revealed that the X and Z1 sequences that facilitate mating-type switching have been extraordinarily well conserved during evolution. Except for a few small insertions and deletions, there is 100% DNA sequence conservation among the five species *S. cerevisiae, S. paradoxus, S. mikatae, S. kudriavzevii* and *S. bayanus* across the entire X region (703 bp) and 150 bp of the Z1 region, whereas interspecies comparisons of Yα or Y**a** regions show much more divergence (79–89% identity among species in Yα, and 84–93% in Y**a**). This phenomenon was first noted by Kellis et al. (2003) who reported that the "MAT**a**2" sequence was completely conserved. ("*MAT***a**2" is a name given to the duplicated copy of the 3′ end of *MAT*α2 which is present in the X region and so is present in *MAT***a** idiomorphs; it is transcribed but there is no evidence that it encodes a functional protein.) However, the 100% conservation extends into the noncoding DNA downstream of *MAT*α2 and *MAT*α1. The X and Z1 regions are among the largest completely conserved DNA sequences in the genomes of these yeasts. Their apparent intolerance of nucleotide substitutions must be due to strong selection for efficient mating-type switching, together with the fact that the X and Z1 sequences each occur in three copies in the genome (at *MAT, HML* and *HMR*) which makes coevolution difficult.

9 Relationship of HO to VDE

The HO protein has strong sequence similarity to inteins, which are unusual selfish genetic elements found primarily in bacteria (see Perler, this Vol.). Only two nuclear genes of eukaryotes are known to contain inteins (Perler 2002), one of which is the vacuolar H^+-ATPase gene *VMA1* of *S. cerevisiae* which contains the VDE intein (Gimble and Thorner 1992). HO has higher sequence similarity to VDE than to any other intein and clusters with it in a phylogenetic tree (Dalgaard et al. 1997; Gogarten et al. 2002). This fact, together with their co-occurrence in *S. cerevisiae* despite the rarity of inteins in eukaryotes, suggests that HO shares a relatively recent common ancestor with VDE and is a sort of renegade intein. Although the *VMA1* ATPase gene is a highly conserved gene with homologues in all eukaryotes, eubacteria and archaea, the distribution of the VDE intein is limited to a few hemiascomycete species closely related to *S. cerevisiae*. Among these species, VDE has a patchy phylogenetic distribution (e.g., it is present in *Candida tropicalis* but not *C. albicans*) and there is evidence that it has been horizontally transferred among yeasts (Koufopanou et al. 2002; Okuda et al. 2003).

Now that the sequences of several HO and VDE proteins are known from different species it is possible to examine patterns of domain conservation, within HO and VDE separately, using the crystal structure of the *S. cerevisiae* VDE protein (also called PI-SceI) to help interpretation (Moure et al. 2002; Bakhrat et al. 2004). Koufopanou et al. (2002) pointed out that mobile inteins are under continual selection for efficient protein splicing to maintain expression of the host gene, but their homing endonuclease activity is only "tested" by natural selection if there are inteinless alleles available for colonization. Thus, once an intein-containing allele has reached a 100% frequency in a population its endonuclease activity can decay. It is still necessary to retain an open reading frame in the endonuclease region of the gene so that the downstream parts of the protein-splicing domain and the host gene are translated, but there will be no selection to preserve amino acid residues needed for the nuclease activity. Indeed, only three of 13 VDE proteins from different yeast species tested by Posey et al. (2004) were active endonucleases. When HO and VDE proteins are compared among a uniform set of four yeast species it is apparent that the endonuclease domain of HO is much better conserved than that of VDE (Fig. 3). However, it should be noted that because VDE can be horizontally transferred among species, it is possible that the amount of evolutionary time involved in these comparisons is not equal for the two proteins. Of the VDE proteins in this alignment, only the *S. cerevisiae* one has been shown to have an active endonuclease; the *Z. rouxii* VDE has no nuclease activity (Posey et al. 2004), and the other two have not been tested.

HO also shows strong conservation of the N-terminal part of the protein-splicing domain, even though *HO* is a free-standing gene, and there is no host gene whose product needs to be spliced. Furthermore, deletion of the N-terminal 112 residues of HO, which correspond to the protein-splicing domain, results in loss of endonuclease activity (Bakhrat et al. 2004). One possible explanation for this conservation is that the protein-splicing domain (properly called domain I) of HO may contain some residues that bind to DNA, similar to the DRR (DNA recognition region) residues in domain I of VDE (Moure et al. 2002). However, the DRR residues are poorly conserved among VDE sequences, let alone HO sequences, so it is impossible to say whether DNA recognition is the sole reason for the conservation of the protein-splicing domain in HO. It is striking that this domain of VDE, which supports protein splicing, has tolerated more insertions and deletions than the equivalent region of HO, which does not (Fig. 3). The C-terminus of HO contains a putative zinc finger domain (Russell et al. 1986), which is an extension relative to VDE. Analysis of the *S. cerevisiae* sequence alone suggested that the domain has three fingers, each with two Cys/Cys or Cys/His pairs (Russell et al. 1986; Bakhrat et al. 2004). However, the last pair (H-X2-C at positions 574–577 of the *S. cerevisiae* protein) is not conserved in other species, and the conserved Cys residues in the first pair have a C-X_3-C rather than the usual C-X_2-C spacing, making it unlikely that this pair is actually part of a zinc finger (Fig. 3). Thus the original proposals for the structure of the zinc finger in HO may not be correct.

One important difference between VDE and HO is that VDE cleavage and homing occurs only during sporulation of diploids, whereas HO is only active

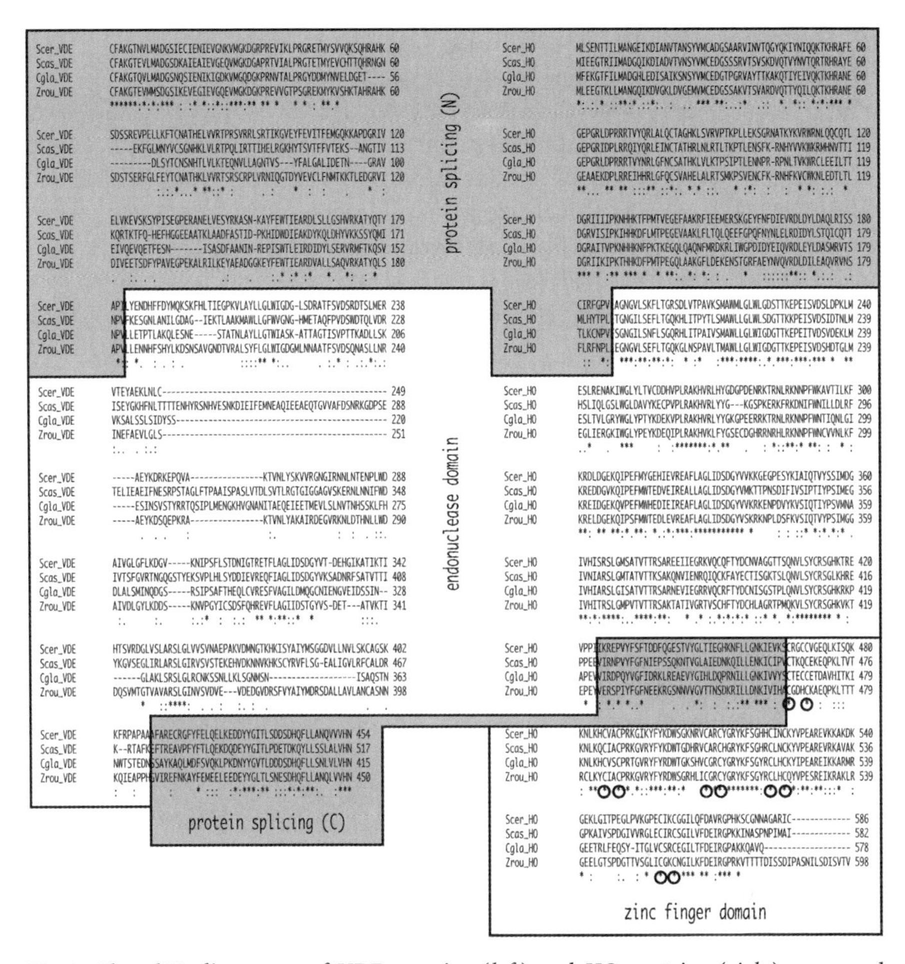

Fig. 3. ClustalW alignments of VDE proteins (*left*) and HO proteins (*right*) compared among *S. cerevisiae, S. castellii, C. glabrata* and *Z. rouxii*. The endonuclease domain of each protein lies between the N- and C-terminal parts of the protein-splicing domain, which are *shaded*. *Circles* mark conserved cysteine residues in the zinc finger of HO

in the late mitotic G1 phase of haploids. The complex way in which HO activity is regulated is reasonably well understood (Cosma 2004), whereas how VDE activity is restricted to meiosis – even though the protein is expressed at high levels in mitotically grown haploid or diploid cells – is unknown (Gimble and Thorner 1992).

10 Hypothesis for the Evolutionary Origin of HO

Where did *HO* come from? It seems very likely that HO and VDE share a recent common ancestor in the hemiascomycetes because their sequences and phylogenetic distributions are similar (Gimble and Thorner 1992; Keeling and Roger 1995; Liu 2000). One scenario is that a mobile intein (presumably from a bacterium) first invaded *VMA1* to form VDE, and has subsequently been horizontally transmitted with frequent gains and losses in various hemiascomycete species to give its current phylogenetic distribution (Koufopanou et al. 2002). The phylogenetic range over which VDE can spread is limited by the requirement for a suitable target site in the *VMA1* sequence of the host species, but this target site can co-evolve with the protein sequence of VDE (Posey et al. 2004). At some point in the cycle of horizontal transfers, gains and losses, the VDE of one species duplicated to give rise to a homing endonuclease gene that was the ancestor of HO. An alternative scenario is that the first mobile intein to invade the hemiascomycetes was located in some other unknown gene, and both VDE and HO genes were derived from it by later duplication.

We speculate that *HO* may have been formed when a mobile intein invaded a gene coding for a protein with a zinc finger domain. The integration site was either very close to the 5′ end of the gene, or else the original 5′ end of the zinc finger protein gene was lost later. Loss of protein-splicing activity led to the formation of a chimeric protein with an endonuclease domain and two DNA-binding regions: one in the zinc finger and one in the N-terminal protein-splicing domain. Somehow this fusion protein began to bind DNA specifically at a site in *MAT*α1, rather than in its own gene, thus creating a DSB that was repaired using the silent cassettes of mating-type information. This step may not be quite as far-fetched as it seems, because VDE recognition sites have been shown to drift (Posey et al. 2004), and there is a small amount of DNA sequence similarity between the recognition sites of HO and *S. cerevisiae* VDE (Gimble and Wang 1996; Bakhrat et al. 2004).

The origin of *HO* may also be connected to the origin of an unlinked gene, *SIR1*. A crucial feature of HO activity is that it cleaves the recognition site present in the active *MAT* locus, but not the sites in the *HM* cassettes which have identical DNA sequences. Cleavage at the silent cassettes is repressed by the presence of heterochromatin, whose formation is directed by the Sir proteins. Whereas Sir2, Sir3 and Sir4 proteins are also involved in transcriptional silencing at rDNA and telomeres, Sir1 is uniquely involved in silencing at the *HM* loci, by recruiting the NAD-dependent histone deacetylase Sir2 (Chien et al. 1993). *SIR1* genes have almost exactly the same phylogenetic distribution as HO genes: present in *Saccharomyces sensu stricto, Saccharomyces castellii* and *Zygosaccharomyces rouxii*, but absent in

K. lactis, K. waltii, S. kluyveri, A. gossypii and more distantly related species. The only discrepancy is that *C. glabrata* has *HO* but not *SIR1*.

The co-option of HO by yeasts led to a dramatic increase in the rate of mating-type switching, from once per million cells to almost one switch per cell generation, resulting in a change from a life cycle where the major growth phase was haploid (like *K. lactis* today) to one where most cells are diploid because budding of a haploid cell is followed almost inevitably by mother–daughter mating. This change was made possible by the fortuitous invasion of a parasitic genetic element but had a profound effect on the biology of a large clade of yeast species.

Acknowledgements. We thank Bernard Dujon, Cécile Fairhead and Gilles Fischer for stimulating conversations. Research in K.H.W.'s laboratory is supported by Science Foundation Ireland. Research in J.E.H.'s laboratory is supported by the US National Institutes of Health.

References

Aylon Y, Kupiec M (2004) DSB repair: the yeast paradigm. DNA Repair (Amst) 3:797–815

Bakhrat A, Jurica MS, Stoddard BL, Raveh D (2004) Homology modeling and mutational analysis of Ho endonuclease of yeast. Genetics 166:721–728

Bressan DA, Vazquez J, Haber JE (2004) Mating type-dependent constraints on the mobility of the left arm of yeast chromosome III. J Cell Biol 164:361–371

Butler G, Kenny C, Fagan A, Kurischko C, Gaillardin C, Wolfe KH (2004) Evolution of the *MAT* locus and its Ho endonuclease in yeast species. Proc Natl Acad Sci USA 101:1632–1637

Chien CT, Buck S, Sternglanz R, Shore D (1993) Targeting of SIR1 protein establishes transcriptional silencing at HM loci and telomeres in yeast. Cell 75:531-541

Cosma MP (2004) Daughter-specific repression of *Saccharomyces cerevisiae* HO: Ash1 is the commander. EMBO Rep 5:953–957

Dalgaard JZ, Klar AJ, Moser MJ, Holley WR, Chatterjee A, Mian IS (1997) Statistical modeling and analysis of the LAGLIDADG family of site-specific endonucleases and identification of an intein that encodes a site-specific endonuclease of the HNH family. Nucleic Acids Res 25:4626–4638

Dietrich FS, Voegeli S, Brachat S, Lerch A, Gates K, Steiner S, Mohr C, Pohlmann R, Luedi P, Choi S, Wing RA, Flavier A, Gaffney TD, Philippsen P (2004) The *Ashbya gossypii* genome as a tool for mapping the ancient *Saccharomyces cerevisiae* genome. Science 304:304–307

Dujon B, Sherman D, Fischer G, Durrens P, Casaregola S, Lafontaine I, de Montigny J, Marck C, Neuvéglise C, Talla E, Goffard N, Frangeul L, Aigle M, Anthouard V, Babour A, Barbe V, Barnay S, Blanchin S, Beckerich J-M, Beyne E, Bleykasten C, Boisramé A, Boyer J, Cattolico L, Confanioleri F, de Daruvar A, Despons L, Fabre E, Fairhead C, Ferry-Dumazet H, Groppi A, Hantraye F, Hennequin C, Jauniaux N, Joyet P, Kachouri R, Kerrest A, Koszul R, Lemaire M, Lesur I, Ma L, Muller H, Nicaud J.-M, Nikolski M, Oztas S, Ozier-Kalogeropoulos O, Pellenz S, Potier S, Richard G.-F, Straub M-L, Suleau A, Swennen D, Tekaia

F, Wésolowski-Louvel M, Westhof E, Wirth B, Zeniou-Meyer M, Zivanovic I, Bolotin-Fukuhara M, Thierry A, Bouchier C, Caudron B, Scarpelli C, Gaillardin C, Weissenbach J, Wincker P, Souciet J-L (2004) Genome evolution in yeasts. Nature 430:35–44

Gimble FS, Thorner J (1992) Homing of a DNA endonuclease gene by meiotic gene conversion in *Saccharomyces cerevisiae.* Nature 357:301–306

Gimble FS, Wang J (1996) Substrate recognition and induced DNA distortion by the PI-SceI endonuclease, an enzyme generated by protein splicing. J Mol Biol 263:163–180

Gogarten JP, Senejani AG, Zhaxybayeva O, Olendzenski L, Hilario E (2002) Inteins: structure, function, and evolution. Annu Rev Microbiol 56:263–287

Haber JE (2000) Lucky breaks: analysis of recombination in *Saccharomyces.* Mutat Res 451:53–69

Haber JE (2002) Switching of *Saccharomyces cerevisiae* mating-type genes. In: Craig RCN, Gellert M, Lambowitz A (eds) Mobile DNA II. ASM Press, Washington, DC, pp 927–952

Herman A, Roman H (1966) Allele specific determinants of homothallism in *Saccharomyces lactis.* Genetics 53:727–740

Jensen RE, Herskowitz I (1984) Directionality and regulation of cassette substitution in yeast. Cold Spring Harbor Symp Quant Biol 49:97–104

Keeling PJ, Roger AJ (1995) The selfish pursuit of sex. Nature 375:283

Kellis M, Patterson N, Endrizzi M, Birren B, Lander ES (2003) Sequencing and comparison of yeast species to identify genes and regulatory elements. Nature 423:241–254

Kellis M, Birren BW, Lander ES (2004) Proof and evolutionary analysis of ancient genome duplication in the yeast *Saccharomyces cerevisiae.* Nature 428:617–624

Koufopanou V, Goddard MR, Burt A (2002) Adaptation for horizontal transfer in a homing endonuclease. Mol Biol Evol 19:239–246

Kurtzman CP, Robnett CJ (2003) Phylogenetic relationships among yeasts of the '*Saccharomyces* complex' determined from multigene sequence analyses. FEMS Yeast Res 3:417–432

Lee SE, Moore JK, Holmes A, Umezu K, Kolodner R, Haber JE (1998) *Saccharomyces* Ku70, Mre11/Rad50 and RPA proteins regulate adaptation to G2/M arrest DNA damage. Cell 94:399–409

Liu XQ (2000) Protein-splicing intein: genetic mobility, origin, and evolution. Annu Rev Genet 34:61–76

MacIntosh GC, Bariola PA, Newbigin E, Green PJ (2001) Characterization of Rny1, the *Saccharomyces cerevisiae* member of the T2 RNase family of RNases: unexpected functions for ancient enzymes? Proc Natl Acad Sci USA 98:1018–1023

Melo J, Toczyski D (2002) A unified view of the DNA-damage checkpoint. Curr Opin Cell Biol 14:237–245

Moore JK, Haber JE (1996a) Capture of retrotransposon DNA at the sites of chromosomal double-strand breaks. Nature 383:644–646

Moore JK (1996b) Cell cycle and genetic requirements of two pathways of nonhomologous end-joining repair of double-strand breaks in *Saccharomyces cerevisiae.* Mol Cell Biol 16:2164–2173

Moure CM, Gimble FS, Quiocho FA (2002) Crystal structure of the intein homing endonuclease PI-SceI bound to its recognition sequence. Nat Struct Biol 9:764–770

Nickoloff JC, Hoekstra MF (1998) Double-strand break and recombinational repair in *Saccharomyces cerevisiae.* In: Hoekstra MF (ed) DNA damage and repair, vol 1. DNA repair in prokaryotes and lower eukaryotes. Humana Press, Totowa, NJ

Okuda Y, Sasaki D, Nogami S, Kaneko Y, Ohya Y, Anraku Y (2003) Occurrence, horizontal transfer and degeneration of VDE intein family in *Saccharomycete* yeasts. Yeast 20:563–573

Perler FB (2002) InBase: the intein database. Nucleic Acids Res 30:383–384

Posey KL, Koufopanou V, Burt A, Gimble FS (2004) Evolution of divergent DNA recognition specificities in VDE homing endonucleases from two yeast species. Nucleic Acids Res 32:3947–3956

Rusche LN, Kirchmaier AL, Rine J (2003) The establishment, inheritance, and function of silenced chromatin in *Saccharomyces cerevisiae*. Annu Rev Biochem 72:481–516

Russell DW, Jensen R, Zoller MJ, Burke J, Errede B, Smith M, Herskowitz I (1986) Structure of the *Saccharomyces cerevisiae* HO gene and analysis of its upstream regulatory region. Mol Cell Biol 6:4281–4294

Sugawara N, Wang X, Haber JE (2003) In vivo roles of Rad52, Rad54, and Rad55 proteins in Rad51-mediated recombination. Mol Cell 12:209–219

Sun K, Coic E, Zhou Z, Durrens P, Haber JE (2002) Saccharomyces forkhead protein Fkh1 regulates donor preference during mating-type switching through the recombination enhancer. Genes Dev 16:2085–2096

Toczyski DP, Galgoczy DJ, Hartwell LH (1997) CDC5 and CKII control adaptation to the yeast DNA damage checkpoint. Cell 90:1097–1106

Wang X, Haber JE (2004) Role of *Saccharomyces* single-stranded DNA-binding protein RPA in the strand invasion step of double-strand break repair. PLoS Biol 2:104–111

Weiler KS, Broach JR (1992) Donor locus selection during *Saccharomyces cerevisiae* mating type interconversion responds to distant regulatory signals. Genetics 132:929–942

Weiler KS, Szeto L, Broach JR (1995) Mutations affecting donor preference during mating type interconversion in *Saccharomyces cerevisiae*. Genetics 139:1495–1510

White CI, Haber JE (1990) Intermediates of recombination during mating type switching in *Saccharomyces cerevisiae*. EMBO J 9:663–673

Wolfe KH, Shields DC (1997) Molecular evidence for an ancient duplication of the entire yeast genome. Nature 387:708–713

Wolner B, van Komen S, Sung P, Peterson CL (2003) Recruitment of the recombinational repair machinery to a DNA double-strand break in yeast. Mol Cell 12:221–232

Wu C, Weiss K, Yang C, Harris MA, Tye BK, Newlon CS, Simpson RT, Haber JE (1998) Mcm1 regulates donor preference controlled by the recombination enhancer in *Saccharomyces* mating-type switching. Genes Dev 12:1726–1737

Wu X, Haber JE (1996) A 700 bp cis-acting region controls mating-type dependent recombination along the entire left arm of yeast chromosome III. Cell 87:277–285

Zhou Z, Sun K, Lipstein EA, Haber JE (2001) A *Saccharomyces servazzii* clone homologous to *Saccharomyces cerevisiae* chromosome III spanning KAR4, ARS 304 and SPB1 lacks the recombination enhancer but contains an unknown ORF. Yeast 18:789–795

Zonneveld BJM, Steensma HY (2003) Mating, sporulation and tetrad analysis in *Kluyveromyces* lactis. In: Barth G (ed) Non-conventional yeasts in genetics, biochemistry and biotechnology. Springer, Berlin Heidelberg New York, pp 151–154

Engineering Homing Endonucleases for Genomic Applications

Frederick S. Gimble

1 Introduction

The rapid progress in molecular biology that has occurred over the last three decades is due in large part to the availability of sequence-specific nucleases. These enzymes have been indispensable tools in recombinant DNA protocols. Most notably, the type II bacterial restriction enzymes have permitted the rapid and low-cost production of recombinant DNAs for cloning and other methods (Pingoud and Jeltsch 2001). One drawback of these enzymes, however, is the small size of their recognition sequences, which range in length from 4–8 base pairs (bp). An enzyme that recognizes a 6-bp target sequence cleaves DNA on average approximately once every 4000 bp. When these enzymes digest DNA from a mammalian genome, which is typically >100 megabases in length, many thousands of DNA fragments are generated, and purification of individual species from this pool is difficult.

The high specificity of homing endonucleases facilitates a myriad of different in vitro and in vivo applications involving complex genomic DNAs that were not previously possible using restriction enzymes. Recognition sequences for these enzymes range between 14–40 bp (Chevalier and Stoddard 2001). However, the utility of homing endonucleases is currently limited for many of the same reasons as for restriction enzymes. First, the number of available homing endonucleases that recognize different target sequences is still relatively small, numbering less than 200. Of those that are available, many do not possess the level of specificity required to target unique loci in DNA from complex genomes. Second, the majority of homing enzymes catalyze a single activity on nucleic acid substrates, namely, they introduce double-strand

F.S. Gimble (e-mail: fgimble@ibt.tamhsc.edu)
Center for Genome Research, Institute of Biosciences and Technology, Texas A&M University System Health Science Center, 2121 W. Holcombe Blvd., Houston, Texas 77096, USA

Nucleic Acids and Molecular Biology, Vol. 16
Marlene Belfort et al. (Eds.)
Homing Endonucleases and Inteins

breaks. Most of these enzymes are not capable of other functions, such as introducing single-strand nicks into DNA or using RNA as a substrate. Third, most of these enzymes are constitutively active in vitro and in vivo. Therefore, a means to regulate their activity in specific tissues or at precise times of development would increase their utility.

The determination of X-ray structures of several homing endonucleases and the biochemical characterization of their reaction mechanisms over the past decade have allowed researchers to turn their attention to engineering these proteins to create reagents with novel activities. This chapter will primarily focus on the LAGLIDADG family of homing endonucleases since these have been the most intensively studied, but many of its conclusions apply equally to members of the other families.

One area of research has concentrated on developing enzymes with altered DNA target specificities through rational design strategies involving computer modeling and domain exchange as well as through genetic selection methods that isolate variants from combinatorial libraries. These efforts initially focused on shifting or subtly altering the specificity of these enzymes but have the eventual goal of engineering proteins that selectively recognize nucleic acid sequences of any desired sequence or length. Other research is directed at regulating homing endonuclease activity in response to external stimuli. Protein derivatives of homing endonucleases possessing novel activities are being developed. This chapter reviews the current and anticipated efforts to engineer homing endonucleases with novel activities.

2 The Modular Organization of Homing Endonucleases and Their Endonucleolytic Activity

Complex macromolecules, including proteins, have the potential to rapidly evolve new functions when they are based on a modular architecture. Diversity arises because each module evolves independently as it acquires point mutations and because combinatorial shuffling of the modules occurs between proteins. The X-ray crystal structures of several divergent intron-encoded (I-CreI (*Chlamydomonas reinhardtii* [Heath et al. 1997; Jurica et al. 1998; Chevalier et al. 2001]), I-DmoI (*Desulfurococcus mobilis* [Silva et al. 1999]), I-MsoI (*Monomastix* [Chevalier et al. 2003]), I-SceI (*Saccharomyces cerevisiae* [Moure et al. 2003]), I-AniI (*Aspergillus nidulans* [Bolduc et al. 2003]) and intein-encoded (PI-SceI [*Saccharomyces cerevisiae* (Duan et al. 1997; Moure et al. 2002)] and PI-PfuI [*Pyrococcus furiosus* (Ichiyanagi et al. 2000)]) LAGLIDADG homing endonucleases suggest that these processes guided their evolution, increasing the complexity of their topology and their target site selectivity. Separate chapters in this volume describe the specific features of the structures.

One group of homing enzymes (i.e. I-CreI and I-MsoI) are symmetric homodimers (Heath et al. 1997; Jurica et al. 1998; Chevalier et al. 2003), a second group (i.e., I-DmoI, I-SceI, and I-AniI) are pseudo-dimeric monomers in which the two symmetry-related domains are tethered (Silva et al. 1999; Bolduc et al. 2003; Moure et al. 2003), and a third group (i.e., PI-SceI and PI-PfuI) consists of the monomeric endonuclease domain connected to a protein-splicing domain (Duan et al. 1997; Ichiyanagi et al. 2000; Moure et al. 2002). The evolution of base-specific and backbone contacts in the PI-SceIsplicing domain increases the total available surface area of the protein that can contact DNA, thereby providing the requisite specificity to cleave a single locus in the yeast genome (Bremer et al. 1992). Mini-inteins (i.e., Mxe GyrA [*Mycobacterium xenopi* gyrase]) consist of the protein-splicing domain but lack an endonuclease domain (Klabunde et al. 1998). A scenario for homing enzyme evolution has been described in which the monomeric intron-encoded enzymes arose following the gene duplication and fusion of a progenitor homodimeric endonuclease (Lykke-Andersen et al. 1996; Silva et al. 1999). Subsequent insertion of a monomeric endonuclease gene into a protein-splicing gene, which may have been mediated by the endonuclease itself, may have produced the bifunctional inteins (Duan et al. 1997).

Other homing endonuclease families and restriction enzymes evolved a modular topology. I-TevI, a member of the GIY-YIG family of homing enzymes, features multiple tethered modules within its C-terminal DNA-binding domain, including a Zn finger, an elongated α-helical peptide segment and a helix-turn-helix subdomain (van Roey et al. 2001). NaeI, a restriction enzyme, consists of an endonuclease domain fused to a topoisomerase domain. Since both domains bind a DNA molecule, NaeI simultaneously contacts two recognition sequences (Huai et al. 2001). Thus, the modular assembly of I-TevI, PI-SceI, and NaeI permits these proteins to increase their overall DNA contact surface area, thereby increasing their specificity.

Homing endonucleases can tolerate significant sequence variation in their homing sites (Chevalier et al., this Vol.; van Roey and Derbyshire, this Vol.). This recognition site flexibility is due in part to the fact that they utilize only a subset of the total available hydrogen-bonding capacity of a DNA sequence to effect specific interaction. This feature may have evolved to enable homing endonucleases to cleave and invade variant sites elsewhere in the host genome or in related species. Only 11 bp within the ~31-bp PI-SceI recognition sequence are important for activity (Gimble and Wang 1996; Moure et al. 2002), and this enzyme cleaves the target sites of 14 other yeast species (Posey et al. 2004). Moreover, it is clear that homing enzymes can evolve different side-chain contacts to the same base pairs, as in the case of the related I-CreI and I-MsoI isoschizomers (Chevalier et al. 2003) and can undergo changes in specificity, such as for PI-SceI and its *Zygosaccharomyces bailii* analog PI-ZbaI

(Posey et al. 2004). The inherent plasticity of the homing enzyme/DNA interface facilitates the engineering of homing enzymes with altered specificity. This has not been the case for restriction enzymes, which make redundant and saturating base-pair contacts, and where there is tight coupling between DNA binding and catalysis (Pingoud and Jeltsch 2001).

Retaining catalytic activity in homing endonucleases during their engineering is complicated by the unique arrangement of the active sites. As described in other chapters in this volume, a distinctive feature of the I-SceI, I-CreI and I-MsoI enzymes is the overlap of the two active sites that results from the sharing of a divalent metal ion (Chevalier et al. 2001, 2003; Moure et al. 2003). Complete uncoupling of the active sites has not been possible (Li and Gimble, unpubl. data). Engineering strategies that involve the exchange or modification of the subunit or domain modules are complicated by the close positioning of the active sites to the dimerization (I-CreI and I-MsoI) or domain (I-SceI) interface at the two-fold symmetry axis.

3 Probing the Modular Structure of Homing Endonucleases by Protein Engineering

Some of the first structure-function analyses of homing endonucleases elucidated the architecture of intein-encoded enzymes. The functional independence of the endonuclease and protein-splicing domains within PI-TliI, an endonuclease encoded by the Tli Pol-2 intein (*Thermococcus litoralis* DNA polymerase), was demonstrated by showing that mutations that abolished the activity of one domain had no effect on the activity of the other (Hodges et al. 1992). Moreover, deletion of the entire endonuclease domain within the bifunctional Sce VMA (*Saccharomyces cerevisiae*) and Mtu RecA (*Mycobacterium tuberculosis*) inteins had no effect on the splicing activity (Chong and Xu 1997; Derbyshire et al. 1997). Thus, some of the first engineering efforts demonstrated that the multi-domain homing enzymes could be disassembled.

Subsequent engineering of homing enzymes showed that protein modules could be recombined into artificial bifunctional inteins. One was assembled by inserting a gene that expresses one of the two I-CreI subunits into the *Mycobacterium xenopi* GyrA mini-intein at the putative site of the missing endonuclease (Fitzsimons Hall et al. 2002). This engineered intein underwent protein splicing to yield a homodimeric site-specific endonuclease activity essentially identical to naturally occurring I-CreI. It is unclear whether this graft of endonuclease and splicing activities mimics an evolutionary "missing link" situated between an intein precursor that lacked any endonuclease activity and the inteins observed today that contain monomeric endonuclease domains. Regardless, this work demonstrates that inteins can be assembled from protein modules in the laboratory, as is thought to have happened during evolution. This construction parallels the assembly of an artificial intron-encoded homing element in which EcoRI supplied the homing endonuclease activity (Eddy and Gold 1992).

Homing endonucleases without protein-splicing domains have been both disassembled and reassembled in order to explore the interactions and relationships between its component modules. The independence of the two domains that comprise monomeric homing enzymes was demonstrated for I-DmoI by showing that they heterodimerize to form an active endonuclease when each is separately expressed in the same cell (Silva and Belfort 2004). Thus, at least for this enzyme, the coupling free energy supplied by tethering of the two domains is not an absolute requirement for the interaction since the two domains heterodimerize in *trans*.

A reciprocal experiment provided evidence for the hypothesis that monomeric endonuclease domains originally evolved from the duplication and fusion of genes that encoded each subunit (Silva et al. 1999). A linker derived from I-DmoI was used to join two copies of the I-CreI gene to generate scI-CreI, which cleaves DNA with the same specificity as native I-CreI in vitro (Epinat et al. 2003). Creation of single-chain (sc) DNA-binding proteins by covalently fusing individual homodimeric subunits had been demonstrated previously for the bacteriophage λ Cro and P22 Arc repressors (Mossing and Sauer 1990; Robinson and Sauer 1996), and the PvuII restriction enzyme (Simoncsits et al. 2001). In principle, DNA binding by a homodimeric protein reflects the coupling of the protein–DNA and protein–protein association equilibria when the protein–protein interaction is weak (Lohman and Bujalowski 1991). Thus, scI-CreI might be expected to be more stable at higher temperatures and more active than the native proteins since it would bind DNA at protein concentrations where I-CreI is predominantly monomeric. In actuality, the apparent catalytic rate of scI-CreI was threefold lower than wild-type I-CreI (Epinat et al. 2003). This may be due to several factors, including a strong I-CreI subunit association constant that negates the benefits of the sc molecule, a suboptimal subunit interface in the engineered scI-CreI and/or a catalytic rather than a DNA-binding step being rate-limiting in the reaction. Regardless of its lower activity in vitro, scI-CreI functions in vivo since it stimulates homologous recombination in both yeast and mammalian cells by cleaving its recognition site (Epinat et al. 2003). Taken together, these works demonstrate that protein subunits or domains can be assembled to yield larger, more complex molecules, as is thought to occur in nature.

4 Engineering Homing Enzymes with Novel Functions

4.1 Changing the Recognition Site Specificity of Homing Endonucleases

Homing endonucleases have become invaluable tools to study complex genomes. By virtue of its ability to introduce unique breaks into complex genomes that mimic lesions made by DNA-damaging agents, I-SceI is commonly used to study DNA repair pathways in organisms ranging from yeast to humans (see, for example, Johnson and Jasin 2001). Moreover, this enzyme stimulates gene targeting over 1000-fold in mammalian systems through homologous recombination (Rouet et al. 1994). However, the current repertoire of homing endonuclease specificities is inadequate to effect gene targeting at desired loci. Gene targeting using the available homing enzymes is limited by the necessity of having to first insert the recognition sites at a locus by conventional targeting methods. The ability to tailor homing endonucleases to effect site-specific cleavage at pre-existing sites would permit repair of defective genes at any given locus. Homing endonucleases could also be engineered to act as transcriptional repressors of genes that are not normally regulated. Uncoupling the DNA-binding and nuclease activities could be accomplished by mutating the acidic residues that bind the metal ion cofactors, leaving intact the site-specific DNA-binding activity (Gimble and Stephens 1995). In addition, site-specific proteins could be developed as therapeutic agents that deliver tethered small molecules to defined regions of the genome.

The approaches used to alter homing endonuclease specificity range from those that target a limited number of contacts to subtly modify DNA recognition to others that generate gross changes in the protein–DNA interaction surface. What also distinguishes these strategies is that some focus on rational design based on high-resolution X-ray structures while others rely more on combinatorial methods coupled with genetic screens and selections. As enumerated below, there are strengths and weaknesses to each approach.

4.1.1 Altering Homing Endonuclease Specificity by Domain Shuffling

One strategy to alter homing endonuclease specificity has been to exchange an entire intein domain or portions of it with domains from other inteins or unrelated DNA-binding proteins. An advantage of working with inteins is that part of their DNA-binding specificity originates in the protein-splicing domain, which, in some cases, can be uncoupled from the endonuclease domain that cleaves the DNA. Some of these engineered proteins have been modeled after chimeric endonucleases created by fusing the non-specific catalytic domain of FokI, a class IIS restriction enzyme, with DNA-binding domains or

DNA-binding proteins (Kim et al. 1996, 1998). For example, the PI-SceI protein splicing domain, which makes independent base-specific contacts to a subset of the total PI-SceI recognition sequence (Grindl et al. 1998; Moure et al. 2002), was fused to the FokI catalytic domain to generate a functional endonuclease (Hu and Gimble, unpubl. data). As expected, this enzyme exhibits less specificity than full-length PI-SceI, which establishes base-specific contacts to the entire recogntion sequence using both protein domains. Furthermore, rather than cleaving at unique sites on each DNA strand like PI-SceI, the PI-SceI/FokI chimera cleaves at multiple neighboring sites (Hu and Gimble, unpubl. data). Regardless, this work demonstrates that the PI-SceI protein-splicing domain can be recruited as a site-specific DNA-binding module to target other protein domains to specific sequences.

Domain shuffling between the PI-SceI and its *Candida tropicalis* analog, PI-CtrIP, was used to design altered specificity proteins (Steuer et al. 2004). Replacement of the PI-SceI protein-splicing domain with that of PI-CtrIP resulted in inactive proteins. Thus, proper register between domains is likely to be critical for domain shuffling to work. However, an active chimera results when the DNA recognition region (DRR) of PI-SceI, which is a subdomain of the protein-splicing domain that makes some of the base-specific DNA contacts, was replaced with its analog from PI-CtrIP. Interestingly, this protein displayed a small preference for the Candida tropicalis site, which differs from the *Saccharomyces cerevisiae* site at six nucleotide positions, suggesting that at least a portion of the specificity has been transferred to PI-SceI (Steuer et al. 2004).

Other strategies to alter specificity have focused on shuffling the domains within the catalytic region of intron-encoded homing endonucleases. Relative to engineering inteins, tinkering with these endonucleases is more ambitious because the DNA-binding determinants are tightly linked to the catalytic center due to their close proximity, making it more difficult to alter binding without also affecting the catalytic machinery (Epinat et al. 2003). Another reason is that the complex hydration shell and shared metal ion cofactor that are key components of the active sites are positioned by residues at the interface of the two domains and can be easily disrupted by alteration of either domain.

Two groups fused the N-terminal domain of I-DmoI to the C-terminal domain of I-CreI to create chimeric endonucleases that cleave hybrid substrates derived from each parent, but not the parental sites (Chevalier et al. 2002; Epinat et al. 2003). Here, the underlying assumption is that each domain would continue to bind its respective native half-site within the composite substrate. Initial structural modeling of the chimeric protein revealed several steric clashes between the domain interface residues. Simply substituting these residues with alanine resulted in insoluble proteins (Chevalier et

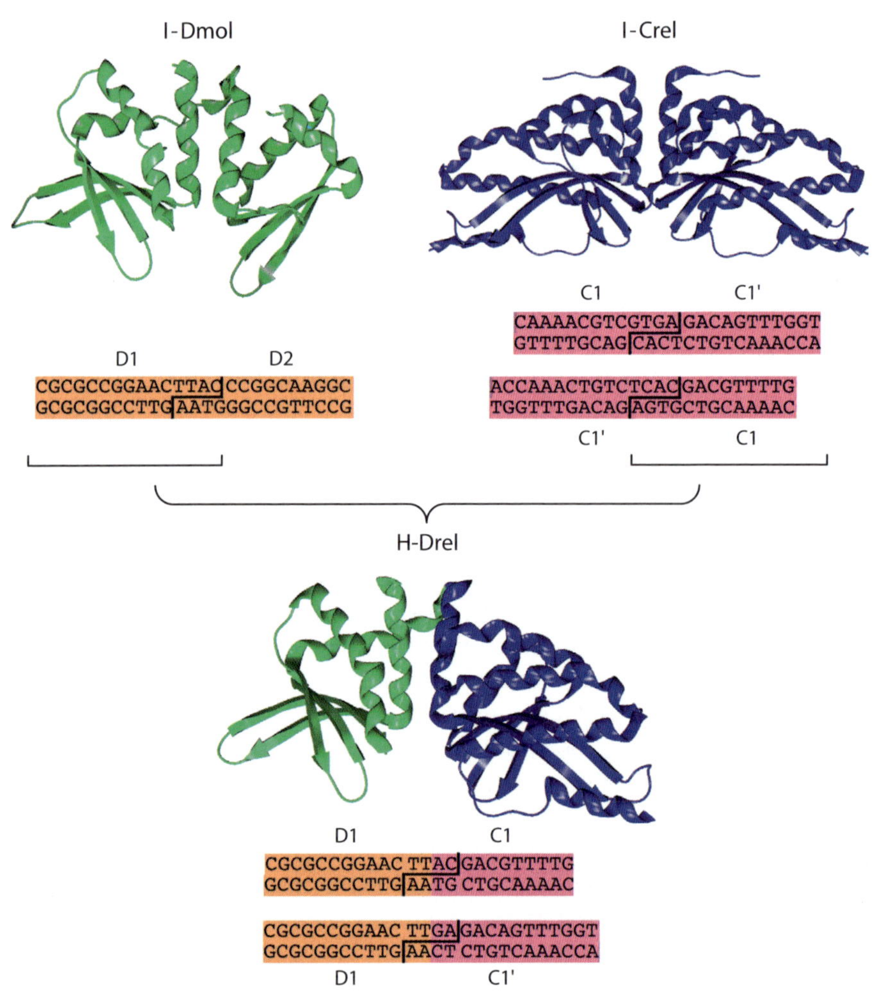

Fig. 1. Engineering of a chimeric homing endonuclease. The X-ray crystal structures of the parent enzymes I-DmoI (1B24 [Silva et al. 1999]) and I-CreI (1G9Z [Chevalier et al. 2001]) are depicted in *green* and *blue*, respectively. Their DNA substrates are shown below each protein, with each half-site labeled separately (I-DmoI substrate half-sites D1 and D2, *orange*; I-CreI substrate half-sites C1 and C1′, *magenta*). I-DmoI, an asymmetric enzyme, binds the substrate only in the depicted orientation, but I-CreI, a symmetric homodimer, binds to its substrate in either orientation. The I-DmoI N-terminal domain (*green*) and the I-CreI C-terminal domain (*blue*) were fused to yield H-DreI (1M0W [Chevalier et al. 2002]). H-DreI only cleaves the two chimeric substrates that are shown of the four that are possible from different combinations of I-DmoI and I-CreI half-sites, and it cleaves neither of the parent enzyme substrates

al. 2002). In order to reduce the number of different combinations of interface residues to be tested of the total sequence space, one group applied computational algorithms to search through 8×10^{17} sequence combinations to identify those with low local free energy amino acid sequences at the interface (Chevalier et al. 2002). Sixteen of these were further screened for solubility using a LacZ-based screen in vivo. One chimeric variant, E-DreI, now termed H-DreI according to convention (Roberts et al. 2003), exhibits a specificity that is distinct from either of the two parent enzymes (Fig. 1). It does not cleave either of the parental substrates and, of the four chimeric substrates with different combinations of I-CreI and I-DreI half-sites, H-DreI only cleaves two, those with one specific dmo half-site fused to either of the cre half-sites. Thus, the I-DmoI N-terminal domain recognizes a specific half-site of its substrate, as expected for an asymmetric enzyme, while the I-CreI domain, which originates from a symmetric homodimer, interacts with either cre half-site. The DNA cleavage geometry is identical to all other characterized LAGLIDADG enzymes, yielding a four base, 3′-overhang, and the cleavage activity is similar to I-CreI. The 2.4 Å X-ray structure of H-DreI closely resembles the computationally redesigned model, confirming the accuracy of the method (Chevalier et al. 2002). A second group engineered a chimera termed *DmoCre* having the same domain architecture as H-DreI (Epinat et al. 2003). The cleavage properties of *DmoCre* and H-DreI were similar even though different mutations were made at the domain interface, indicating that there are multiple solutions to the problem of engineering the LAGLIDADG α-helical interface. A constraint of domain shuffling is that the specificity of the resulting chimera is always directly related to the specificities of the parent enzymes. These studies illustrate that the partial independence of LAGLIDADG domains permits them to be recombined to yield novel reagents.

4.1.2 Altering Homing Endonuclease Specificity Using Genetic Screens and Selections

Complementary strategies to obtain altered specificity enzymes have used genetic selections or screens in bacteria to isolate variants from complex libraries. In theory, if the entire sequence space of a protein could be screened, variants with any desired specificity could be obtained without prior structural or biochemical knowledge of the system. In reality, however, the complexity of the library that can be probed, which increases exponentially with the linear increase in the number of randomized residue positions, is limited by the number of library members that can be transformed into E. coli and assayed. Since the upper limit on the number of independent transformants for a given library is $\sim10^7$–10^9, the size of the sequence that can be completely randomized (e.g. 20 different codons at n different positions) with 99% confidence that all variants will be obtained is limited to n=6–8 (Lowman and Wells 1991). Thus, mutagenizing an entire protein domain has not been possible, and it has been necessary, instead, to use existing structural information to target specific interactions for analysis.

The genetic methods that have been used to alter homing endonuclease specificity can be broadly divided into those that select/screen for DNA-binding activity and those that select/screen for catalytic activity. Selections for DNA-cleavage activity are inherently preferable because variants are obtained with the desired property. Variations that contribute to cleavage activity by affecting any of the steps of the reaction pathway, including ligand binding, metal ion cofactor binding, coupling between DNA binding and catalysis, the catalytic reaction, or product release, will be selected. By contrast, variants that are selected for DNA binding are not guaranteed to be catalytically active. If the DNA-binding site is tightly coupled to the catalytic site, it may be impossible to perturb the DNA-binding determinants without also decreasing the DNA-cleavage activity. Moreover, since these methods select for binding of the protein in the ground state, it is expected that isolated variants will have low turnover numbers. This latter constraint may not be an issue for homing endonucleases, however, since these evolved to have extreme specificity, rather than high turnover numbers, in order to cleave at single genomic target sites. In general, genetic selections are preferred to genetic screens since they generally permit a larger part of the total sequence space to be sampled in a single experiment.

An adaptation of a bacterial two-hybrid strategy (Joung et al. 2000) selected altered specificity variants of the intein-encoded PI-SceI homing endonuclease (Gimble et al. 2003). This DNA-binding selection consists of three components (Fig. 2A); a protein fusion between the yeast Gal11P protein and PI-SceI, a second protein fusion between the *E. coli* RNA polymerase α subunit (RNAPα) and the yeast Gal4 protein, and a DNA target that includes a mutant PI-SceI recognition sequence located upstream of a weak *lac* promoter that controls the selectable *HIS3* and *aadA* genes (Gimble et al. 2003). Gal11P/PI-SceI variants expressed from a plasmid library that bind to the mutant target site can be selected because they recruit the RNAPα/Gal4 fusion protein to the promoter through the Gal11P–Gal4 interaction, thereby conferring bacterial growth on histidine- and spectinomycin-selective media. The study targeted a protein/DNA interaction within the PI-SceI protein-splicing domain because its distance from the endonuclease active sites minimizes the deleterious effects of the mutations on catalysis. Whether this DNA-binding selection can be successfully applied to intron-encoded homing endonucleases, where the DNA-binding determinants and the catalytic residues are immediately adjacent and presumably tightly coupled, is unclear. A residue that makes a critical

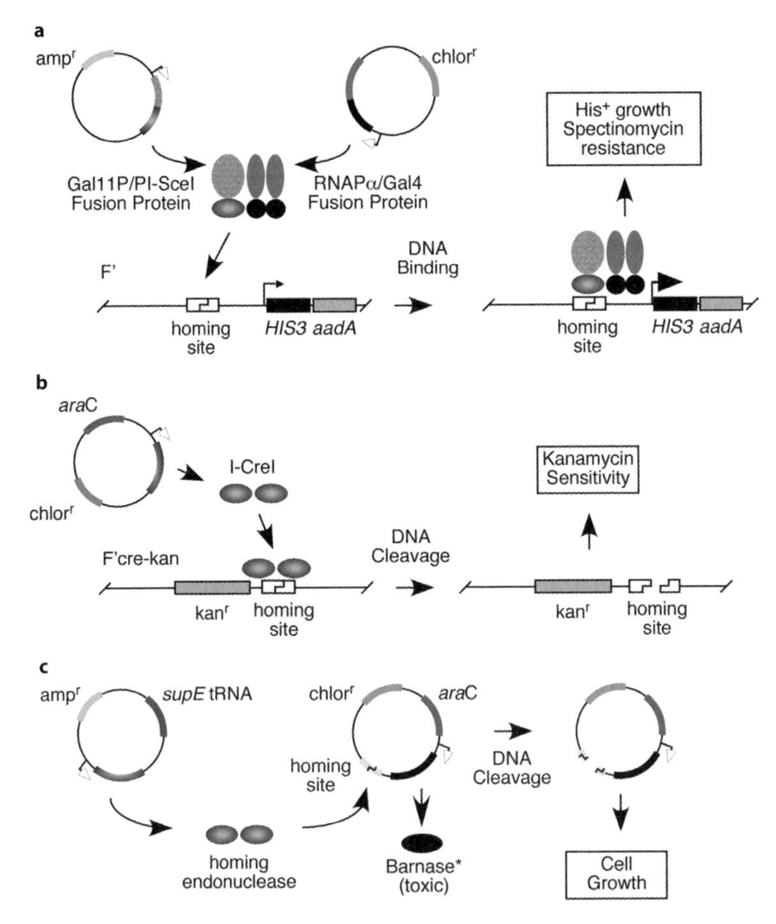

Fig. 2. Bacterial selections and screens for altered specificity homing endonucleases. **a** Bacterial two-hybrid binding selection (Gimble et al. 2003). Gal11P/PI-SceI fusion variants expressed from a plasmid library associate with a RNAPα/Gal4 fusion protein encoded by a separate plasmid. If the PI-SceI variant binds to a homing site located on an F′, it recruits the RNAPα subunit proximal to the weak Plac promoter, leading to an increase in *HIS3* and *aadA* expression and growth on histidine-selective and spectinomycin-selective media. **b** Screen for I-CreI cleavage activity (Seligman et al. 2002). Plasmid-encoded I-CreI derivatives that bind and cleave an I-CreI homing site located on an F′ lead to its elimination. Concomitant loss of an adjacent kanamycin antibiotic marker yields kanamycin-sensitive cells. **c** Selection for homing endonuclease activity (Gruen et al. 2002). Co-existence of two plasmids kills bacterial cells because one expresses an amber nonsense allele of a toxic gene product, barnase, and the other expresses an amber tRNA suppressor. However, when the tRNA supressor plasmid also expresses a homing enzyme that cleaves the homing site on the barnase expression plasmid, the cells survive due to elimination of the barnase gene

contact, Arg-94, and four neighboring residues were randomized to generate a plasmid library for the selection. Altered specificity proteins were selected that gain the ability to bind to substrates containing mutations at two nucleotides contacted by Arg-94 in wild-type PI-SceI. A powerful feature of the method is that the binding stringency of the isolates could be increased by including 3-aminotriazole, a competitive inhibitor of HIS3, during the selection. The DNA-binding specificities of the selected variants ranged from being relaxed (i.e. able to cleave the wild-type and mutant targets equally) to being dramatically shifted to preferring the selection targets. None of the variants displayed the same degree of specificity as wild-type PI-SceI (Gimble et al. 2003). To achieve this level of specificity, a negative selection against binding to the wild-type recognition sequence may need to be applied concomitant with the positive selection for binding to the new target site.

Genetic screens and selections for homing endonucleases that cleave altered recognition sequences have been developed. In a study of I-CreI, proteins containing each of the 19 possible non-wild-type amino acid substitutions for a residue that makes a critical DNA contact were expressed and screened in vivo for their ability to cleave and eliminate a reporter plasmid containing a mutant I-CreI recognition sequence (Seligman et al. 2002). Plasmid loss following DNA cleavage at the mutant site results in loss of kanamycin resistance or β-galactosidase activity, depending on whether the plasmid contains the kan^R or *lacZ* genes (Fig. 2B). Shifted, rather than completely altered, specificity proteins were obtained, perhaps due to the lack of a negative selection/screen against cleavage of other recognition sequences. Determination of the crystal structures of these constructs revealed that the new contacts do not significantly perturb the local protein conformation or surrounding DNA contacts (Sussman et al. 2004). The observed independence of the contacts will simplify the engineering of homing endonucleases. A similar method was set up as a genetic selection (Fig. 2C) in which a toxic gene product, barnase, expressed by a reporter plasmid, kills all of the bacteria except those that also encode a homing endonuclease variant that cleaves a homing site within the barnase plasmid (Gruen et al. 2002). Whether the background level of false positive isolates is sufficiently low to obtain altered specificity variants by this method is still unclear. An ingenious approach using phage display, which has been applied to staphylococcal nuclease but not to homing enzymes, permits selection for catalytic activity as well as rapid, iterative rounds of directed evolution to refine endonuclease specificity. Here, M13 phage that express library candidates that are catalytically active cleave a neighboring DNA substrate on the phage surface, thereby causing the selected phage particle, but no others, to be released from a solid support (Pedersen et al. 1998).

4.2 Introducing Molecular Switches into Homing Endonucleases

The ability to temporally or spatially control homing endonuclease activity would increase their utility in vitro and in vivo. Two different approaches can be used to modulate activity, by incorporating a trigger that activates or inactivates the enzyme irreversibly or by inserting a switch that allows the endonuclease to be reversibly turned on and off. One approach to tightly regulate homing enzymes that requires no engineering is to simply use chelating agents to alter the concentration of the divalent metal ion, which is essential to the reaction. However, this strategy is impractical in vivo since it is difficult to modulate metal concentrations within the cell, and because numerous other metabolic activities would be affected by changes in the metal concentration. Another approach, which has been applied to restriction endonucleases (Muir et al. 1997), is to isolate temperature-sensitive variants that become inactivated at high temperature. Altering the temperature in vivo, however, is likely to adversely affect other cellular functions.

Inserting switches or triggers into homing endonucleases that regulate only these proteins would be the most effective means to control their activity. Progress has been made in this respect by showing that PI-SceI can be reversibly regulated using a redox switch (Posey and Gimble 2002). To create the switch, one of two cysteines was inserted into a flexible loop that undergoes a hinge–flap motion in the presence of DNA to make minor-groove contacts, while the second was inserted into an adjacent β-hairpin loop that contacts the DNA in the major groove. In the reduced state, the loops freely undergo the conformational changes that permit them to contact DNA, and the protein is nearly as active as wild-type PI-SceI. However, under oxidizing conditions, formation of a disulfide bond between the cysteines constrains the loops, effectively trapping the protein in a non-productive conformation that binds DNA poorly and is catalytically inactive. Successive cycles of reducing and oxidizing treatments can be used to switch PI-SceI activity on and off (Posey and Gimble 2002). The major drawback of the switch is that it is currently limited to in vitro applications since changing the intracellular redox potential is untenable. The next step will be to develop switches in homing endonucleases that are activated using small molecules that are easily transported into cells and that do not negatively affect cellular metabolism. An alternative strategy may involve "caged" homing enzymes that have been chemically modified at inserted cysteine residues with different ortho-nitrobenzyl moieties that inactive the proteins (Bayley et al. 1998). After exposure to UV-visible radiation, the release of the covalent modification triggers the activation of the protein.

5 Conclusions and Future Prospects

Protein engineering of homing endonucleases has helped elucidate their domain architecture and has generated enzymes with novel functions. Endonucleases with shifted sequence specificities have been engineered by interchanging protein domains and by coupling combinatorial methods with genetic selections or screens. Iterative rounds of selection or screening and the application of negative selections may be required to further hone the specificity of these endonucleases. Homing endonucleases have been engineered that can be activated or deactivated using a redox switch. Further development may involve introducing switches that can be selectively activated by small molecule ligands or by other external stimuli. Homing endonucleases may be engineered to acquire novel functions, such as the ability to selectively nick DNA or to act on RNA substrates. These reagents are likely to be useful tools for genomic engineering applications, including gene targeting.

Acknowledgements. We are grateful for financial support for this work from the National Science Foundation (MCB-0321550) and the Welch Foundation (BE-1452).

References

Bayley H, Chang CY, Miller WT, Niblack B, Pan P (1998) Caged peptides and proteins by targeted chemical modification. Methods Enzymol 291:117–135

Bolduc JM, Spiegel PC, Chatterjee P, et al (2003) Structural and biochemical analyses of DNA and RNA binding by a bifunctional homing endonuclease and group I intron splicing factor. Genes Dev 17:2875–2888

Bremer MCD, Gimble FS, Thorner J, Smith CL (1992) VDE endonuclease cleaves *Saccharomyces cerevisiae* genomic DNA at a single site: physical mapping of the VMA1 gene. Nucleic Acids Res 20:5484

Chevalier B, Stoddard BL (2001) Homing endonucleases: structural and functional insight into the catalysts of intron/intein mobility. Nucleic Acids Res 29:3757–3774

Chevalier B, Monnat RJ Jr, Stoddard BL (2001) The homing endonuclease I-CreI uses three metals, one of which is shared between the two active sites. Nat Struct Biol 8:312–316

Chevalier B, Kortemme T, Chadsey MS, Baker D, Monnat RJ, Stoddard BL (2002) Design, activity, and structure of a highly specific artificial endonuclease. Mol Cell 10:895–905

Chevalier B, Turmel M, Lemieux C, Monnat RJ, Stoddard BL (2003) Flexible DNA target site recognition by divergent homing endonuclease isoschizomers I-CreI and I-MsoI. J Mol Biol 329:253–269

Chong S, Xu MQ (1997) Protein splicing of the *Saccharomyces cerevisiae* VMA intein without the endonuclease motifs. J Biol Chem 272:15587–15590

Derbyshire V, Wood DW, Wu W, Dansereau JT, Dalgaard JZ, Belfort M (1997) Genetic definition of a protein-splicing domain: functional mini-inteins support structure predictions and a model for intein evolution. Proc Natl Acad Sci USA 94:11466–11471

Duan X, Gimble FS, Quiocho FA (1997) Crystal structure of PI-SceI, a homing endonuclease with protein splicing activity. Cell 89:555–564
Eddy SR, Gold L (1992) Artificial mobile DNA element constructed from the EcoRI endonuclease gene. Proc Natl Acad Sci USA 89:1544–1547
Epinat JC, Arnould S, Chames P, et al (2003) A novel engineered meganuclease induces homologous recombination in yeast and mammalian cells. Nucleic Acids Res 31:2952–2962
Fitzsimons Hall M, Noren CJ, Perler FB, Schildkraut I (2002) Creation of an artificial bifunctional intein by grafting a homing endonuclease into a mini-intein. J Mol Biol 323:173–179
Gimble FS, Stephens BW (1995) Substitutions in conserved dodecapeptide motifs that uncouple the DNA binding and DNA cleavage activities of PI-SceI endonuclease. J Biol Chem 270:5849–5856
Gimble FS, Wang J (1996) Substrate recognition and induced DNA distortion by the PI-SceI endonuclease, an enzyme generated by protein splicing. J Mol Biol 263:163–180
Gimble FS, Moure CM, Posey KL (2003) Assessing the plasticity of DNA target site recognition of the PI-SceI homing endonuclease using a bacterial two-hybrid selection system. J Mol Biol 334:993–1008
Grindl W, Wende W, Pingoud V, Pingoud A (1998) The protein splicing domain of the homing endonuclease PI-SceI is responsible for specific DNA binding. Nucleic Acids Res 26:1857–1862
Gruen M, Chang K, Serbanescu I, Liu DR (2002) An in vivo selection system for homing endonuclease activity. Nucleic Acids Res 30:e29
Heath PJ, Stephens KM, Monnat RJ Jr, Stoddard BL (1997) The structure of I-CreI, a group I intron-encoded homing endonuclease. Nat Struct Biol 4:468–476
Hodges RA, Perler FB, Noren CJ, Jack WE (1992) Protein splicing removes intervening sequences in an archaea DNA polymerase. Nucleic Acids Res 20:6153–6157
Huai Q, Colandene JD, Topal MD, Ke H (2001) Structure of NaeI-DNA complex reveals dual-mode DNA recognition and complete dimer rearrangement. Nat Struct Biol 8:665–669
Ichiyanagi K, Ishino Y, Ariyoshi M, Komori K, Morikawa K (2000) Crystal structure of an archaeal intein-encoded homing endonuclease PI-PfuI. J Mol Biol 300:889–901
Johnson RD, Jasin M (2001) Double-strand-break-induced homologous recombination in mammalian cells. Biochem Soc Trans 29:196–201
Joung JK, Ramm EI, Pabo CO (2000) A bacterial two-hybrid selection system for studying protein-DNA and protein–protein interactions. Proc Natl Acad Sci USA 97:7382–7387
Jurica MS, Monnat RJ Jr, Stoddard BL (1998) DNA recognition and cleavage by the LAGLIDADG homing endonuclease I-CreI. Mol Cell 2:469–476
Kim Y-G, Cha J, Chandrasegaran S (1996) Hybrid restriction enzymes: zinc fusion to FokI cleavage domain. Proc Natl Acad Sci USA 93:1156–1160
Kim YG, Smith J, Durgesha M, Chandrasegaran S (1998) Chimeric restriction enzyme: Gal4 fusion to FokI cleavage domain. Biol Chem 379:489–495
Klabunde T, Sharma S, Telenti A, Jacobs WR, Sacchettini JC (1998) Crystal structure of *gyrA* intein from *Mycobacterium xenopi* reveals structural basis of protein splicing. Nat Struct Biol 5:31–36
Lohman TM, Bujalowski W (1991) Thermodynamic methods for model-independent determination of equilibrium binding isotherms for protein–DNA interactions: spectroscopic approaches to monitor binding. Methods Enzymol 208:258–290
Lowman HB, Wells JA (1991) Monovalent phage display: a method for selecting variant proteins from random libraries. Methods 3:205–216

Lykke-Andersen J, Garrett RA, Kjems J (1996) Protein footprinting approach to mapping DNA binding sites of two archaeal homing enzymes: evidence for a two-domain protein structure. Nucleic Acids Res 24:3982–3989

Mossing MC, Sauer RT (1990) Stable, monomeric variants of lambda Cro obtained by insertion of a designed β-hairpin sequence. Science 250:1712–1715

Moure CM, Gimble FS, Quiocho FA (2002) Crystal structure of the intein homing endonuclease PI-SceI bound to its recognition sequence. Nat Struct Biol 9:764–770

Moure CM, Gimble FS, Quiocho FA (2003) The crystal structure of the gene targeting homing endonuclease I-SceI reveals the origins of its target site specificity. J Mol Biol 334:685–695

Muir RS, Flores H, Zinder ND, Model P, Soberon X, Heitman J (1997) Temperature-sensitive mutants of the *Eco*RI endonuclease. J Mol Biol 274:722–737

Pedersen H, Holder S, Sutherlin DP, Schwitter U, King DS, Schultz PG (1998) A method for directed evolution and functional cloning of enzymes. Proc Natl Acad Sci USA 95:10523–10528

Pingoud A, Jeltsch A (2001) Structure and function of type II restriction endonucleases. Nucleic Acids Res 29:3705–3727

Posey KL, Gimble FS (2002) Insertion of a reversible redox switch in a rare-cutting DNA endonuclease. Biochemistry 41:2184–2190

Posey KL, Koufopanou V, Burt A, Gimble FS (2004) Evolution of divergent DNA recognition specificities in VDE homing endonucleases from two yeast species. Nucleic Acids Res 32:3947–3956

Roberts RJ, Belfort M, Bestor T, et al (2003) A nomenclature for restriction enzymes, DNA methyltransferases, homing endonucleases and their genes. Nucleic Acids Res 31:1805–1812

Robinson CR, Sauer RT (1996) Covalent attachment of Arc repressor subunits by a peptide linker enhances affinity for operator DNA. Biochemistry 35:109–116

Rouet P, Smih F, Jasin M (1994) Expression of a site-specific endonuclease stimulates homologous recombination in mammalian cells. Proc Natl Acad Sci USA 91:6064–6068

Seligman LM, Chisholm KM, Chevalier BS, et al (2002) Mutations altering the cleavage specificity of a homing endonuclease. Nucleic Acids Res 30:3870–3879

Silva GH, Belfort M (2004) Analysis of the LAGLIDADG interface of the monomeric homing endonuclease I-DmoI. Nucleic Acids Res 32:3156–3168

Silva GH, Dalgaard JZ, Belfort M, van Roey P (1999) Crystal structure of the thermostable archaeal intron-encoded endonuclease I-DmoI. J Mol Biol 286:1123–1136

Simoncsits A, Tjornhammar ML, Rasko T, Kiss A, Pongor S (2001) Covalent joining of the subunits of a homodimeric type II restriction endonuclease: single-chain PvuII endonuclease. J Mol Biol 309:89–97

Steuer S, Pingoud V, Pingoud A, Wende W (2004) Chimeras of the homing endonuclease PI-SceI and the homologous *Candida tropicalis* intein: a study to explore the possibility of exchanging DNA-binding modules to obtain highly specific endonucleases with altered specificity. Chembiochem 5:206–213

Sussman D, Chadsey M, Fauce S, Engel A, Bruett A, Monnat R Jr, Stoddard BL, Seligman LM (2004) Isolation and characterization of new homing endonuclease specificities at individual target site positions. J Mol Biol 342:31–42

Van Roey P, Waddling CA, Fox KM, Belfort M, Derbyshire V (2001) Intertwined structure of the DNA-binding domain of intron endonuclease I-TevI with its substrate. EMBO J 20:3631–3637

Inteins – A Historical Perspective

FRANCINE B. PERLER

Abstract

Protein splicing elements, termed inteins, were first identified in 1990. Since then, post-translational protein splicing has been demonstrated and the self-catalytic mechanism deciphered. The robust nature of these single turnover enzymes is evidenced by the expanding list of naturally occurring variations in the protein splicing mechanism. Protein splicing must be efficient and neutral, and must not cause detrimental effects to the spliced extein; otherwise, selective pressure would lead to intein loss. Inteins are probably ancient elements, but their original function can only be speculated upon, because invasion by homing endonucleases mobilized them into new locations and converted them into selfish DNA. To date, there is no evidence of regulation of protein splicing in native systems. The sporadic distribution of inteins may relate more to the types of genes found in mobile elements capable of spreading inteins, than to the function of those genes. Inteins tend to be found in conserved host protein motifs, which may be due to conservation of homing endonuclease recognition sites, difficulty in removing inteins from essential regions or the ease of accepting an insertion sequence in a conserved substrate or cofactor binding site designed to interact with the environment. The ability to cleave peptide bonds, to ligate protein fragments and to generate carboxy-terminal alpha-thioesters have made inteins the fastest growing tool for protein engineering and biotechnology.

F.B. Perler (e-mail: perler@neb.com)
New England Biolabs, Inc., 32 Tozer Road, Beverly, Massachusetts 01915, USA

Nucleic Acids and Molecular Biology, Vol. 16
Marlene Belfort et al. (Eds.)
Homing Endonucleases and Inteins

1 Introduction

It is hard to imagine that only 15 years ago protein splicing was unheard of and the commonly accepted view of gene expression involved one gene yielding one protein, with processing of polyproteins being the rare exception. Between 1990 and 2004, over 200 putative protein-splicing elements were identified and over 500 protein-splicing papers were published [see InBase (Perler 2002), the on-line protein-splicing database at www.neb.com/neb/inteins.html]. Between 1993 and 1995, more reviews of protein splicing were published than experimental studies.

In 1990, the groups of Stevens, Neff and Hirata independently found that a *Saccharomyces cerevisiae* ATPase gene (Sce VMA) contained an insertion sequence unrelated to other ATPases, which they predicted was removed by protein splicing (Anraku et al. 1990; Kane et al. 1990). Amazingly, expression of the Sce VMA gene yielded two vastly different proteins, the mature ATPase and the VMA homing endonuclease. The homing endonuclease contained amino- and carboxy-terminal regions absent in previously described intron-encoded LAGLIDADG family homing endonucleases. These regions were later shown to form the protein-splicing domain. Proving that splicing was at the protein level and not at the RNA level was extremely difficult, because the precursor protein was not observed in native systems.

2 Efforts To Prove That Splicing Is Post-translational

The fact that the intervening sequence in the Sce VMA gene was in the same open reading frame (ORF) as the surrounding sequences suggested that it could be removed from a protein precursor, rather than by traditional RNA splicing (Shih et al. 1988). The challenge to prove post-translational processing was compounded by the need to provide more than circumstantial evidence to change such a fundamental aspect of gene expression. The inability to detect spliced mRNA by any technique was inconclusive, since spliced mRNA could be rapidly degraded after translation (Hirata et al. 1990; Kane et al. 1990; Davis et al. 1992; Hodges et al. 1992; Perler et al. 1992; Gu et al. 1993). Silent substitutions that changed splice junction proximal nucleotide sequences, but not amino acids (aa), had no effect on splicing (Davis et al. 1992; Perler et al. 1992), arguing against RNA splicing since introns have essential sequences at the intron–exon junctions. Frameshift mutations generating translational stops were introduced into the intervening sequence (Kane et al. 1990; Davis et al. 1992; Perler et al. 1992; Gu et al. 1993). Mutations that created a stop codon in the intervening sequence blocked synthesis of the mature protein. Splicing was rescued with a second frameshift mutation that restored the original ORF. Although this supported protein splicing, it was possible that these mutations were at positions critical to intron function. Next, pulse-chase labeling experiments were performed to establish a precursor–product relationship (Kane et al. 1990; Davis et al. 1992; Hodges et al. 1992). The analysis was complicated by extra labeled proteins, which we now know are products of single splice junction cleavage. Finally, temperature-dependent splicing of a purified engineered precursor protein proved beyond a reasonable doubt that protein splicing was post-translational (Xu et al. 1993).

In 1994, many of the workers in the field recognized the need to adopt a uniform nomenclature for protein splicing (Perler et al. 1994). The primary translation product is called the **precursor protein**. The intervening protein sequence that is removed by protein splicing is called the **intein** (***in****ternal pro****tein***). The regions flanking the intein are called the **exteins** (***ex****ternal pro****tein***). These terms apply to both genes and proteins. **Protein splicing** is defined as the excision of an intein coupled with ligation of the exteins. A native peptide bond exists between the ligated exteins (Cooper et al. 1993). **Excision** refers to the production of free intein from a precursor due to cleavage at both sides of the intein, irrespective of extein ligation. **Cleavage** usually refers to the breakage of the bond at a single intein/extein splice junction. In analyzing splicing data, it is important not to rely on excision as a measure of protein splicing, since excision is frequently not accompanied by ligation in heterologous exteins.

3 Intein Motifs and Conserved Residues

No set convention has been adopted for numbering of residues. One method (Fig. 1) that simplifies comparison of inteins in different host proteins (Noren et al. 2000) consists of separately numbering each part of the precursor: (1) **intein residues** are numbered sequentially from amino to carboxy terminus of the intein beginning with 1; (2) **N-extein residues** are negative numbers starting with –1 at the carboxy terminus of the N-extein, counting towards the precursor amino terminus; (3) **C-extein residues** include a plus sign beginning with +1 at the amino terminus of the C-extein, counting towards the precursor carboxy terminus.

Analysis of even the first few inteins indicated that sequences surrounding the splice sites were conserved. Three inteins were flanked by cysteines, making it unclear whether the intein began or ended with cysteine (Hirata et al. 1990; Kane et al. 1990; Davis et al. 1991, 1992). The *Thermococcus litoralis* DNA polymerase second intein (Tli Pol-2) was predicted to begin with serine and have a threonine at the +1 position based on the conserved DNA polymerase motif into which the intein was inserted (Perler et al. 1992). This

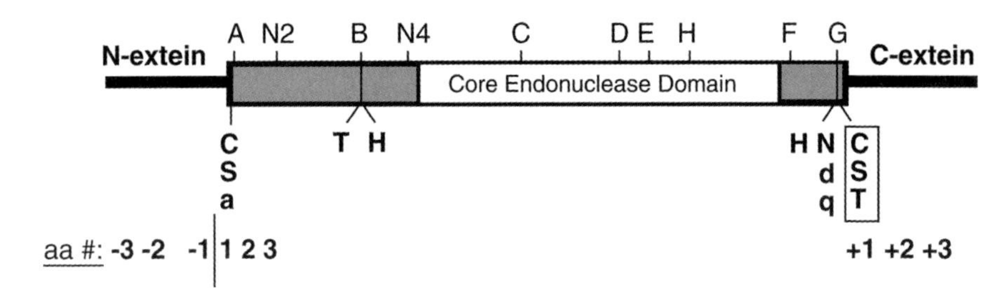

Fig. 1. Organization of a protein-splicing precursor. Intein motifs are depicted *above* and conserved amino acids are depicted *below*, as are amino acid numbers. In this precursor, a LAGLIDADG family homing endonuclease is shown with its standard motifs (C, D, E, and H). The amino-terminal splicing region (motifs A, N2, B, and N4) associates with the carboxy-terminal splicing region (motifs F and G) to form the intein-splicing domain. The conserved nucleophiles and assisting groups are in *uppercase*, while residues present in non-canonical inteins are in *lowercase*

was confirmed by sequencing the excised Tli Pol-2 intein (Perler et al. 1992). Similar results were obtained with the Sce VMA intein, which began with cysteine (Cooper et al. 1993). Although threonine has not been observed at the intein amino terminus, substitution of serine by threonine in the Tli Pol-2 intein yielded a functional intein (Hodges et al. 1992). If the intein began with cysteine or serine, then it must end with the conserved asparagine, since comparison to intein-less homologues required one of the cysteine, serine or threonine residues flanking the intein to be part of the extein.

Intein motifs contain groups of similar amino acids at specific positions interspersed with non-conserved positions, making it more difficult to identify inteins by simple sequence comparison. Sophisticated sequence comparison systems including a hidden Markov model have been devised to find new inteins (Dalgaard et al. 1997b; Pietrokovski 1998a). Two motif nomenclatures are widely used. The earlier system used blocks A–H (Fig. 1), where only blocks A, B, F, and G are in the splicing domain (Pietrokovski 1994; Perler et al. 1997a). A second method uses `N´ for N-extein motifs, `EN´ for homing endonuclease motifs, and `C´ for C-extein motifs (Pietrokovski 1998a). Equivalent motifs are: A=N1, B=N3, C=EN1, D=EN2, E=EN3, H=EN4, F=C2, and G=C1. Motifs N2 and N4 are characterized by an acidic residue (Pietrokovski 1998a). Motif sequences are listed in InBase. Block A begins with the intein amino terminus. Block B is usually 60–90 aa from the intein amino terminus and often contains a Thr-x-x-His motif. This histidine is the most conserved intein residue and is only absent from one putative intein. A similar motif is also found in proteases. Block F directly precedes block G, which contains the intein carboxy-terminal dipeptide His-Asn and the C-extein +1 nucleophile.

The serine, cysteine and threonine at the carboxy-terminal side of both splice junctions are nucleophiles in the protein-splicing reaction. The third nucleophile is the intein carboxy-terminal asparagine. The two conserved histidines assist these nucleophiles. The remaining conserved residues either directly assist the chemical reactions or are important for folding of the intein. A pattern of residues with similar chemical functionalities instead of conservation of a particular amino acid is a hallmark of inteins. This may reflect the fact that many different amino acids can increase nucleophilicity and electrophilicity. However, each intein appears to have evolved a network of residues to facilitate the nucleophilic displacements mediated by the specific combination of nucleophiles present in that intein, since splicing is often reduced or blocked by conservative substitution of any of these three nucleophiles.

4 Intein Polymorphisms and Non-canonical Inteins

Over 200 inteins have been listed in InBase as of 2004. Unlike other enzymes, instead of honing the consensus sequences with the discovery of more members of the class, the identification of more inteins has led to the recognition of more polymorphisms. Mini-inteins have a small linker in place of a homing endonuclease domain. The 10–20% of inteins that lack a penultimate histidine are at different points in the evolutionary path towards efficient splicing after loss of this histidine (Wang and Liu 1997; Chen et al. 2000). Inteins have been identified with an amino-terminal alanine in Snf2, KlbA and DnaB helicase exteins (Dalgaard et al. 1997a; Gorbalenya 1998; Pietrokovski 1998a; Southworth et al. 2000; Yamamoto et al. 2001; Southworth and Perler 2002). These Ala1 inteins were originally thought to be inactive, but were later shown to splice using a slightly different mechanism (Southworth et al. 2000; Yamamoto et al. 2001). Aspartic acid and glutamine have been found at the carboxy terminus of some inteins, which appear to use yet another variation of the standard protein-splicing mechanism (Pietrokovski 1998b; Amitai et al. 2004; Mills et al. 2004). Finally, some inteins lack one or more block G essential residues. Based on phylogeny, location and additional motifs, these elements have been termed bacterial intein-like sequences (BILs; Amitai et al. 2003). Splicing of a *Magnetospirillum magnetotacticum* BIL in Magn8951 was restored when the +1 tyrosine was 'reverted' to cysteine (Southworth et al. 2004), indicating that this BIL is probably a mutated intein. *Trichodesmium erythraeum* inteins in DnaE (Liu and Yang 2003) and DnaB (Yang et al. 2004) have multiple repeat sequences. Is this a property of *T. erythraeum* or a new type of sequence that has invaded inteins?

In summary, the six most prevalent intein polymorphisms involve the presence of an endonuclease domain, the presence of a penultimate histidine, the presence of multiple repeat units in the linker region, inteins beginning with alanine, inteins ending with glutamine or aspartic acid, and BILs.

5 Criteria for Intein Designation

Many types of protein rearrangements occur besides intein-mediated protein splicing. Some are autocatalytic, while others require processing enzymes. These include processing of polyproteins, the N-terminal nucleophile amidohydrolase precursors, pyruvoyl enzymes, Hedgehog embryonic signaling protein, LexA and lambda repressor (Beachy et al. 1997; Perler et al. 1997b; Perler 1998; Noren et al. 2000). Recently, a proteasome-mediated `protein-splicing´ process involved in antigen presentation in lymphocytes was described in humans; however, this process is unrelated to inteins and the use of the term protein splicing has led to confusion with inteins (Hanada et al. 2004; Vigneron et al. 2004). It would be clearer if the term protein splicing was limited to inteins and protein rearrangement (the traditional term) or protein editing was used for these other processes.

A combination of criteria differentiates protein splicing from other types of protein editing. The first is sequence comparison with homologues lacking the in-frame insertion. Second, the size of the mature protein is similar to homologues without the extra sequence. Third, proof of extein ligation by Western blot analysis, amino acid sequencing or by identifying peptides spanning the ligation point by mass spectrometry. Fourth, the intervening sequence should contain intein motifs (Fig. 1).

6 The Minimal Splicing Element

After proving that splicing was at the protein level, the next priority was to determine the minimal splicing element. Did the intein mediate splicing? Did the extein contribute to protein splicing? Several groups cloned inteins with fewer and fewer native extein residues into model host proteins until it became clear that the intein plus the first C-extein amino acid (the +1 residue) were sufficient for protein splicing (Davis et al. 1992; Cooper et al. 1993; Xu et al. 1993). However, splicing generally improves when an intein is inserted at a site that mimics its native insertion site. An in vitro protein-splicing system (Xu et al. 1993) demonstrated that no accessory factors, sources of energy or additional enzymes were required.

Several lines of evidence suggest that the protein-splicing domain was split by insertion of a homing endonuclease or small linker domain. Two different families of homing endonucleases (HNH and LAGLIDADG) are present in the intein central region. Protein-splicing and homing endonuclease functionalities have separate active sites. Mutation of intein active site residues do not block endonuclease function and vice versa (Hodges et al. 1992; Kawasaki et al. 1997). Phylogenetic analysis suggests that the splicing and endonuclease domains may have evolved independently (Dalgaard et al. 1997a,b). A key insight was provided by a structure of the *Drosophila* Hedgehog protein autoprocessing domain, which was found to share a `fold´ (a three-dimensional structure) with the splicing regions of the Sce VMA intein (Hall et al. 1997). Many residues in the two structures had superimposable backbone positions, despite the fact that there was little sequence identity except in intein motifs A and B (Koonin 1995). The shared structural fold was called a HINT module, for *H*edgehog and *INT*ein. In all probability, homing endonuclease domains have been gained and lost during evolution. In some cases, some homing endonuclease motifs are missing and the intein appears to be in the process of deleting the endonuclease. Several mini-inteins have alleles with homing endonuclease domains, such as the *Porphyra purpurea* DnaB helicase intein or the *Mycobacterium xenopi* DNA gyrase subunit A (Mxe GyrA) intein. Several groups independently generated artificial mini-inteins (Chong and Xu 1997; Derbyshire et al. 1997; Mills et al. 1998; Shingledecker et al. 1998; Wu et al. 1998b; Mathys et al. 1999). The endonuclease domain is dispensable for protein splicing, which further defines the minimal splicing element as the amino-terminal and carboxy-terminal splicing regions of the intein, plus the C-extein +1 nucleophile.

7 Splicing of Split Inteins: *Trans*-Splicing

A major challenge to studying inteins and using them for biotechnology is the difficulty in controlling the reaction so that precursors can be isolated and splicing initiated at will. Several workers in the field realized that splicing could be controlled if the precursor was split and the fragments reassembled to form an active intein (Lew et al. 1998; Mills et al. 1998; Shingledecker et al. 1998; Southworth et al. 1998; Wu et al. 1998b). The homing endonuclease region was chosen as the best split site in these systems. Since the carboxy-terminal splicing region is so small, synthetic peptides could be used. As with many other types of proteins, split precursor fragments can spontaneously interact to reconstitute an active precursor. In fact, fragments of the Psp Pol intein still splice with deletions of the endonuclease domain or with 80% overlap of the intein in the two fragments. The biggest problem with engineered split precursors was solubility and aggregation of protein fragments, which usually required denaturation and renaturation for splicing to occur. In the case of the split Psp-GBD Pol intein, *trans*-splicing was as efficient in 6 M urea as in aqueous solutions. A further level of splicing control was achieved with a split mini-intein derivative of the *Mycobacterium tuberculosis* (Mtu) RecA intein, which formed a disulfide bond upon reassembly and required thiols to initiate splicing. A precursor with two inteins was split in each intein and only fragments from the same intein associated, yielding a correctly spliced mature product from three fragments (Otomo et al. 1999).

The discovery of a naturally split DnaE gene in *Synechocystis* sp. strain PCC6803 (Ssp) with intein fragments attached to each DnaE extein fragment (Gorbalenya 1998; Pietrokovski 1998a; Wu et al. 1998a) solved many of the problems associated with the artificially split precursors. Ssp DnaE fragments are soluble and readily associate (Martin et al. 2001; Nichols et al. 2003). A second naturally split intein with Ser1 and Thr+1 nucleophiles is present in the DNA polymerase from *Nanoarchaeum equitans.*

8 The Influence of Exteins and Insertion Site Characteristics

The ability to splice in different host proteins initially led researchers to hypothesize that the extein sequence was irrelevant, except for the +1 residue. We now know that this is untrue. Inteins and their native extein hosts have evolved to optimize splicing while minimizing dead-end cleavage reactions. Inteins can be considered single turnover enzymes, with the proximal extein sequences being the substrate of the intein enzymatic activities. Like any other enzyme family, different inteins have different degrees of substrate specificity. Some inteins are very permissive, while other inteins are very restrictive. The Mxe GyrA intein splices efficiently with 3 of 20 aa at the –1 position (Southworth et al. 1999), while the Sce VMA intein splices with most –1 residues (Chong et al. 1998). Several groups have examined the ability of inteins to splice or cleave when a non-native residue precedes or follows the intein (Chong et al. 1997, 1998; Evans et al. 1999; Mathys et al. 1999; Southworth et al. 1999).

The number of native extein residues surrounding one or both sides of an intein that leads to the highest percentage of spliced product varies with each intein and, in some cases, with each extein. This extein effect may be due to substrate specificity issues, global folding of the precursor and/or the local architecture of the intein active site in different precursors. These requirements may explain the observation that splicing of various inteins in certain foreign precursors is sometimes temperature dependent. To optimize splicing, many researchers routinely include one to five native extein residues on both sides of an intein. Experiments with model precursors versus native precursors clearly indicate that the intein and native extein have evolved to permit facile splicing with minimal perturbation of extein function and that we have yet to understand these requirements.

Intein insertion sites tend to be conserved extein motifs (Perler et al. 1992; Dalgaard et al. 1997b; Pietrokovski 1998a), leading some to suggest that finding the location of the extein active site is as simple as probing the position around an intein. Why are inteins found in conserved motifs? Several theories exist. First, selective pressure on the extein conserves the homing endonuclease recognition site, facilitating horizontal transfer. This is supported by the observation that intein-less extein alleles are not cut (Fsihi et al. 1996; Gimble 2000; Saves et al. 2000). Second, in the absence of spliced mRNA for retrotransposition back into the genome, the intein can only be removed by deletion or recombination. When the intein is in an essential region, deletion has to be precise in order to maintain extein function. Loss of an intein by recombination is unlikely if the intein has endonuclease activity because the endonuclease initiates homing, which favors transfer of the intein gene into the intein-less extein, rather than loss of the intein-plus gene by recombination (Gimble and Thorner 1992).

Finally, I favor biochemical reasons that involve protein folding and flexibility. The intein and extein must coexist. The intein must fold properly in the precursor for splicing to occur and splicing cannot yield a misfolded extein. An intein in the extein hydrophobic core might disrupt the extein-folding pathway, stimulate aggregation, or require refolding for extein function. The intein must be able to undergo subtle conformational changes thought to occur at its active site during splicing (Poland et al. 2000; Southworth et al. 2000; Wood et al. 2000). Conserved motifs have properties that make them ideal intein insertion sites. They are often surface-associated docking regions designed to interact with substrates and cofactors. They thus provide an accessible area where insertion of more than 130 aa might not globally disrupt folding of the extein. Active sites and binding regions tend to be loops between α-helices or β-strands, which could provide the flexibility mandated by conformational changes at the intein active site (Poland et al. 2000). This does not conflict with the proposal that proximal extein residues interact with intein residues to form a β-sheet (Kawasaki et al. 1997; Nogami et al. 1997), since a loop region might provide structural flexibility to assume a β-strand conformation in the precursor without affecting the final local structure in the spliced extein.

9 The Challenge of Deciphering the Protein-Splicing Mechanism

It was evident at the start that inteins had chemically reactive amino acids at both splice junctions – amino acids normally found in enzyme active sites.

This led to the hypothesis that folding of the intein brought the conserved residues at both splice junctions into close proximity to form the protein-splicing active site and also brought together the exteins for ligation. We still do not know whether there is a single protein-splicing active site or the relationship between the potentially different active sites for each step in the protein-splicing pathway. All experiments with split inteins fail to detect cleavage in single precursor fragments, confirming that the both ends of the intein are necessary for activity at either splice junction (Lew et al. 1998; Mills et al. 1998; Shingledecker et al. 1998; Southworth et al. 1998; Wu et al. 1998a,b; Martin et al. 2001; Nichols et al. 2003).

Although we know the major players in the splicing reaction (the nucleophiles), the assisting groups have yet to be fully identified, partly because different amino acids perform similar functions in different inteins. The block B histidine activates the amino-terminal splice junction (Kawasaki et al. 1997) and the penultimate histidine activates the carboxy-terminal splice junction (Cooper et al. 1993; Chong et al. 1996, Chong et al. 1998; Xu and Perler 1996; Kawasaki et al. 1997; Wang and Liu 1997; Chen et al. 2000), although the block B histidine may sometimes play a role in carboxy-terminal cleavage (Southworth et al. 2000). Mutagenesis and structural data suggest that block F plays a role in activating the carboxy-terminal splice junction (Ghosh et al. 2001; Ding et al. 2003).

The protein-splicing mechanism was deciphered using a combination of techniques. It did not take long before each group was mutating the intein amino-terminal residue, the conserved histidines, the carboxy-terminal asparagine and the +1 residue (Davis et al. 1992; Hirata and Anraku 1992; Hodges et al. 1992; Cooper et al. 1993; Chong et al. 1996; Xu and Perler 1996; Kawasaki et al. 1997; Wang and Liu 1997; Evans et al. 1999; Mathys et al. 1999; Chen et al. 2000). Results varied slightly with each system. Some inteins were able to splice if conservative mutations (serine, threonine, cysteine) were made at the carboxy-terminal side of either splice junction, but usually not efficiently. Mutation of the Tli Pol-2 intein Ser1 to threonine in the natural extein resulted in about 10% DNA polymerase activity compared to wild type, while mutation to cysteine yielded only minor amounts of spliced DNA polymerase and mutation to alanine blocked splicing (Hodges et al. 1992). Mutation of Cys1 to Ser1 or Cys+1 to Ser+1 in the Sce VMA intein blocked splicing in yeast cells when assayed by Western blot, although enough spliced product was made to rescue growth (Hirata and Anraku 1992; Cooper et al. 1993). This highlights a significant difference between a biochemical and a biological assay: cells usually require far fewer properly spliced molecules than needed for detection in an in vitro assay. Conservative mutation of the carboxy-terminal asparagine blocked splicing.

Many protein-splicing mechanisms were proposed, which served as guides for experimentation. All of the early models were discarded when the organization of the branched intermediate was determined (Xu et al. 1994), because they were inconsistent with this intermediate. The first proposed model for protein splicing consisted of three steps (Wallace 1993; Xu et al. 1993): an N/O acyl shift of both Ser1 and Ser+1, followed by attack on the upstream ester by the downstream primary amine, yielding ligated exteins with the carboxy terminus of the intein connected to the side chain of Ser+1. These models did not make strong predictions for how the branch was resolved, although hydrolysis was suggested. Also in 1993, Cooper et al. suggested a protein-splicing mechanism based on their mutational analysis of the Sce VMA intein (Cooper et al. 1993). According to this model, the peptide bond between the intein and the C-extein is cleaved during cyclization of the intein carboxy-terminal asparagine (succinimide formation). The resultant amino terminus of the C-extein cleaves the peptide bond at the amino-terminal splice junction, releasing the intein and ligating the exteins. In 1994 another mechanism involving two cyclizations of the intein carboxy-terminal asparagine was suggested (Clarke 1994).

In 1993, Xu et al. developed an in vitro protein-splicing system where the intein in the DNA polymerase gene from the extreme thermophilic archaeon *Pyrococcus* sp. strain GB-D (Psp-GBD Pol) was cloned between the *Escherichia coli* maltose-binding protein and the delta Sal fragment of *Dirofilaria immitis* paramyosin, for the first time allowing purification of an active precursor (Xu et al. 1993). This in vitro splicing system was critical to defining the protein-splicing mechanism (Xu et al. 1993, 1994; Xu and Perler 1996) in combination with studies of model peptides (Shao et al. 1995, 1996). The key discovery was a branched intermediate, initially identified as a slowly migrating protein in sodium dodecyl sulfate/polyacrylamide gel electrophoresis (SDS-PAGE). Protein sequencing revealed two amino termini (the intein and the N-extein). The branched intermediate reverted to a linear molecule at pH 10. It was alkali labile, yielding the N-extein and a fusion of the intein with the C-extein. Mutation of either Ser1 or Ser+1 prevented branch formation. These data suggested an ester linkage between the N-extein and the side chain of Ser+1. The observation that mutation of Ser1 also blocked branched intermediate formation suggested that an acyl shift of this residue to form a linear ester intermediate might be necessary for branch formation, which was later verified (Shao et al. 1996; Xu and Perler 1996). Mutagenesis data indicated that the intein carboxy-terminal asparagine was necessary for resolution of the branched intermediate. A less common form of side-chain cyclization of asparagine was proposed. Analysis of the carboxy-terminal peptide produced by cyanogen bromide cleavage at an engineered methionine in the Psp-GBD Pol intein proved that the intein ended in a cyclized form of asparagine (a carboxy-terminal succinimide; Xu et al. 1994; Shao et al. 1995). Similar experiments performed with the Sce VMA intein confirmed this mechanism (Chong et al. 1996, 1998).

These data support the accepted four-step protein-splicing pathway: (1) an acyl shift of Ser1 or Cys1 yields an ester or thioester linkage at the amino-terminal splice junction; (2) this (thio)ester linkage is cleaved and the branched intermediate is formed by transesterification resulting from attack by the C-extein Ser+1, Cys+1 or Thr+1; (3) the branch is resolved by cyclization of the conserved asparagine, which cleaves the carboxy-terminal splice junction and releases the ligated exteins; and (4) a peptide bond is formed between the exteins by a spontaneous acyl shift of the +1 residue. The details of this process are described by Mills and Paulus (this Vol.).

10 Splicing of Non-canonical Inteins

Mutation of Ser1 or Cys1 to alanine abolishes protein splicing in all standard inteins tested. How then do inteins splice when they naturally contain Ala1? The answer lies in one of the earlier models of protein splicing, which suggested that the +1 residue could directly attack a peptide bond at the amino-terminal splice junction, rather than a (thio)ester bond (Xu et al. 1994). This turns out to be the case, with the remainder of the mechanism (branch resolution and peptide bond formation) being identical to the standard protein-splicing pathway. The question of how natural Ala1 inteins have overcome the barrier to direct attack of the peptide bond at the amino-terminal splice junction remains unanswered. In standard inteins, a conformational change (global or proximal to the active site) triggered by linear (thio)ester formation has been proposed, which allows alignment of essential nucleophiles or accessory groups for the transesterification step. This hypothesis is supported by the crystal structure of a Sce VMA precursor, where the mutated Ser+1 residue is 9 Å away from the scissile bond (Poland et al. 2000). Ala1 inteins may always be in this second conformation or they can shift to it without prior ester formation. The final answer may have to await structural analysis of a precursor containing a natural Ala1 intein. Another possibility is that residues within Ala1 inteins make the scissile bond more electrophilic than in standard inteins, allowing cleavage of a peptide bond by the +1 nucleophile.

Inteins with carboxy-terminal glutamine or aspartic acid have been shown to splice (Pietrokovski 1998b; Amitai et al. 2004; Mills et al. 2004). These mechanisms will be reviewed in other chapters of this volume. In the Psp-GBD Pol intein, mutation of the carboxy-terminal asparagine to glutamine or aspartic acid yielded branched intermediate and amino-terminal cleavage products (Xu and Perler 1996).

11 Why Are Inteins So Robust?

The nature and number of allowable variations in the standard protein-splicing mechanism is remarkable. This versatility may be directly due to the fact that protein splicing is a unimolecular reaction, which keeps the extein `substrate´ locked in the intein active site allowing reactions to proceed suboptimally or even in altered order. New nucleophiles and variant mechanisms may be easier to accommodate in inteins because of (1) the ability of intein active sites to function with similar amino acid nucleophiles and (2) the putative structural flexibility of the intein active site required by the proposed conformational changes associated with different steps of the splicing pathway. The strength of inteins may turn out to be the supporting cast, the network(s) of facilitating groups that make the scissile bonds so susceptible to attack by a variety of nucleophiles. This, combined with being a unimolecular reaction, has probably allowed inteins to adapt to various chemical mechanisms dictated by mutated catalytic groups.

12 Control, Control, Control

Controlling splicing and cleavage led to understanding protein splicing. Several chapters of this volume will describe applications based on protein splicing. These applications are predicated on the ability to turn inteins on and off by mutation, temperature-dependent or pH-sensitive splicing, placing inteins in suboptimal insertion sites, chemically caging nucleophiles, thiols, zinc, splitting inteins, or activating split inteins with small molecules.

Most applications involve using inteins to purify proteins and for protein semisynthesis. Protein purification is usually accomplished by mutating the intein so that it will cleave on demand the splice junction between the intein and the protein of interest. A purification tag or a folding reporter is often added to one end of the intein or in place of the homing endonuclease domain. Protein semisynthesis can be performed directly with inteins using trans-splicing or it can be indirectly performed by taking advantage of the carboxy-terminal α-thioester produced by thiol cleavage of a target protein linked to the intein amino terminus after Cys1 undergoes its acyl shift. In the presence of thiol reagents, proteins or peptides with carboxy-terminal α-thioesters will spontaneously ligate to polypeptides beginning with cysteine. For synthetic peptides, this process is called native chemical ligation (Dawson et al. 1994); when inteins are involved, the process is called expressed protein ligation (EPL; Muir et al. 1998) or intein-mediated protein ligation (IPL; Evans et al. 1998). Using these techniques, proteins can be stitched together using synthetic or biosynthetic building blocks. Specific regions of proteins can be modified and then ligated to the remaining unmodified portion of the protein. Proteins can be segmentally labeled, phosphorylated, tagged, or synthetic moieties (biosensors, cross-linking reagents, etc.) added, etc. Protein and peptide backbones can be cyclized for increased stability. Cytotoxic proteins can be expressed (Wu et al. 2002). *trans*-Splicing may reduce the possibility of unwanted gene transfer in transgenic plants (Chin et al. 2003; Yang et al. 2003). Bringing control to a new level, temperature-sensitive inteins were cloned into the transcription regulators Gal4 and Gal80, controlling expression of any gene activated by Gal4 (Zeidler et al. 2004). This system was then used to turn on and off expression of genes regulated by Gal4 in a spatial and temporal fashion in *Drosophila* embryos by combining temperature-dependent splicing of Gal4 or Gal80 precursors with enhancer-driven tissue-specific expression.

13 Perspectives for the Future

We have learned a lot about inteins, but there is much more to discover. Mechanisms for splicing of non-canonical inteins still need to be refined. How inteins control the order of reactions in the splicing pathway is still to be determined. The active site of inteins needs to be further analyzed to understand the mechanism of splicing at a more detailed level. Structures of intermediates and active precursors are needed. Many basic questions about evolution of inteins are still unanswered, including the role of inteins in early evolution, how inteins became associated with homing endonucleases, how intein insertion sites and homing endonuclease recognition sites merged, or how inteins spread across phylogenetic domains. Is the absence of inteins in multicellular organisms due to barriers preventing lateral transmission to germ cells? Although inteins have not been found in higher organisms, hedgehog autoprocessing domains are present. The diversity of intein applications is growing exponentially. The future is only limited by our imagination.

Acknowledgements. I thank Marlene Belfort for initiating the intein nomenclature project that led to InBase as a resource for those interested in inteins and the many protein splicers that continue to contribute to InBase. I pay homage to the creativity displayed by researchers developing intein applications whose contributions are too numerous to include herein, especially Tom Muir. I apologize to all those whom I have not mentioned due to space limitations. I thank Ming Xu who came to my laboratory as a postdoc with an enthusiasm for determining the protein-splicing mechanism and Maurice Southworth who has been part of my group since the time before inteins and who has enjoyed playing with them as much as I have. I thank Donald Comb for supporting my research.

References

Amitai G, Belenkiy O, Dassa B, Shainskaya A, Pietrokovski S (2003) Distribution and function of new bacterial intein-like protein domains. Mol Microbiol 47:61–73

Amitai G, Dassa B, Pietrokovski S (2004) Protein splicing of inteins with atypical glutamine and aspartate C-terminal residues. J Biol Chem 279:3121–3131

Beachy PA, Cooper MK, Young KE, von Kessler DP, Park W, Hall TMT, Leahy DJ, Porter JA (1997) Multiple roles of cholesterol in hedgehog protein biogenesis and signaling. Cold Spring Harbor Symp Quant Biol 62:191–204

Chen L, Benner J, Perler FB (2000) Protein splicing in the absence of an intein penultimate histidine. J Biol Chem 275:20431–20435

Chin HG, Kim GD, Marin I, Mersha F, Evans TC Jr, Chen L, Xu MQ, Pradhan S (2003) Protein trans-splicing in transgenic plant chloroplast: reconstruction of herbicide resistance from split genes. Proc Natl Acad Sci USA 100:4510–4515

Chong S, Xu MQ (1997) Protein splicing of the *Saccharomyces cerevisiae* VMA intein without the endonuclease motifs. J Biol Chem 272:15587–15590

Chong S, Shao Y, Paulus H, Benner J, Perler FB, Xu MQ (1996) Protein splicing involving the *Saccharomyces cerevisiae* VMA intein. The steps in the splicing pathway, side reactions leading to protein cleavage, and establishment of an in vitro splicing system. J Biol Chem 271:22159–22168

Chong S, Mersha FB, Comb DG, Scott ME, Landry D, Vence LM, Perler FB, Benner J, Kucera RB, Hirvonen CA, Pelletier JJ, Paulus H, Xu MQ (1997) Single-column purification of free recombinant proteins using a self-cleavable affinity tag derived from a protein splicing element. Gene 192:271–281

Chong S, Williams KS, Wotkowicz C, Xu MQ (1998) Modulation of protein splicing of the *Saccharomyces cerevisiae* vacuolar membrane ATPase intein. J Biol Chem 273:10567–10577

Clarke ND (1994) A proposed mechanism for the self-splicing of proteins. Proc Natl Acad Sci USA 91:11084–11088

Cooper AA, Chen Y, Lindorfer MA, Stevens TH (1993) Protein splicing of the yeast *tfp1* intervening protein sequence: a model for self-excision. EMBO J 12:2575–2583

Dalgaard JZ, Klar AJ, Moser MJ, Holley WR, Chatterjee A, Mian IS (1997a) Statistical modeling and analysis of the LAGLIDADG family of site-specific endonucleases and identification of an intein that encodes a site-specific endonuclease of the HNH family. Nucleic Acids Res 25:4626–4638

Dalgaard JZ, Moser MJ, Hughey R, Mian IS (1997b) Statistical modeling, phylogenetic analysis and structure prediction of a protein splicing domain common to inteins and hedgehog proteins. J Comput Biol 4:193–214

Davis EO, Sedgwick SG, Colston MJ (1991) Novel structure of the *rec*A locus of *Mycobacterium tuberculosis* implies processing of the gene product. J Bacteriol 173:5653–5662

Davis EO, Jenner PJ, Brooks PC, Colston MJ, Sedgwick SG (1992) Protein splicing in the maturation of *M. tuberculosis* RecA protein: a mechanism for tolerating a novel class of intervening sequence. Cell 71:201–210

Dawson PE, Muir TW, Clark-Lewis I, Kent SB (1994) Synthesis of proteins by native chemical ligation. Science 266:776–779

Derbyshire V, Wood DW, Wu W, Dansereau JT, Dalgaard JZ, Belfort M (1997) Genetic definition of a protein-splicing domain: functional mini-inteins support structure predictions and a model for intein evolution. Proc Natl Acad Sci USA 94:11466–11471

Ding Y, Xu M-Q, Ghosh I, Chen X, Ferrandon S, Lesage G, Rao Z (2003) Crystal structure of a mini-intein reveals a conserved catalytic module involved in side chain cyclization of asparagine during protein splicing. J Biol Chem 278:39133–39142

Evans TC, Benner J, Xu M-Q (1998) Semisynthesis of cytotoxic proteins using a modified protein splicing element. Protein Sci 7:2256–2264

Evans TC Jr, Benner J, Xu M-Q (1999) The in vitro ligation of bacterially expressed proteins using an intein from *Methanobacterium thermoautotrophicum.* J Biol Chem 274:3923–3926

Fsihi H, Vincent V, Cole ST (1996) Homing events in the GyrA gene of some mycobacteria. Proc Natl Acad Sci USA 93:3410–3415

Ghosh I, Sun L, Xu MQ (2001) Zinc inhibition of protein trans-splicing and identification of regions essential for splicing and association of a split intein. J Biol Chem 276:24051–24058

Gimble FS (2000) Invasion of a multitude of genetic niches by mobile endonuclease genes. FEMS Microbiol Lett 185:99–107

Gimble FS, Thorner J (1992) Homing of a DNA endonuclease gene by meiotic gene conversion in *Saccharomyces cerevisiae*. Nature 357:301–306

Gorbalenya AE (1998) Non-canonical inteins. Nucleic Acids Res 26:1741–1748

Gu HH, Xu J, Gallagher M, Dean GE (1993) Peptide splicing in the vacuolar ATPase subunit a from *Candida tropicalis.* J Biol Chem 268:7372–7381

Hall TM, Porter JA, Young KE, Koonin EV, Beachy PA, Leahy DJ (1997) Crystal structure of a hedgehog autoprocessing domain: homology between hedgehog and self-splicing proteins. Cell 91:85–97

Hanada K, Yewdell JW, Yang JC (2004) Immune recognition of a human renal cancer antigen through post-translational protein splicing. Nature 427:252–256

Hirata R, Anraku Y (1992) Mutations at the putative junction sites of the yeast *vma1* protein, the catalytic subunit of the vacuolar membrane H+-ATPase, inhibit its processing by protein splicing. Biochem Biophys Res Commun 188:40–47

Hirata R, Ohsumi Y, Nakano A, Kawasaki H, Suzuki K, Anraku Y (1990) Molecular structure of a gene, *vma1*, encoding the catalytic subunit of H+-translocating adenosine triphosphatase from vacuolar membranes of *Saccharomyces cerevisiae.* J Biol Chem 265:6726–6733

Hodges RA, Perler FB, Noren CJ, Jack WE (1992) Protein splicing removes intervening sequences in an archaea DNA polymerase. Nucleic Acids Res 20:6153–6157

Kane PM, Yamashiro CT, Wolczyk DF, Neff N, Goebl M, Stevens TH (1990) Protein splicing converts the yeast *tfp1* gene product to the 69-kD subunit of the vacuolar H+-adenosine triphosphatase. Science 250:651–657

Kawasaki M, Nogami S, Satow Y, Ohya Y, Anraku Y (1997) Identification of three core regions essential for protein splicing of the yeast VMA1 protozyme. J Biol Chem 272:15668–15674

Koonin EV (1995) A protein splice-junction motif in hedgehog family proteins. Trends Biochem Sci 20:141–142

Lew BM, Mills KV, Paulus H (1998) Protein splicing in vitro with a semisynthetic two-component minimal intein. J Biol Chem 273:15887–15890

Liu XQ, Yang J (2003) Split-DnaE genes encoding multiple novel inteins in *Trichodesmium erythraeum.* J Biol Chem 24:24

Martin DD, Xu M-Q, Evans TC (2001) Characterization of a naturally occurring trans-splicing intein from *Synechocystis* sp. PCC6803. Biochemistry 40:1393–1402

Mathys S, Evans TC, Chute IC, Wu H, Chong S, Benner J, Liu XQ, Xu M-Q (1999) Characterization of a self-splicing mini-intein and its conversion into autocatalytic N- and C-terminal cleavage elements: facile production of protein building blocks for protein ligation. Gene 231:1–13

Mills KV, Lew BM, Jiang S, Paulus H (1998) Protein splicing in trans by purified N- and C-terminal fragments of the mycobacterium tuberculosis RecA intein. Proc Natl Acad Sci USA 95:3543–3548

Mills KV, Manning JS, Garcia AM, Wuerdeman LA (2004) Protein splicing of a *Pyrococcus abyssi* intein with a C-terminal glutamine. J Biol Chem 279:20685–20691

Muir TW, Sondhi D, Cole PA (1998) Expressed protein ligation: a general method for protein engineering. Proc Natl Acad Sci USA 95:6705–6710

Nichols NM, Benner JS, Martin DD, Evans TC Jr (2003) Zinc ion effects on individual Ssp DnaE intein splicing steps: regulating pathway progression. Biochemistry 42:5301–5311

Nogami S, Satow Y, Ohya Y, Anraku Y (1997) Probing novel elements for protein splicing in the yeast VMA1 protozyme: a study of replacement mutagenesis and intragenic suppression. Genetics 147:73–85

Noren CJ, Wang J, Perler FB (2000) Dissecting the chemistry of protein splicing and its applications. Angew Chem Int Ed 39:450–466

Otomo T, Ito N, Kyogoku Y, Yamazaki T (1999) NMR observation of selected segments in a larger protein: central-segment isotope labeling through intein-mediated ligation. Biochemistry 38:16040–16044

Perler FB (1998) Breaking up is easy with esters. Nat Struct Biol 5:249–252

Perler FB (2002) InBase: the intein database. Nucleic Acids Res 30:383–384

Perler FB, Comb DG, Jack WE, Moran LS, Qiang B, Kucera RB, Benner J, Slatko BE, Nwankwo DO, Hempstead SK, Carlow CKS, Jannasch H (1992) Intervening sequences in an archaea DNA polymerase gene. Proc Natl Acad Sci USA 89:5577-5581

Perler FB, Davis EO, Dean GE, Gimble FS, Jack WE, Neff N, Noren CJ, Thorner J, Belfort M (1994) Protein splicing elements: inteins and exteins – a definition of terms and recommended nomenclature. Nucleic Acids Res 22:1125–1127

Perler FB, Olsen GJ, Adam E (1997a) Compilation and analysis of intein sequences. Nucleic Acids Res 25:1087–1093

Perler FB, Xu M-Q, Paulus H (1997b) Protein splicing and autoproteolysis mechanisms. Curr Opin Chem Biol 1:292–299

Pietrokovski S (1994) Conserved sequence features of inteins (protein introns) and their use in identifying new inteins and related proteins. Protein Sci 3:2340–2350

Pietrokovski S (1998a) Modular organization of inteins and C-terminal autocatalytic domains. Protein Sci 7:64–71

Pietrokovski S (1998b) Identification of a virus intein and a possible variation in the protein-splicing reaction. Curr Biol 8:R634–R635

Poland BW, Xu M-Q, Quiocho FA (2000) Structural insights into the protein splicing mechanism of PI-SceI. J Biol Chem 275:16408–16413

Saves I, Eleaume H, Dietrich J, Masson JM (2000) The Thy pol-2 intein of *Thermococcus hydrothermalis* is an isoschizomer of PI-TliI and PI-TfuI endonucleases. Nucleic Acids Res 28:4391–4396

Shao Y, Xu MQ, Paulus H (1995) Protein splicing: characterization of the aminosuccinimide residue at the carboxyl terminus of the excised intervening sequence. Biochemistry 34:10844–10850

Shao Y, Xu M-Q, Paulus H (1996) Protein splicing: evidence for an N-O acyl rearrangement as the initial step in the splicing process. Biochemistry 35:3810–3815

Shih C, Wagner R, Feinstein S, Kanik-Ennulat C, Neff N (1988) A dominant trifluoperazine resistance gene from *Saccharomyces cerevisiae* has homology with fof1 ATP synthase and confers calcium-sensitive growth. Mol Cell Biol 8:3094–3103

Shingledecker K, Jiang S, Paulus H (1998) Molecular dissection of the *Mycobacterium tuberculosis* RecA intein: design of a minimal intein and of a trans-splicing system involving two intein fragments. Gene 207:187–195

Southworth MW, Perler FB (2002) Protein splicing of the *Deinococcus radiodurans* strain R1 Snf2 intein. J Bacteriol 184:6387–6388

Southworth MW, Adam E, Panne D, Byer R, Kautz R, Perler FB (1998) Control of protein splicing by intein fragment reassembly. EMBO J 17:918–926

Southworth MW, Amaya K, Evans TC, Xu M-Q, Perler FB (1999) Purification of proteins fused to either the amino or carboxy terminus of the *Mycobacterium xenopi* gyrase A intein. BioTechniques 27:110–120

Southworth MW, Benner J, Perler FB (2000) An alternative protein splicing mechanism for inteins lacking an N-terminal nucleophile. EMBO J 19:5019–5026

Southworth MW, Yin J, Perler FB (2004) Rescue of protein splicing activity from a *Magnetospirillum magnetotacticum* intein-like element. Biochem Soc Trans 32:250–254

Vigneron N, Stroobant V, Chapiro J, Ooms A, Degiovanni G, Morel S, van der Bruggen P, Boon T, van den Eynde B J (2004) An antigenic peptide produced by peptide splicing in the proteasome. Science 304:587–590

Wallace CJ (1993) The curious case of protein splicing: mechanistic insights suggested by protein semisynthesis. Protein Sci 2:697–705

Wang S, Liu XQ (1997) Identification of an unusual intein in chloroplast clpp protease of *Chlamydomonas eugametos.* J Biol Chem 272:11869–11873

Wood DW, Derbyshire V, Wu W, Chartrain M, Belfort M, Belfort G (2000) Optimized single-step affinity purification with a self-cleaving intein applied to human acidic fibroblast growth factor. Biotechnol Prog 16:1055–1063

Wu H, Hu Z, Liu XQ (1998a) Protein trans-splicing by a split intein encoded in a split DnaE gene of *Synechocystis* sp. PCC6803. Proc Natl Acad Sci USA 95:9226–9231

Wu H, Xu M-Q, Liu XQ (1998b) Protein trans-splicing and functional mini-inteins of a cyanobacterial DnaB intein. Biochem Biophys Acta 1387:422–432

Wu W, Wood DW, Belfort G, Derbyshire V, Belfort M (2002) Intein-mediated purification of cytotoxic endonuclease I-TevI by insertional inactivation and pH-controllable splicing. Nucleic Acids Res 30:47864–47871

Xu M, Perler FB (1996) The mechanism of protein splicing and its modulation by mutation. EMBO J 15:5146–5153

Xu M, Southworth MW, Mersha FB, Hornstra LJ, Perler FB (1993) In vitro protein splicing of purified precursor and the identification of a branched intermediate. Cell 75:1371–1377

Xu M, Comb DG, Paulus H, Noren CJ, Shao Y, Perler FB (1994) Protein splicing: an analysis of the branched intermediate and its resolution by succinimide formation. EMBO J 13:5517–5522

Yamamoto K, Low B, Rutherford SA, Rajagopalan M, Madiraju MV (2001) The *Mycobacterium avium-intracellulare* complex DnaB locus and protein intein splicing. Biochem Biophys Res Commun 280:898–903

Yang J, Fox GC Jr, Henry-Smith TV (2003) Intein-mediated assembly of a functional beta-glucuronidase in transgenic plants. Proc Natl Acad Sci USA 100:3513–3518

Yang J, Meng Q, Liu XQ (2004) Intein harbouring large tandem repeats in replicative DNA helicase of *Trichodesmium erythraeum.* Mol Microbiol 51:1185–1192

Zeidler MP, Tan C, Bellaiche Y, Cherry S, Hader S, Gayko U, Perrimon N (2004) Temperature-sensitive control of protein activity by conditionally splicing inteins. Nat Biotechnol 22:871–876

Origin and Evolution of Inteins and Other Hint Domains

Bareket Dassa, Shmuel Pietrokovski

Abstract

Intein protein-splicing domains are part of the Hint superfamily. This superfamily includes three other characterized families: Hog-Hint and two types of Bacterial intein-like (BIL) domains. Hint domains share the same structure fold and common sequence features, and have similar biochemical activities. They post-translationally auto-process the proteins in which they are present by protein-splicing, self-cleavage or ligation activities. Yet, each Hint family apparently has its own distinct biological role. We discuss the evolution of the different Hint families, the origin of primordial Hint domains themselves, and their possible activities and biological functions.

1 Introduction

The intein protein family is part of the Hint superfamily, named after the characteristic structure fold first identified in Hedgehog and intein protein domains (Hall et al. 1997). Four characterized Hint domain families are currently known: Hog–Hint, intein, and two types of bacterial intein-like (BIL) domains (Fig. 1). Together with sharing the same structural fold and common sequence features, Hint domains have similar biochemical activities. The domains post-translationally process the proteins in which they are present by protein-splicing, self-cleavage or ligation activities (Paulus 2000; Dassa et al. 2004a).

B. Dassa, S. Pietrokovski (e-mail: shmuel.pietrokovski@weizmann.ac.il)
Department of Molecular Genetics, Weizmann Institute of Science, Rehovot 76100, Israel

Nucleic Acids and Molecular Biology, Vol. 16
Marlene Belfort et al. (Eds.)
Homing Endonucleases and Inteins

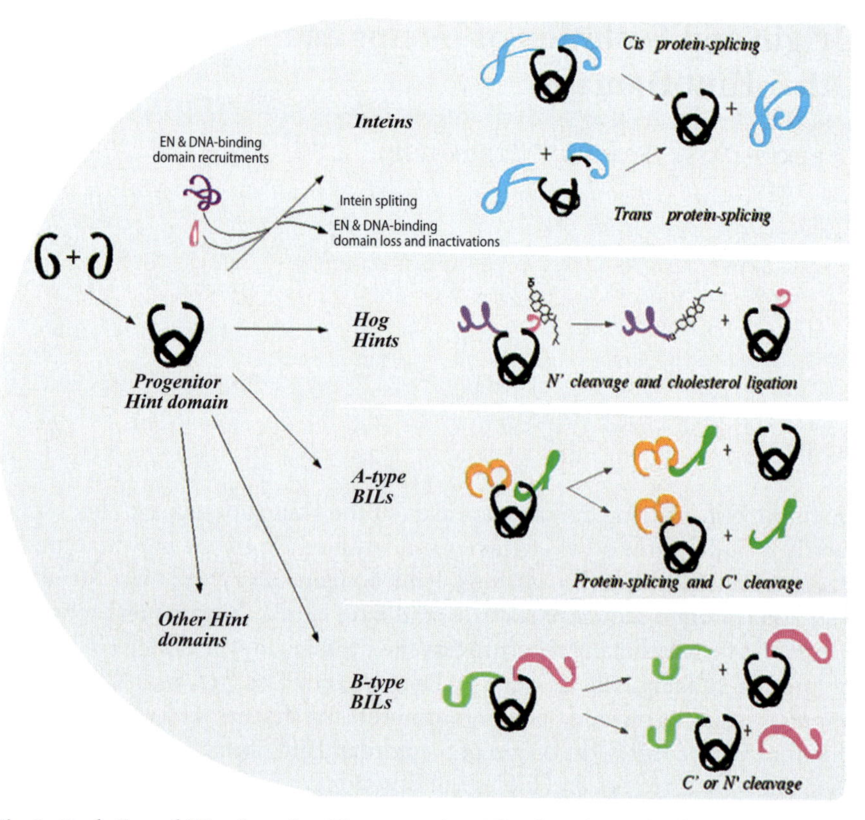

Fig. 1. Evolution of Hint domains. The progenitor Hint domain evolved into at least four currently known protein families: intein, Hog–Hint, A-type BIL and B-type BIL domains. The progenitor domain itself probably emerged by fusion of two duplicated and symmetrical subdomains. Inteins acquire and lose EN and DNA-binding domains during their evolution and some have irreversibly split. More information on Hint domain families can be found at http://bioinfo.weizmann.ac.il/~pietro/Hints

Hint domains are 130 to 160 amino acids long, sharing 4–6 conserved sequence motifs (Amitai et al. 2003). The Hint protein fold is a compact, relatively flat, symmetrical structure, mainly composed of β-strands, with its N- and C- termini close together (Duan et al. 1997; Hall et al. 1997). Inteins usually include additional homing-endonuclease (EN) and DNA-binding domains, not necessary for protein-splicing, which mediate the homing of the intein gene.

Species from the three domains of life, *Eukarya*, *Bacteria* and *Archaea*, include Hint domains. While inteins are apparently limited to unicellular eukaryotes and prokaryotes, Hog–Hint domains are present in multicellular animals. BIL domains and inteins overlap in their phylogenetic distributions, both are present in bacteria, but in different types of proteins.

Data gathered on Hint domains since their discovery in 1990 have identified their biochemical activity, genetics and evolution. Apparently, each Hint family has its own distinct biological role. Nevertheless, many facets of Hint domains are still unknown or debated.

This chapter describes each of the known Hint families, and evidence for the existence of other Hint families. The data available for each family are used to illuminate its evolution. We then discuss hypotheses for the evolutionary history of the Hint families, the possible activities and biological roles of the progenitor Hint domains, and how these progenitor domains themselves originated.

2 Hint Domain Families

2.1 Inteins

Inteins are selfish genetic elements. They are not known to confer any advantage to their host proteins and species. Inteins survive (1) by having a negligible impact on the fitness of their host genes, due to effectively removing themselves from the precursor proteins; (2) by being integrated in conserved positions of essential genes, where they are difficult to remove without compromising their host genes; and (3) by frequently being able to mediate the insertion of their gene into unoccupied integration points (homing; Belfort and Roberts 1997), counteracting removal of their genes. This survival strategy can account for the phylogenetic distribution of inteins. They are found across all known types of unicellular organisms, but in a highly sporadic manner.

2.1.1 Inteins Include Different Domains

All inteins include a protein-splicing Hint domain. Efficient self-catalyzed protein-splicing minimizes the intein's effect on its host protein, and hence on its host cell. Proper folding of the Hint domain seems all that is required for accurate splicing of the intein flanks (exteins). No additional proteins or energy sources are required. Protein-splicing allows a gene with an intein to function with no apparent difference from its intein-less version. Nevertheless, without contributing to the fitness of the host gene the intein gene is liable to eventually be lost by genomic events that will remove it. However, incomplete deletions and changes in the intein gene, that abolish or reduce its protein-splicing activity, will leave its host protein with an "inert" inserted domain. Removal of the intein gene together with some flanking sequences will also be deleterious. This is probably the major selection force for the presence of inteins in conserved sequence motifs of essential proteins. Most changes in such sequence motifs of these proteins will harm the cells. Intein genes are thus most difficult to remove from conserved protein-coding regions and these form genomic niches for their survival (Pietrokovski 2001).

Many inteins have homing-endonuclease and DNA-binding domains integrated in their Hint domain. Most of the domains are of the LAGLIDADG type, and a few are of the HNH type (Hirata et al. 1990; Pietrokovski 1994; Dalgaard et al. 1997). Several of the remaining inteins seem to have lost their EN domains. Inactivating deletions and mutations are also noticeable in several EN domains. Inteins apparently undergo cycles of EN-activity gain and loss. EN domains are not necessary for protein-splicing and are also found in other proteins. Since inteins can also survive just by efficient protein-splicing, it is most likely that the earliest intein only consisted of a Hint domain. Current inteins with no EN domains either lost or never acquired them.

2.1.2 Intein Protein Hosts and Insertion Points

Inteins are present in ancient protein types, ones that are found in prokaryotes and eukaryotes or in bacteria and archaea (Pietrokovski 2001). Most of these are also essential proteins (e.g., Table 1). A specific explanation for the abundance of inteins in proteins which are involved in DNA and RNA metabolism was previously suggested (Liu 2000). Liu noted that these types of host proteins can enhance the repair of DNA lesions generated by intein EN domains during their homing process and possible non-specific activity.

Inteins are often present at the same insertion points in homologous genes (Table 1). Such inteins are particularly similar to each other and are termed alleles. Some proteins include more than one intein, but these are no more related to each other than to other inteins (Perler et al. 1997). The relatively high similarity between intein alleles is attributed to customary vertical gene transmission during speciation events and to homing of intein genes between strains and species. Non-allelic inteins have only weak sequence similarity to each other. Trees created from intein multiple sequence alignments are thus characterized by very deep branches between many inteins, indicating distant uncertain relations, and by clusters of intein alleles.

Different, non-allelic, intein insertion points are each differently conserved and share only one common sequence feature. The residue C-terminal to the intein (+1 residue) is typically Cys, Ser or Thr. Its side chain is crucial for ligating the exteins in canonical intein protein-splicing. However, some inteins utilize alternative protein-splicing mechanisms (Southworth et al. 2000; Amitai et al. 2004; Mills et al. 2004).

2.1.3 *Inteins Are Sporadically Distributed*

Inteins are sporadically present in diverse bacteria, archaea, fungi, phages and viruses (see http://bioinfo.weizmann.ac.il/~pietro/inteins for updated details). By examining the few hundred fully and almost fully sequenced genomes of microbes and higher organisms, it seems that inteins are absent from multicellular organisms (higher plants and animals) and that closely related species and orthologous proteins often have different inteins (Table 1).

Inteins are thus sporadically present in conserved motifs of essential proteins across diverse organisms. This could result from independent loss of inteins in different lineages (Pietrokovski 2001) and from horizontal interspecies transfer of inteins (Gogarten et al. 2002). While the two hypotheses are not exclusive, we believe the first one is the main explanation for current intein distribution.

Horizontal transfer readily accounts for the discontinuous intein dispersion in corresponding insertion points (e.g. Fsihi et al. 1996; Saves et al. 2000; Isabelle et al. 2001; Koufopanou et al. 2002; Okuda et al. 2003) since inteins have an established homing mechanism. However, there are no known specific mechanisms for intein gene "invasion" into *heterologous* insertion points. Intein homing ability is irrelevant for this process since copying of the intein gene relies on specific sequence similarity between the intein gene flanks and the insertion point. However, constraints on loss of intein genes from conserved motifs of essential genes are relevant for their establishment in such new points. The ends of the copied region must exactly correspond to the intein gene, not including any additional flanks or missing the intein ends. One property that greatly facilitates such exact copying is the occurrence of the mobile gene region by itself as a free molecule. This is found in introns, as spliced-out RNA molecules, but not in inteins, that are removed at the protein level.

Independent loss of inteins from different lineages assumes that while at present each species only has a fraction of intein insertion points occupied, past species had inteins in all insertion points. Inteins were thus more common in the past, and we need to account for their gradual loss. Either inteins had some advantage for the cells that was lost, or new mechanisms selecting against inteins arose. Possible lost advantages include control of protein activity and combinatorial *trans*-protein-splicing (see Sect. 3). These functions could be made redundant with the rise of current transcriptional and post-translational control mechanisms, and of larger genomes affording the direct coding of diverse proteins. New mechanisms selecting against inteins might have included improved barriers against invasion of foreign DNA into the genome, recognition and elimination of cells with invaded genomes, and additional repair mechanisms of DNA lesions. All of these would make positions that lost their inteins less likely to be reinvaded.

Table 1. Intein distribution in Archaea

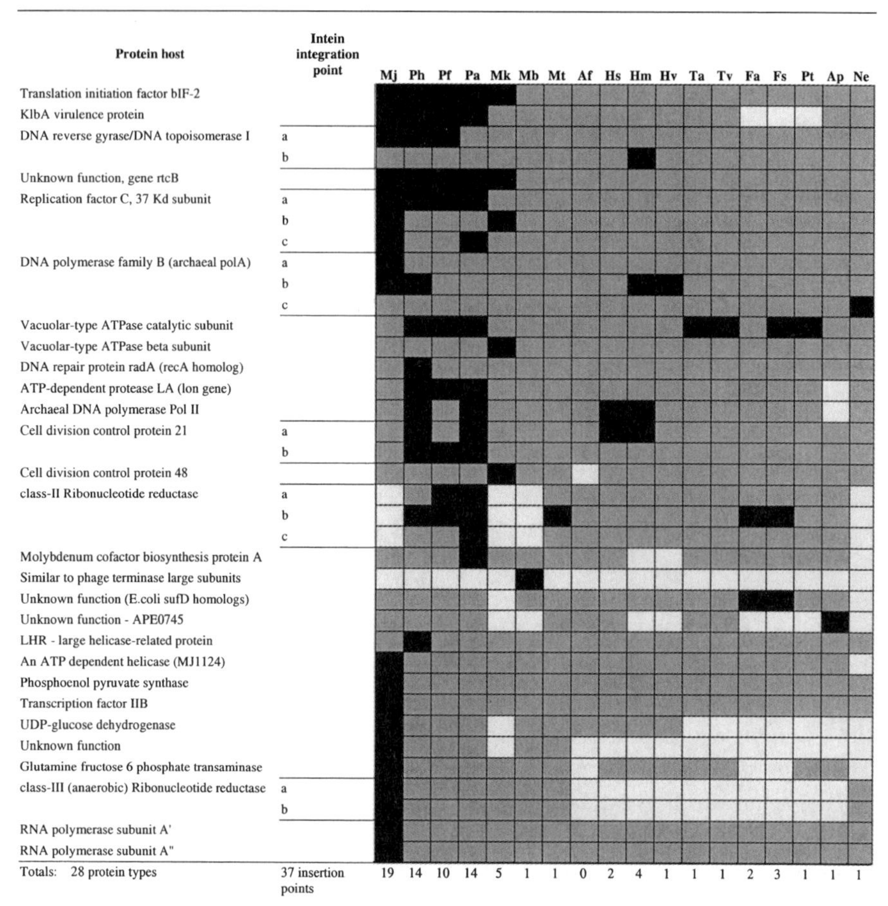

Protein host	Intein integration point	Mj	Ph	Pf	Pa	Mk	Mb	Mt	Af	Hs	Hm	Hv	Ta	Tv	Fa	Fs	Pt	Ap	Ne
Translation initiation factor bIF-2																			
KlbA virulence protein																			
DNA reverse gyrase/DNA topoisomerase I	a																		
	b																		
Unknown function, gene rtcB																			
Replication factor C, 37 Kd subunit	a																		
	b																		
	c																		
DNA polymerase family B (archaeal polA)	a																		
	b																		
	c																		
Vacuolar-type ATPase catalytic subunit																			
Vacuolar-type ATPase beta subunit																			
DNA repair protein radA (recA homolog)																			
ATP-dependent protease LA (lon gene)																			
Archaeal DNA polymerase Pol II																			
Cell division control protein 21	a																		
	b																		
Cell division control protein 48																			
class-II Ribonucleotide reductase	a																		
	b																		
	c																		
Molybdenum cofactor biosynthesis protein A																			
Similar to phage terminase large subunits																			
Unknown function (E.coli sufD homologs)																			
Unknown function - APE0745																			
LHR - large helicase-related protein																			
An ATP dependent helicase (MJ1124)																			
Phosphoenol pyruvate synthase																			
Transcription factor IIB																			
UDP-glucose dehydrogenase																			
Unknown function																			
Glutamine fructose 6 phosphate transaminase																			
class-III (anaerobic) Ribonucleotide reductase	a																		
	b																		
RNA polymerase subunit A'																			
RNA polymerase subunit A"																			
Totals: 28 protein types	37 insertion points	19	14	10	14	5	1	1	0	2	4	1	1	1	2	3	1	1	1

Intein insertion points in genes from complete or mostly complete archeal genomes. Intein presence (black), absence (gray) or no corresponding ortholog gene in the species (white) is marked for these species and strains: Mj - *Methanocaldococcus jannaschii* DSM 2661, Ph - *Pyrococcus horikoshii* OT3, Pf - *Pyrococcus furiosus*, Pa - *Pyrococcus abyssi* GE5, Mk - *Methanopyrus kandleri* AV19, Mb - *Methanococcoides burtonii* DSM 6242, Mt - *Methanothermobacter thermautotrophicus* DH, Af - *Archaeoglobus fulgidus* DSM 4304, Hs - *Halobacterium* sp. NRC-1, Hm - *Haloarcula marismortui* ATCC43049, Hv - *Haloferax volcanii* DS2 ATCC29605, Ta - *Thermoplasma acidophilum* DSM 1728, Tv - *Thermoplasma volcanium* GSS1, Fa - *Ferroplasma acidiphilum* fer1, Fs - *Ferroplasma* species type II, Pt - *Picrophilus torridus* DSM 9790, Ap - *Aeropyrum pernix* K1, Ne - *Nanoarchaeum equitans* Kin4-M.
More information on intein protein host types can be found at http://bioinfo.weizmann.ac.il/~pietro/inteins .

2.1.4 A Non-selfish Intein-Derived Protein

No example of inteins with a beneficial cellular role is known. However, one intein-derived protein is crucial for an important cellular process. Mating-type switches in *Saccharomyces* yeasts are initiated with a specific double-strand cleavage of the MAT locus by the HO endonuclease (Haber 1998). The full switching system evolved gradually and is only present in *Saccharomyces sensu-stricto* and related yeasts. The final step in the development of the system was acquisition of the HO endonuclease (Butler et al. 2004). HO is derived from a VDE intein present in *Saccharomyces* yeasts, as shown by sequence similarity and phylogenetic analysis (Hirata et al. 1990; Pietrokovski 1994; Dalgaard et al. 1997). HO includes the VDE protein-splicing, homing-endonuclease and DNA-binding domains. It has no N-terminal flanking region upstream of the protein-splicing domain and includes a C-terminal Zn-finger DNA-binding domain. It is not known why the protein-splicing domain was retained and its motifs conserved when it seems that HO only needed to retain the endonuclease activity of the VDE intein.

2.1.5 Split Inteins

Inteins can be split into two or more parts which associate and ligate their exteins in a *trans* protein-splicing reaction. Split inteins naturally occur in *Cyanobacteria* and *Nanoarchaea* (Wu et al. 1998; Caspi et al. 2003; Waters et al. 2003), and can be engineered by splitting contiguous inteins (Shingledecker et al. 1998; Southworth et al. 1998; Sun et al. 2004). Naturally split inteins include all the typical intein Hint sequence features. They probably resulted from genomic DNA rearrangements that split contiguous inteins and their host genes. These rearrangements might not be rare, but probably only very few produced active split-intein genes. These genes had to code for intein parts that could *trans*-splice, and produce the right amounts of the split products at the right time, to generate active, mature proteins. The two types of naturally split inteins are in catalytic subunits of DNA polymerases. These proteins are crucial in every cell generation. It is thus remarkable that the cells could survive the splitting event, since the gene's function was immediately needed. If an alternative mechanism for the gene's activity was present, it was either transient or inferior to the function of the split genes that thus managed to survive.

Once established in a conserved motif of an essential gene a split intein is very difficult to lose. Both intein parts must be precisely excised, and the split host part rejoined to recreate a functional intein-less gene. One way to remove the split intein is to acquire a gene that can replace the split one. However, this replacement is difficult for genes such as DNA polymerases that interact with many other proteins. Allelic split DnaE inteins were found in diverse cyanobacteria. This led us to propose that the split dnaE gene is fixed in a number of large cyanobacterial lineages (Caspi et al. 2003). The *Nanoarchaeum equitans* split pol intein is an allele of several archaeal type-B DNA polymerase contiguous inteins. No other homologous DNA polymerases are known from this phylum, and thus we do not know if this split intein is fixed in it.

We do not know how split intein genes managed to establish themselves in cyanobacteria and *Nanoarchaeum*. The genomic event that created them might have generated other advantages to the cell, ensuring its survival and fixing the split intein gene.

2.2 Hog–Hint Domains

Hedgehog developmental proteins of animals are composed of three protein domains. The N-terminal domain (Hedge) is a secreted developmental signal. It is first cleaved off from the precursor protein, then modified by covalent attachment of lipids at both ends and secreted from the cell. The C-terminal part of the protein (Hog) is composed of two domains, a Hint domain and a sterol-recognition region (SRR). The Hint domain shifts the peptide bond, which attaches it to the Hedge domain, to a thioester bond. This bond is attacked by the hydroxyl group of a cholesterol molecule bound by the SRR. This cleaves off the Hedge domain while attaching a cholesterol molecule to the resulting C-terminus by an ester bond (Mann and Beachy 2004).

Hog–Hint domains have the same structural fold as intein Hint domains and also share their sequence motifs, only missing the C1 motif of the intein C-terminus (Hall et al. 1997). Hog–Hint domains also have a few additional motifs including one with an active site Asp or His residue responsible for activation of the cholesterol molecule. These specific sequence features of the Hint motifs allow us to distinguish Hog–Hint from the other Hint domains (Pietrokovski 1998; Amitai et al. 2003).

2.2.1 Phylogenetic Distribution of Hog–Hint Domains

Arthropods apparently have single hedgehog genes, while vertebrates have three types of such genes: Sonic, Desert and Indian hedgehogs (Hammerschmidt et al. 1997). Other chordate subphyla, urochordates and cephalochordates, have fewer hedgehog genes. Earlier diverged animal groups, echinoderms, mollusks, and annelids (segmented worms), each have at least one hedgehog gene. Nematodes have a larger number of genes with Hog domains (Table 2). *C. elegans* has ten genes belonging to three families. While their biological role is as yet unclear, these nematode proteins seem to be distant homologues of Hedgehog proteins. Their N-terminal domains correspond to the Hedge domain, and the region downstream of their Hint domains corre-

Table 2. Phylogenetic distribution of Hog–Hint domains

Phylogenetic group	Genes with Hog–Hint domains	Reference or examples
Vertebrates (mammals, amphibians and fish)	Sonic, Desert and Indian hedgehogs	Hammerschmidt et al. (1997)
Urochordates (*Ciona*)	hh1 and hh2 hedgehogs	Takatori et al. (2002)
Cephalochordates (*Amphioxus*)	Amphioxus hedgehog	Shimeld (1999)
Arthropods (insects, arachnids, millipedes)	Hedgehog	Hammerschmidt et al. (1997)
		Janssen et al. (2004)
		GenBank accessions AAP38182, BAD01490
Echinoderms (sea urchin)	Hedgehog	GenBank accession AAC15065
Mollusks (snail)	Hedgehog	Nederbragt et al. (2002)
Annelids – segmented worms (leach)	Hedgehog	Kang et al. (2003)
Nematodes – round worms (*C. elegans*)	Warthogs, groundhogs and M110	Aspock et al. (1999)
Rhodophyta – red algae (*Cyanidioschyzon porphyra*)	Rhodhogs	Matsuzaki et al. (2004)
Cryptosporidium parvum – apicomplexan protist	Gene with Hog–Hint domain	Abrahamsen et al. (2004)

sponds to the Hedgehog SRR and is termed ARR, for adduct recognition region (Aspock et al. 1999; Mann and Beachy 2004). A nematode Hog protein region was also shown to have the cholesterol-mediated autocleavage activity of Hedgehog Hog regions (Porter et al. 1996). However, in their natural context, nematode Hog proteins might use other molecules to cleave the ester bond to their N-terminal domains (Mann and Beachy 2004).

No Hog–Hint domains were identified in the few very different protists that have been completely and almost completely sequenced, except for *Cryptosporidium parvum*. This apicomplexan intracellular parasite of mammals includes a single gene with a Hog–Hint domain. It is highly expressed during in vitro development but no other experimental data are currently available on it (Abrahamsen et al. 2004).

Different subclasses of red algae (*Rhodophyta*) include genes with Hog domains, which we term Rhodhog genes. These include *Cyanidioschyzon merolae* and *Porphyra yezoensis* red algae, that each has a multi-gene Rhodhog family (Matsuzaki et al. 2004).

Hog–Hint domains are present in animals and red algae, but are not found in fungi and plants. The most parsimonious explanation is the presence of a single hedgehog gene in the animal progenitor that diversified in different phyla. The *C. parvum* Hog–Hint domain is probably the result of horizontal transfer, since no other Hog–Hint domain is known in other protists. The red algae Rhodhog domains might also result from an ancient horizontal transfer, but this depends on the contested phylogenetic position of this kingdom relative to green plants and animals.

2.3 BIL Hint Domains

Two new types of Hint domains were recently identified and termed A- and B-type BIL (for bacterial intein-like) domains (Amitai et al. 2003). Members of each BIL type are more similar by sequence to each other than to other types of Hint domains. The two types are also as different from one another as they are from intein and Hog–Hint domains. While inteins are integrated in highly conserved sites of essential proteins, both A- and B-type BIL domains are integrated in variable regions of non-conserved and diverse bacterial proteins, some of which are extracellular. Similar to inteins, BIL domains can auto-cleave at either their N- or C-termini. A-type BIL domains can also protein splice, but not by the canonical intein mechanism, since the C-terminal flanking residue of the A-type BIL does not always have a nucleophilic side chain. This and other features suggest that the biological role of BIL domains differs from that of inteins (Amitai et al. 2003; Dassa et al. 2004a). BIL domains may contribute to the variability of their flanking protein by protein-splicing, cleavage and ligation. Studying BIL domains may reveal new ways of protein maturation and control, and enhance our understanding of other Hint domains.

2.3.1 Phylogenetic Distribution of BIL Domains

The first 100 BIL domains were identified exclusively in bacterial genomes, hence their name, bacterial intein-like domains (Amitai et al. 2003; Dassa et al. 2004a). They are present in more than 25 taxonomically diverse bacterial species, including *Proteobacteria, Cyanobacteria, Planctomycetes, Clostridia,* and others (see http://bioinfo.weizmann.ac.il/~pietro/BILs for updated details). Some BIL domains are found in pathogens of humans and plants, such as *Neisseria meningitidis* and *Pseudomonas syringae*. A-type BIL domains were recently found in ciliates, thus extending their phylogenetic distribution to eukaryotic protists.

The genomic distribution of BIL domains is variable and highly dynamic, with relatively fast gene duplications and losses. While some species include more than 20 BIL domains, other related species have none. Sequence similarity between BIL domains shows that multiple BIL domains occurring in the same or related species are the result of gene expansions within each lineage (Amitai et al. 2003). This includes pseudo-genes, apparent by deletions, and non-sense mutations within BIL domains and their protein flanks.

BIL domains usually appear once in each protein. However, the ciliate *Tetrahymena thermophila* includes two genes with the same domain organization, each containing a pair of A-type BIL domains interspersed with ubiquitin-like (Ubl) domains. These genes are termed BUBL, for BIL-Ubl domains. Another BUBL gene with poly-Ubl domains and a single A-type BIL domain is found in the ciliate *Paramecium tetraurelia* (Dassa et al. 2004b; Fig. 2). The original BUBL gene probably included one BIL and poly-Ubl domains, and, after the divergence of *Tetrahymena* from *Paramecium*, the *Tetrahymena*

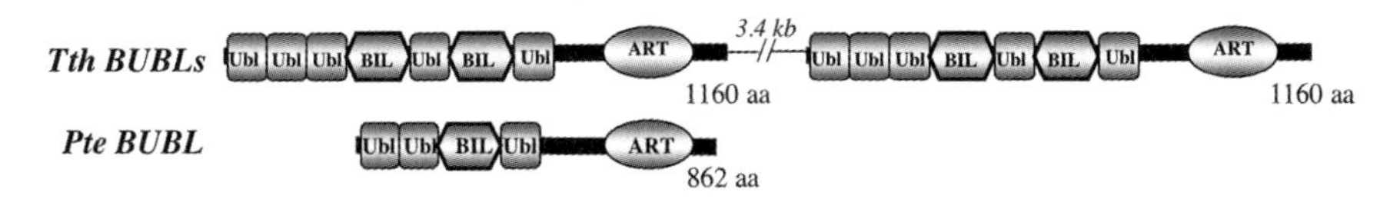

Fig. 2. Ciliate BUBL genes. Schemes of the *Tetrahymena thermophila (Tth)* and *Paramecium tetraurelia (Pte)* BUBL genes with ubiquitin-like *(Ubl)* domains shown as *round boxes*, A-type BIL domains shown as *hexagons*, and the ADP-ribosyltransferase *(ART)* domains as *ovals*

BUBL gene underwent two duplication events. The first one was intra-genic, duplicating the BIL domain, followed by a tandem duplication of the entire BUBL gene.

2.3.2 Protein Distribution of BIL Domains

BIL domains are integrated into hyper-variable positions of non-conserved proteins and are not always flanked at their C-terminus by nucleophilic residues. This is in sharp contrast to inteins, which are present in conserved positions of essential proteins. BIL domains are frequently found in the C-terminal regions of proteins, which are sometimes unstable coding regions that change by microevolutionary processes. Many BIL domains are flanked by N-terminal regions with characteristic motifs of extracellular proteins (Amitai et al. 2003).

A-type BIL domains are found in the 3′ region of Neisserial mafB adhesins or in short coding regions downstream of these genes. These regions can recombine with *mafB* gene 3′ ends, creating hyper-variable C-terminal regions. Thus, BIL domains might offer a post-translational mechanism for generating protein variability, together with genetic rearrangements in microevolution.

Ciliate A-type BIL domains are found between conserved residues of ubiquitin-like (Ubl) domains in BUBL genes (Dassa et al. 2004b). This probably relates to the role of the BUBL BIL domains as discussed below.

2.3.3 Biochemical Activity and Biological Roles of BIL Domains

BIL domains can protein splice and auto-cleave at either N- or C-termini. Activity of BIL domains was extrapolated from that of inteins and Hog–Hint domains by comparing corresponding active-site positions (Fig. 4A). Similar to intein and Hog–Hint domains, the N-terminus of BIL domains is conserved, typically being Cys or Ser. The domains also include a highly conserved His in a motif corresponding to the N3 motif of inteins and Hog–Hint domains. The N-terminal and N3 (His) residues are responsible for the N/S or N/O acyl shifts of inteins and Hog–Hint domains (Paulus 2000; Romanelli et al. 2004). Thus, it is assumed that this reaction also occurs in BIL domains (Dassa et al. 2004a).

A-type BIL domains have highly conserved His-Asn residues at their C-termini (like inteins), followed by a diverse position (unlike inteins). Asn is necessary for cleavage of the C-termini of inteins, and is followed by a conserved Cys, Ser or Thr +1 residue to allow ligation of the two intein flanks. According to intein protein-splicing mechanisms, A-type BIL domains can cleave their C-termini but not ligate their flanking sequences. Nevertheless, A-type BIL domains can protein splice, even with a non-nucleophilic +1 residue (Ala) (Dassa et al. 2004a).

In the suggested mechanism for protein-splicing of A-type BIL domains, the C-terminus of the domain is cleaved by Asn cyclization and the resulting free amino group of the C-terminal flank attacks the previously formed thioester/ester bond at the N-terminus of the domain by an aminolysis reaction. This ligates the two flanks and completes the release of the BIL domain. Nevertheless, A-type BIL protein-splicing is not as efficient as that of inteins, also producing large amounts of cleavage products (Dassa et al. 2004a). This could be the reason why inteins do not utilize such a mechanism. Cleavage products of essential proteins, like those flanking inteins, may be deleterious to the cell in a dominant-negative manner – partial products inactivating the properly processed products.

Aminolysis was also suggested for proteosomal peptide ligation (Vigneron et al. 2004). There, an ester bond between a cleaved peptide and the proteosomal Thr active site is attacked by the amino group of another cleaved peptide. This ligates two peptides, which are then displayed on the cell surface. This is an example of natural convergence, since there are no apparent evolutionary or structural commonalities between Hint protein-splicing and proteosomal peptide ligation. Thus, features and chemical reactions of Hint domains may illuminate other, seemingly unrelated, biological systems.

B-type BIL domains share all intein protein-splicing motifs except for the C-terminal motif (Amitai et al. 2003), yet the domains can cleave their N- or C-termini (Dassa et al. 2004a). While N-terminal cleavage may resemble the intein reaction, C-terminal cleavage may proceed as in atypical reactions of inteins, where the N-terminus of the domain was suggested to attack its C-terminus (Amitai et al. 2003).

The expansion of BIL genes in different lineages indicates that they benefit the cell. Additionally, unlike inteins, both BIL domains are typically present in hyper-variable sites of non-conserved proteins, and the domains do not have any evident means for horizontal transfer. At present, there is only scant circumstantial data for the biological roles of BIL domains. We thus raise some possible hypotheses addressing both BIL types together.

We suggest that BIL domains contribute to their host protein by post-translational modifications. This can increase the host protein variability by generating alternative products from a single protein precursor: the precursor itself, N- or C-terminal- cleavage products, both N- and C-terminal flanks, and the splicing product.

Ligation of diverse molecules to the host protein is another means to increase its variability. This provides single-cell organisms with an advantageous adaptation to dynamically changing environments (Ziebuhr et al. 1999). One well-studied example is evading the immune system of multicellular organisms infected by pathogens. Another example is adhering the cell to new targets. Activity of BIL domains may directly assist bacteria in binding to their target hosts, living environment, or other bacteria (forming biofilms). Hog-Hint domains efficiently ligate their N-terminal flank to cholesterol. Similarly, BIL domains might ligate their host proteins to other molecules. This can covalently attach bacteria to their adherence surface. The preceding hypotheses concur with finding BIL domains in proteins that undergo microevolution (varying proteins by rapid genetic changes), and in extracellular proteins.

BIL domains may control the expression of their host proteins and modulate their activity by external signals. Splicing or autocleavage can be triggered by changes in redox environment or by conformational changes induced by an allosteric modification to the BIL domain or its host protein.

Ciliate BUBL A-type BIL domains are present together with Ubl domains. Ubiquitin (Ub) and Ubl domains are translated as inactive pre-protein precursors that need to be proteolytically cleaved to release the functional Ub or Ubl monomers. These are then conjugated by a series of activating (E1), conjugating (E2) and ligating (E3) enzymes (Jentsch and Pyrowolakis 2000). We suggest that BIL domains in BUBL proteins release the Ubl domains from the pre-protein in a self-processing reaction, and may also conjugate them to their targets (Dassa et al. 2004b).

2.4 Other Hint Domains

Advances in sequence analysis methods and the accumulation of sequence data enabled the definition of the Hint superfamily (Koonin 1995; Hall et al. 1997; Amitai et al. 2003). However, as more Hint families are found it is difficult to decide if a specific domain, or a whole group, belongs to a new or to an already known Hint family. The recently sequenced genome of the enterobacterium *Photorhabdus luminescens* strain TT01 (Duchaud et al. 2003) includes an example of a probable new family of Hint domains. A 595-residue open reading frame (ORF) of unknown function includes a region with significant similarities to the N-terminus of Hint domains (Fig. 3). The regions flanking the Hint domain are not similar to known proteins, and the domain itself does not have any of the motifs (or EN and DNA-binding domains) characteristic of known Hint domains.

```
     Pl 448 CIAEGTLIDMADGSKKKVEDIRSGDKVLTKQGGVLQVKSR-----IVGHDTEFVDLIYNNDEK 505
            C   GT+I  A G    +E+I++GDKV+       +V  +      V   TE + L     E
BIL4 Ct 316 CFVAGTMILTATGLVA-IENIKAGDKVIATNPETFEVAEKTVLETYVRETTELLHLTIGG-EV 376

     Pl 506 VSLTPTHPVATLR-GIVKADELKIGDTIY-TRDGQTTLSSVKLRNTD-PLNVYNFVLEKTDD 564
            +  T  HP      G V+A +L++GD +  +R     +   KL   D P+ VYNF   K DD
BIL4 Ct 377 IKTTFDHPFYVKDVGFVEAGKLQVGDKLLDSRGNVLVVEEKKLEIADKPVKVYNF---KVDD 435
```

Fig. 3. *Photorhabdus luminescens* new type of Hint domain. BLASTP alignment of *Photorhabdus luminescens* locus 1731 (Pl, NCBI gi code 36785096) Hint domain and *Clostridium thermocellum* BIL4 domain, BIL4 Ct (Dassa et al. 2004a). Putative Hint domain active-site residues are *highlighted*

3 Origin of the Hint Domains

3.1 Features of the Progenitor Hint Domain

Hint families diverged from each other very early in evolution. This is evident from their sequence diversity and phylogenetic dispersion. The presence of Hint families in different phylogenetic domains is a very strong indicator that the Hint progenitor was present before the split between prokaryotes and eukaryotes (Pietrokovski 2001). In the simplest scenario, the progenitor Hint domain did not have additional domains inserted in it, being similar in size to current Hog–Hint and BIL domains (Fig. 4A).

All characterized Hint families apparently catalyze an N-S/O acyl shift of the peptide bond between their N-terminus and host flank (Paulus 2000; Dassa et al. 2004a). This is supported by the presence of the N-terminal sequence motifs (N1 and N3) that are necessary for this reaction in all Hint families (Amitai et al. 2003; Fig. 4A). We thus assume that the progenitor Hint domain also had these sequence features and activity. However, C-termini of current Hint families are diverse by sequence and seem responsible for different reactions. Hence, it is unknown which, if any, of these reactions were present in the progenitor domain.

Current Hint domains have very different known biological roles. Best characterized are inteins, whose role seems limited to selfish propagation of their genes, and Hog–Hint domains that perform a crucial role in the maturation of their host proteins. Was the progenitor Hint domain a selfish element or did it have some beneficial role for its host cell? The Hint progenitor domain was probably already a complex fold, catalyzing several reactions. This domain was thus more likely to have arisen by a positive selection for some advantageous cellular role, rather than by the non-selective (spontaneous) appearance of a selfish function.

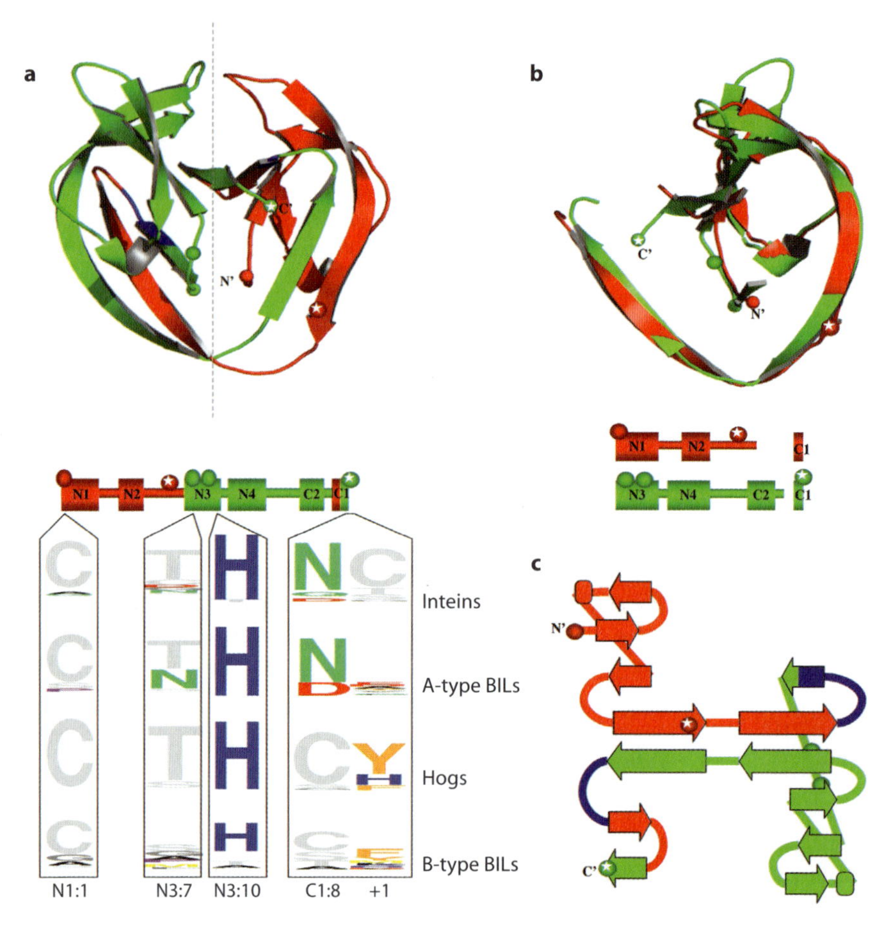

Fig. 4. Structure and active sites of Hint domains. **a** Two symmetrical subdomains of a Hog–Hint domain (Hall et al. 1997) are shown in red and green, with common Hint active-site residues as *balls*, and *balls with stars* marking specific active sites: the Hog–Hint sterol/adduct activating site [1AT0 Asp303], and the intein's C-terminal residue. A linear scheme of a Hint domain shows the subdomains below, with intein motifs as *boxes*. Beneath the scheme are amino-acid distributions in active sites and corresponding positions of the four Hint families. Each *column* points to its position in the scheme. **b** Two Hint subdomains are superimposed (*top*) and aligned in the linear scheme (*bottom*). **c** A representative secondary structure elements diagram of Hint domains, colored by subdomains with active sites indicated by *balls*

We present two possible beneficial roles for Hint progenitor domains, combinatorial *trans*-protein-splicing and the generation of protein variability. These roles are non-exclusive and other roles are possible too.

Trans-protein-splicing by split inteins was suggested to confer selective advantage to early organisms (Perler 1999; Pietrokovski 2001). A protein domain having one split intein portion could be ligated to either of several other proteins that include the complementary second portion of the split intein. Moreover, a few short protein regions, each including both a C-terminal intein part on the N-terminus and an N-terminal intein part on the C-terminus, would be able to ligate to each other in many different combinations, producing a large mixture of diverse linear and cyclic protein products. These processes would allow a primitive cell, with a small genome coding for relatively few protein domains, to produce many complex proteins without intricate enzymatic systems. Such proteins would enable the emergence of intricate cells that could support larger genomes that could directly code for large proteins.

The progenitor Hint domain could also enhance protein variability by the processes we suggest for present-day BIL domains. Each protein precursor could be in several states depending whether its Hint domain spliced, cleaved either end, cleaved both ends, ligated a molecule to one of its ends, or was inert. The different activities could be modulated by some signal or even be stochastic. Once more this could be of great benefit for early primitive cells that could only code for a limited number of proteins and did not possess complex systems for protein processing and degradation.

3.2 Emergence of the Progenitor Hint Domain

Hint domains have a pseudo-two-fold symmetry, being made up of two tandem subdomains that can be structurally superimposed (Hall et al. 1997). This is also apparent from subtle sequence similarity between the N2 and N4 motifs, present in all Hint families (Pietrokovski 1998). The originally described structural symmetry can be extended to the strand-loop-strand-loop C-terminal region of Hint domains (Amitai and Pietrokovski, unpubl. research). The progenitor Hint domain itself thus arose from a tandem duplication of a primordial subdomain, about 60 residues long, that exchanged short regions within the Hint fold by rearrangements of its gene (Hall et al. 1997). The active-site residues of both inteins and Hog–Hint domains are positioned asymmetrically on the two subdomains (Fig. 4B). For example, the catalytic N-terminal and N3 His residues are present in different subdomains, and each of their corresponding residues on the other subdomain is not a catalytic residue. This is notable since both these and adjacent residues are necessary for catalyzing the N-terminal acyl shift probably present in the progenitor Hint domain. For the primordial Hint subdomain to catalyze the acyl shift reaction, by itself or as a homodimer, the subdomain needed to include all active-site residues. This is not the current situation and thus either the active-site residues were differently distributed in the primordial Hint subdomain, or it was unlikely to catalyze the activities of present Hint domains.

Figure 1 summarizes the likely evolution of Hint domain families from a progenitor Hint subdomain, through the primordial Hint domain, to current Hint families. Each family adopted its activity according to its biological role. This led to the differences in activity, structure and sequence we can observe at present.

Many inteins include an EN domain, that seems to be accompanied by a DNA-binding domain. The relationship between the Hint domain and the EN and DNA-binding domains is dynamic. During evolution intein genes acquire, inactivate, lose and probably reacquire EN and DNA-binding coding regions (Gimble 2000, 2001).

Inteins and A-type BIL domains have a C-terminal motif that allows them to cleave that end and ligate their flanks. However, the ligation reaction and its temporal position relative to cleavage of the C-terminus differ in the two families. This is probably related to the role of protein-splicing in each family and to its required fidelity. Inteins need to protein splice very efficiently to avoid negative selection. Cys/Ser/Thr residues at the +1 position of intein integration points allow such efficient splicing. Intein integration points also need to be in conserved positions of essential proteins, reducing the loss of intein genes due to genomic rearrangements and mutations.

Hog–Hint domains include a conserved Asp/His active-site residue and downstream domains that bind sterols and perhaps other adducts. This bi-domain Hog unit might have the general role of processing its N-terminal flanking domain by cleaving it off while adding an adduct to the resulting C-terminus of the flank. Animal, nematode and red algae Hog domains diverged independently from a common ancestral Hog domain. The biological roles of Hog domains in nematode and red algae are not yet known.

BIL domains mainly differ in the sequence of their C-termini. A-type BIL domains have the same motif present in inteins but are flanked by diverse C-terminal flanking residues. The C-terminus motif of type-B BIL domains is different from that of inteins and A-type BIL domains, and of Hog–Hint domains. The domains are present in hyper-variable positions of non-conserved proteins in bacteria, and in polyubiquitin-like coding genes of ciliates. The role of both types of domain is hypothesized to benefit their hosts by various post-translational modifications.

Acknowledgements. This research was supported by The Israel Science Foundation, founded by The Israel Academy of Sciences and Humanities. S. Pietrokovski holds the Ronson and Harris Career Development Chair.

References

Abrahamsen M, Templeton T, Enomoto S, Abrahante J, Zhu G, Lancto C, Deng M, Liu C, Widmer G, Tzipori S, Buck G, Xu P, Bankier A, Dear P, Konfortov B, Spriggs H, Iyer L, Anantharaman V, Aravind L, Kapur V (2004) Complete genome sequence of the apicomplexan, *Cryptosporidium parvum*. Science 304:441–445

Amitai G, Belenkiy O, Pietrokovski S (2003) Distribution and function of new bacterial intein-like protein domains. Mol Microbiol 47:61–73

Amitai G, Dassa D, Pietrokovski S (2004) Protein-splicing of inteins with atypical glutamine and aspartate C-terminal residues. J Biol Chem 279:3121–3131

Aspock G, Kagoshima H, Niklaus G, Burglin TR (1999) *Caenorhabditis elegans* has scores of hedgehog-related genes: sequence and expression analysis. Genome Res 9:909–923

Belfort M, Roberts RJ (1997) Homing endonucleases: keeping the house in order. Nucleic Acids Res 25:3379–3388

Butler G, Kenny C, Fagan A, Kurischko C, Gaillardin C, Wolfe KH (2004) Evolution of the MAT locus and its Ho endonuclease in yeast species. Proc Natl Acad Sci USA 101:1632–1637

Caspi J, Amitai G, Belenkiy O, Pietrokovski S (2003) Distribution of split DnaE inteins in cyanobacteria. Mol Microbiol 50:1569–1577

Dalgaard JZ, Klar AJ, Moser MJ, Holley WR, Chatterjee A, Mian IS (1997) Statistical modeling and analysis of the LAGLIDADG family of site-specific endonucleases and identification of an intein that encodes a site-specific endonuclease of the HNH family. Nucleic Acids Res 25:4626–4638

Dassa B, Haviv H, Amitai G, Pietrokovski S (2004a) Protein-splicing and auto-cleavage of bacterial intein-like domains lacking a C′-flanking nucleophilic residue. J Biol Chem 279:32001–32007

Dassa B, Yanai I, Pietrokovski S (2004b) New type of poly ubiquitin-like genes with intein-like autoprocessing domains. Trends Genet 20:538–542

Duan X, Gimble FS, Quiocho FA (1997) Crystal structure of PI-SceI, a homing endonuclease with protein-splicing activity. Cell 89:555–564

Duchaud E, Rusniok C, Frangeul L, Buchrieser C, Givaudan A, Taourit S, Bocs S, Boursaux-Eude C, Chandler M, Charles J, Dassa E, Derose R, Derzelle S, Freyssinet G, Gaudriault S, Medigue C, Lanois A, Powell K, Siguier P, Vincent R, Wingate V, Zouine M, Glaser P, Boemare N, Danchin A, Kunst F (2003) The genome sequence of the entomopathogenic bacterium *Photorhabdus luminescens*. Nat Biotechnol 21:1307–1313

Fsihi H, Vincent V, Cole ST (1996) Homing events in the gyrA gene of some mycobacteria. Proc Natl Acad Sci USA 93:3410–3415

Gimble FS (2000) Invasion of a multitude of genetic niches by mobile endonuclease genes. FEMS Microbiol Lett 185:99–107

Gimble FS (2001) Degeneration of a homing endonuclease and its target sequence in a wild yeast strain. Nucleic Acids Res 29(20):4215–4223

Gogarten JP, Senejani AG, Zhaxybayeva O, Olendzenski L, Hilario E (2002) Inteins: structure, function, and evolution. Annu Rev Microbiol 56:263–287

Haber J (1998) Mating-type gene switching in *Saccharomyces cerevisiae*. Annu Rev Genet 32:561–599

Hall TM, Porter JA, Young KE, Koonin EV, Beachy PA, Leahy DJ (1997) Crystal structure of a hedgehog autoprocessing domain: homology between hedgehog and self-splicing proteins. Cell 91:85–97

Hammerschmidt M, Brook A, McMahon AP (1997) The world according to *hedgehog*. Trends Genet 13:14–21

Hirata R, Ohsumk Y, Nakano A, Kawasaki H, Suzuki K, Anraku Y (1990) Molecular structure of a gene, VMA1, encoding the catalytic subunit of H(+)-translocating adenosine triphosphatase from vacuolar membranes of *Saccharomyces cerevisiae.* J Biol Chem 265:6726–6733

Isabelle S, Mamadou D, Jean-Michel M (2001) Distribution of GyrA intein in non-tuberculous mycobacteria and genomic heterogeneity of Mycobacterium gastri. FEBS Lett 508:121–125

Janssen R, Prpic NM, Damen WG (2004) Gene expression suggests decoupled dorsal and ventral segmentation in the millipede *Glomeris marginata (Myriapoda: Diplopoda).* Dev Biol 268:89–104

Jentsch S, Pyrowolakis G (2000) Ubiquitin and its kin: how close are the family ties? Trends Cell Biol 10:335–342

Kang D, Huang F, Li D, Shankland M, Gaffield W, Weisblat DA (2003) A hedgehog homolog regulates gut formation in leech (*Helobdella*). Development 130:1645–1657

Koonin EV (1995) A protein splice-junction motif in hedgehog family proteins. Trends Biochem Sci 20:141–142

Koufopanou V, Goddard MR, Burt A (2002) Adaptation for horizontal transfer in a homing endonuclease. Mol Biol Evol 19:239–246

Liu XQ (2000) Protein-splicing intein: genetic mobility, origin, and evolution. Annu Rev Genet 34:61–76

Mann R, Beachy P (2004) Novel lipid modifications of secreted protein signals. Annu Rev Biochem 73:891–923

Matsuzaki M, Misumi O, Shin IT, Maruyama S, Takahara M, Miyagishima SY, Mori T, Nishida K, Yagisawa F, Yoshida Y, Nishimura Y, Nakao S, Kobayashi T, Momoyama Y, Higashiyama T, Minoda A, Sano M, Nomoto H, Oishi K, Hayashi H, Ohta F, Nishizaka S, Haga S, Miura S, Morishita T, Kabeya Y, Terasawa K, Suzuki Y, Ishii Y, Asakawa S, Takano H, Ohta N, Kuroiwa H, Tanaka K, Shimizu N, Sugano S, Sato N, Nozaki H, Ogasawara N, Kohara Y, Kuroiwa T (2004) Genome sequence of the ultrasmall unicellular red alga *Cyanidioschyzon merolae* 10D. Nature 428:653–657

Mills KV, Manning JS, Garcia AM, Wuerdeman LA (2004) Protein-splicing of a *Pyrococcus abyssi* intein with a C-terminal glutamine. J Biol Chem 279:20685–20691

Nederbragt AJ, van Loon AE, Dictus WJ (2002) Evolutionary biology: hedgehog crosses the snail's midline. Nature 417:811–812

Okuda Y, Sasaki D, Nogami S, Kaneko Y, Ohya Y, Anraku Y (2003) Occurrence, horizontal transfer and degeneration of VDE intein family in *Saccharomycete yeasts.* Yeast 20:563–573

Paulus H (2000) Protein-splicing and related forms of protein autoprocessing. Annu Rev Biochem 69:447–496

Perler FB (1999) A natural example of protein trans-splicing. Trends Biochem Sci 24:209–211

Perler FB, Olsen GJ, Adam E (1997) Compilation and analysis of intein sequences. Nucleic Acids Res 25:1087–1093

Pietrokovski S (1994) Conserved sequence features of inteins (protein introns) and their use in identifying new inteins and related proteins. Protein Sci 3:2340–2350

Pietrokovski S (1998) Modular organization of inteins and C-terminal autocatalytic domains. Protein Sci 7:64–71

Pietrokovski S (2001) Intein spread and extinction in evolution. Trends Genet 17:465–472

Porter JA, Ekker SC, Park WJ, von Kessler DP, Young KE, Chen CH, Ma Y, Woods AS, Cotter RJ, Koonin EV, Beachy PA (1996) Hedgehog patterning activity: role of a lipophilic modification mediated by the carboxy-terminal autoprocessing domain. Cell 86:21–34

Romanelli A, Shekhtman A, Cowburn D, Muir TW (2004) Semisynthesis of a segmental isotopically labeled protein-splicing precursor: NMR evidence for an unusual peptide bond at the N-extein-intein junction. Proc Natl Acad Sci USA 101:6397–6402

Saves I, Laneelle MA, Daffe M, Masson JM (2000) Inteins invading mycobacterial RecA proteins. FEBS Lett 480:221–225

Shimeld S (1999) The evolution of the hedgehog gene family in chordates: insights from amphioxus hedgehog. Dev Genes Evol 209:40–47

Shingledecker K, Jiang S-Q, Paulus H (1998) Molecular dissection of the *Mycobacterium tuberculosis* RecA intein: design of a minimal intein and of a trans-splicing system involving two intein fragments. Gene 207:187–195

Southworth MW, Adam E, Panne D, Byer R, Kautz R, Perler FB (1998) Control of protein-splicing by intein fragment reassembly. EMBO J 17:918–926

Southworth MW, Benner J, Perler FB (2000) An alternative protein-splicing mechanism for inteins lacking an N-terminal nucleophile. EMBO J 19(18):5019–5026

Sun W, Yang J, Liu X-Q (2004) Synthetic two-piece and three-piece split inteins for protein trans-splicing. J Biol Chem 279:35281–35286

Takatori N, Satou Y, Satoh N (2002) Expression of hedgehog genes in *Ciona intestinalis* embryos. Mech Dev 116:235–238

Vigneron N, Stroobant V, Chapiro J, Ooms A, Degiovanni G, Morel S, van der Bruggen P, Boon T, van den Eynde B (2004) An antigenic peptide produced by peptide splicing in the proteasome. Science 304:587–590

Waters E, Hohn MJ, Ahel I, Graham DE, Adams MD, Barnstead M, Beeson KY, Bibbs L, Bolanos R, Keller M, Kretz K, Lin X, Mathur E, Ni J, Podar M, Richardson T, Sutton GG, Simon M, Soll D, Stetter KO, Short JM, Noordewier M (2003) The genome of *Nanoarchaeum equitans*: insights into early archaeal evolution and derived parasitism. Proc Natl Acad Sci USA 100:12984–12988

Wu H, Hu Z, Liu X-Q (1998) Protein trans-splicing by a split intein encoded in a split DnaE gene of *Synechocystis* sp. PCC6803. Proc Natl Acad Sci USA 95:9226–9231

Ziebuhr W, Ohlsen K, Karch H, Korhonen T, Hacker J (1999) Evolution of bacterial pathogenesis. Cell Mol Life Sci 56:719–728

Biochemical Mechanisms of Intein-Mediated Protein Splicing

KENNETH V. MILLS, HENRY PAULUS

Abstract

This chapter discusses the mechanism of the self-catalyzed process by which inteins promote both their own excision from a host protein and the direct linkage of the flanking host protein segments, the N- and C-exteins, by a peptide bond. The majority of inteins have a nucleophilic amino acid at their N-terminus and asparagine at their C-terminus and are linked to a C-extein with an N-terminal nucleophilic amino acid. These canonical inteins promote protein splicing by a four-step mechanism of sequential acyl rearrangements. Non-canonical inteins, which lack either the N-terminal nucleophile or the C-terminal asparagine, promote protein splicing by a variant of this mechanism or promote protein cleavage rather than splicing. A remarkable feature of the protein splicing process is that it involves multiple steps that are chemically autonomous yet proceed in a highly coordinated manner without side reactions unless perturbed by mutation, unnatural exteins, or non-physiological conditions. The factors that may serve to integrate protein splicing into a system that ordinarily operates efficiently without side reactions are discussed.

K.V. Mills (e-mail: kmills@holycross.edu)
College of the Holy Cross, Department of Chemistry, 1 College Street, Worcester, Massachusetts 01610, USA

H. Paulus (e-mail: paulus@bbri.org)
Boston Biomedical Research Institute, 64 Grove Street, Watertown, Massachusetts 02472, USA

Nucleic Acids and Molecular Biology, Vol. 16
Marlene Belfort et al. (Eds.)
Homing Endonucleases and Inteins

1 Conserved Features of Intein Structure

Nearly 200 putative intein sequences have been identified to date (Perler 2002). They interrupt about 30 types of proteins, usually in a conserved region, and range in size from 134 to more than 600 residues (Pietrokovski 1994; Dalgaard et al. 1997; Perler 2002). Within the 100–150 N-terminal and approximately 50 C-terminal intein residues, six conserved sequence motifs have been identified that probably play a role in protein splicing, as illustrated schematically in Fig. 1. The larger inteins harbor homing endonucleases and have four additional conserved motifs (Pietrokovski 1998). The amino acid residues indicated in Fig. 1 play an essential role in the protein-splicing reaction of most inteins, which we refer to as the "canonical" mechanism. These canonical residues consist of the nucleophilic residue at the intein N-terminus, the His in motif N3, the Asn at the intein C-terminus, and the nucleophilic residue at the N-terminus of the C-extein[1]. The canonical mechanism of protein splicing is described in the next section, whereas alternate protein-splicing mechanisms of inteins or intein-like proteins that lack some of the canonical residues are discussed in Section 3.

2 The Canonical Protein-Splicing Mechanism

In the protein-splicing reaction, the intein is excised and the N- and C-exteins are linked directly by a peptide bond. Early in vitro studies of protein splicing using inteins from extreme thermophiles expressed as fusion proteins in *E. coli* led to two general conclusions about the nature of protein splicing: (1) protein splicing is mediated entirely by amino acid residues within the intein and by the adjacent C-extein residue and requires no additional protein catalysts, and (2) protein splicing requires no coenzymes or sources of metabolic energy, suggesting that it occurs by bond rearrangements rather than by the cleavage and synthesis of peptide bonds. The nature of the amino acids that flank the scissile bonds in canonical inteins (Fig. 1) suggests a possible mechanism. The nucleophilic amino acids at the N-termini of the intein and the C-extein could function in nucleophilic attacks on amide and ester bonds, including attacks on amide bonds involving their own amino groups, and the

[1] Throughout this chapter, we will refer to the nucleophilic residues at the N-terminus of the intein and the N-terminus of the C-extein as the upstream and downstream nucleophilic residues, respectively. The intein residues are numbered starting with the upstream nucleophilic residue as 1; the C-extein residues are numbered starting with the downstream nucleophilic residue as +1.

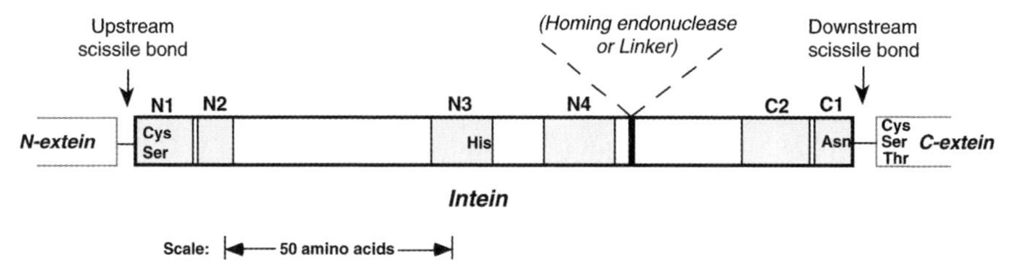

Fig. 1. Conserved elements in a canonical intein. The *shaded areas* are conserved intein motifs, identified by the revised nomenclature of Pietrokovski (1998). Motifs *N1, N3, C1* and *C2* correspond to motifs A, B, G, and F of the older nomenclature, respectively (Pietrokovski 1994; Perler 2002). Amino acid residues that play an essential role in the canonical protein-splicing mechanism are indicated. The conserved motifs and other intein sequences are to scale, based on the *M. xenopi* GyrA intein. The site of insertion of the homing endonuclease domain or linker regions is indicated by the *dark vertical line*. In the *S. cerevisiae* VMA intein, a DNA recognition region (DRR) is inserted just before motif N4

Asn residue at the intein C-terminus could undergo spontaneous cyclization coupled to peptide bond cleavage.

Detailed biochemical studies on protein splicing during the period 1993–1996 succeeded in defining the chemical reactions that underlie the canonical protein-splicing mechanism (reviewed in Perler et al. 1997; Paulus 1998, 2000; Noren et al. 2000; Evans and Xu 2002). These reactions, which consist of the four steps illustrated in Fig. 2, are described in detail in the sections that follow.

2.1 First Step of Protein Splicing – N/O or N/S Acyl Shift

Protein splicing is initiated by attack of the nucleophilic side chain of the Ser or Cys residue at the intein N-terminus on its upstream peptide bond to generate an ester intermediate (Fig. 2, step 1). Spontaneous N/O and N/S acyl shifts have not been observed under physiological conditions, owing in part to the very unfavorable equilibrium position at neutral pH, but can be induced under strongly acidic conditions, and are thought to proceed through oxyoxazolidine or oxythiazolidine anion intermediates, respectively (Fig. 3; reviewed in Iwai and Ando 1967).

Evidence for N/O or N/S Acyl Shift. The occurrence of N/O or N/S acyl shifts in protein splicing was demonstrated in vitro with a modified version of the *Pyrococcus* sp. GB-D (*Psp*) Pol intein in which the N-terminal Ser was replaced with Cys (Shao et al. 1996). In the presence of high concentrations of hydroxylamine or ethylenediamine, fusion proteins containing the modified intein undergo cleavage at the upstream scissile bond, consistent with the aminolysis of a thioester. The site of hydroxylaminolysis was identified by analysis of the C-terminus of the polypeptide cleavage product. Similar studies were performed using the *S. cerevisiae (Sce)* VMA intein, in which Cys occurs naturally at the upstream splice junction (Chong et al. 1996). Studies of the N/S acyl

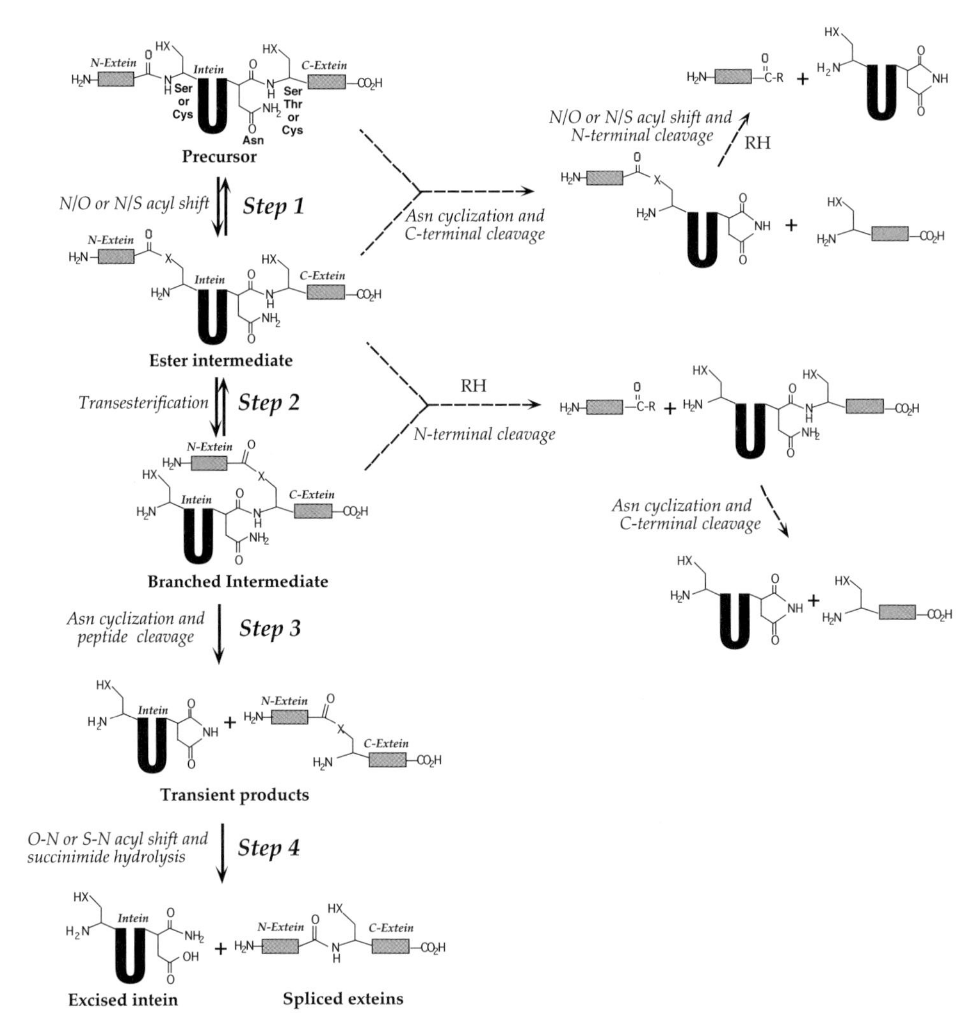

Fig. 2. Canonical protein-splicing mechanism. The canonical protein-splicing pathway is shown on the *left* and possible side reactions are indicated by the *dotted arrows* on the *right*. The amino acid residues that participate directly in the chemical reactions are shown (X=O or S). The remaining portions of the intein and the exteins are indicated schematically and are not to scale. *RH* indicates a nucleophile, which may be an organic thiol such as DTT or cysteine, an amine such as hydroxylamine or ethylenediamine, or water

shifts of these inteins showed that the thioester intermediate can be cleaved by transesterification with a large excess of low molecular weight thiols such as cysteine or 1,4-dithiothreitol (DTT), which led to the development of a protein expression/purification system with self-cleavable affinity tags (Chong et al. 1997).

The Amide–Thioester Equilibrium in Protein Splicing. The equilibrium position of the N/S acyl shift involving the *Sce* VMA intein was measured after arresting further amide–ester interconversion by adding 8 M urea and then estimating the amount of ester intermediate by reaction with hydroxylamine (Chong et al. 1998b). Depending on the amino acid at the C-terminus of the N-extein, the K_{eq} value for the N/S acyl shift ranges from less than 0.05 to as high as 10. A K_{eq} value of about 1 was also observed with the *Synechocystis* sp. PCC6803 (*Ssp*) DnaB intein (Mathys et al. 1999). The large equilibrium constants for the N/S acyl shifts of the scissile bond are unexpected because S/N acyl shifts of typical peptide bonds at neutral pH are essentially irreversible (Iwai and Ando 1967; Shao and Paulus 1997).

Potential Use of Catalytic Strain To Promote N/S Acyl Shifts. Intein-promoted destabilization of the upstream scissile amide bond is a possible driving force for the first step of protein splicing and would also explain the large K_{eq} value of the N/S acyl shift. The destabilization of this bond toward the oxythiazolidine transition state would both enhance the reaction rate and make the equilibrium of the acyl shift more favorable in the direction of the thioester. Evidence for strain in the scissile bond was provided by analysis of the crystal structure of the *Mycobacterium xenopi (Mxe)* GyrA intein, which shows the peptide bond in question to be in the uncommon *cis* configuration (Klabunde et al. 1998), as well as one crystal structure of the Sce VMA intein, which reports main-chain distortion (Poland et al. 2000). The energy of *cis*-peptide bonds is about 5 kcal mol^{-1} higher than that of *trans*-peptide bonds (Ramachandran and Mitra 1976), making the K_{eq} value about 4000 times larger for N/S acyl shifts in which a *cis*-peptide bond is rearranged to an ester.

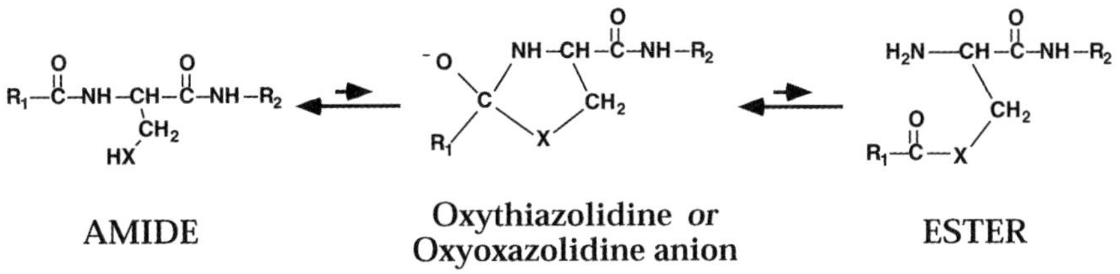

AMIDE

Oxythiazolidine *or* Oxyoxazolidine anion

ESTER

Fig. 3. N/O or N/S acyl shift, with the postulated oxyoxazolidine or oxythiazolidine anion intermediate. X=O or S

A more definitive examination of the conformation of the upstream scissile bond made use of nuclear magnetic resonance (NMR), which can be carried out on a much shorter time scale than X-ray crystallography and therefore permits the use of functional inteins (Romanelli et al. 2004). Using protein semisynthesis (Muir 2003), a segmental isotopically labeled *Mxe* GyrA intein was constructed, which allows for unequivocal assignment of the resonances of the scissile amide bond. The wild-type intein is able to promote thiol-induced N-terminal cleavage and hence promote the N/S acyl shift, but a His^{75}Ala intein can promote only trace amounts of N-terminal cleavage. The possibility of a hydrogen bond between His^{75} and the scissile amide is suggested by an unusually large upfield chemical shift for the amide proton of the wild-type intein, together with a downfield shift in the His^{75}Ala mutant. In addition, a possible change in the polarization of the scissile bond, perhaps due to a loss of amide resonance due to bond rotation, is suggested by the complementary observation that the one-bond dipolar coupling constant ($^{1}J_{NC'}$) for the scissile bond is abnormally low in the wild-type intein, but not in the His^{75}Ala mutant. However, the spectral overlap of the $^{1}H\{^{15}N\}$ HSQC spectra suggests that the His^{75}Ala mutation does not significantly alter the overall structure of the intein. These data could be interpreted to suggest that His^{75} stabilizes a locally distorted scissile amide bond conformation.

2.2 Second Step of Protein Splicing – Transesterification

Intramolecular transesterifications are freely reversible and relatively rapid reactions that require good nucleophiles. In protein splicing, transesterification occurs either between a thioester and a thiol or alcohol, or between an oxygen ester and an alcohol, but never between an oxygen ester and a thiol. In the first two cases, the equilibrium constant at neutral pH is about 1; in the thioester–alcohol case it is about 50 (Jencks et al. 1960). The transesterification step involves attack by the nucleophilic side chain of the N-terminal amino acid of the C-extein on the ester linking the N-extein and intein, yielding a branched ester intermediate (Fig. 2, step 2).

Evidence for Branched Intermediate Formation. A branched intermediate was first detected in the in vitro splicing reaction involving the *Psp*-GBD Pol intein, which has Ser residues at both splice junctions (Xu et al. 1993). The intermediate was detected by its unusually slow migration upon sodium dodecyl sulfate/polyacrylamide gel electrophoresis (SDS-PAGE) and by the detection of two N-termini, corresponding to those of the N-extein and the intein. Mild alkaline hydrolysis of the branched intermediate trapped with 6 M guanidinium chloride shows that the N-extein is esterified to the downstream

Ser hydroxyl group (see Fig. 2; Xu et al. 1994). Transesterification can be reversed by shifting the equilibrium of the N/O acyl shift to the amide by raising the pH (Xu et al. 1993), and the branched ester can be restored by lowering the pH, indicating full reversibility (Xu et al. 1994). Branched intermediates were also characterized in protein-splicing systems involving a thioester rather than an oxygen ester (Chong and Xu 1997; Chen et al. 2000), and the equivalent structure has tentatively been identified in the naturally split *Synechocystis* sp. (*Ssp*) DnaE intein (Nichols et al. 2003). On the other hand, the occurrence of branched intermediates has not been reported in every extensively studied protein-splicing system.

Mechanistic Aspects of Transesterification. The first intein crystal structures lack C-exteins (Duan et al. 1997; Klabunde et al. 1998), although modeling of their N-terminal Cys residues suggests that the nucleophilic side chains of the N-terminal residues of the intein and the C-extein are closely juxtaposed to allow facile transesterification. This was confirmed by the crystal structure of the *Sce* VMA intein with adjacent extein residues, in which the two nucleophilic side chains are 3.6 Å apart (Mizutani et al. 2002). It is likely that this intein is functional because the crystals are isomorphic with those of a mutant *Sce* VMA intein that can undergo splicing in the crystal lattice (Mizutani et al. 2002, 2004). This contrasts with the structures of the inactive Zn^{2+}-complex of the *Sce* VMA intein (Poland et al. 2000) and of the *Ssp* DnaB intein (Ding et al. 2003), in which the two nucleophilic side chains are separated by a distance of 9 and 8.5 Å, respectively, implying either that a conformational change has to occur before transesterification can proceed or that these crystal structures correspond to inactive forms of the intein.

A general base catalyst may serve to increase the nucleophilicity of the side chain that initiates transesterification, especially when a hydroxyl group is involved, but the intein crystal structures available at this time provide no insights into this matter. Modeling of the linear ester intermediate, based on a crystal structure of the *Sce* VMA intein, suggests that the neutral amino group of the N-terminal Cys produced in the N/S acyl shift can activate the N-terminal thiol of the C-extein for nucleophilic attack (Mizutani et al. 2002, 2004). In the *Mycobacterium tuberculosis (Mtu)* RecA intein, the apparent pK_a of the intein N-terminal Cys, which initiates the N/S acyl shift, is approximately 8.2, whereas that of the C-extein N-terminal Cys, which initiates the transesterification reaction, is about 5.8 (Shingledecker et al. 2000). This suggests that transesterification may be facilitated by the induction of an unusually low pK_a for the attacking thiol group by the unique electrostatic and dielectric environment of the protein-splicing active center rather than by general base catalysis.

2.3 Third Step of Protein Splicing – Asparagine Cyclization

Certain Asn residues in peptides and proteins can undergo slow cyclization to aminosuccinimide (reviewed in Clarke 1987). This reaction usually results in deamidation of the side chain by attack of the peptide bond nitrogen on the carbonyl carbon of the asparagine β-amide. Another mode of asparagine cyclization, the attack of the β-amide nitrogen on the carbonyl carbon of the peptide bond, leads to peptide bond cleavage and can occur at high temperatures (Geiger and Clarke 1987) or in unusually long-lived proteins such as the lens crystallins (Voorter et al. 1988). It is this second type of asparagine cyclization that occurs in the course of protein splicing and leads to the concomitant cleavage of the downstream scissile bond (Fig. 2, step 3).

Evidence for Asparagine Cyclization. The occurrence of asparagine cyclization in protein splicing was demonstrated by the identification of an aminosuccinimide residue at the C-terminus of the excised *Psp*-GBD Pol intein by high-performance liquid chromatography (HPLC), mass spectrometry and colorimetric analysis (Xu et al. 1994; Shao et al. 1995). The excised *Sce* VMA intein also has a C-terminal aminosuccinimide residue, suggesting that the same cleavage mechanism occurs in the inteins of hyperthermophiles and mesophiles, regardless of whether protein splicing involves oxygen or thioesters (Chong et al. 1996). The essential role of asparagine cyclization in protein splicing was shown by the replacement of the C-terminal Asn by Asp, which completely blocks splicing and leads to the accumulation of the branched intermediate (Xu and Perler 1996).

Mechanistic Aspects of Asparagine Cyclization. The mechanism of asparagine cyclization and the attendant cleavage of the peptide bond linking the intein to the C-extein is not yet clear. Some biochemical evidence suggests that the adjacent conserved His residue plays a role in Asn cyclization. For instance, its replacement with other amino acids can inhibit protein splicing (Cooper et al. 1993) and lead to the accumulation of the branched ester intermediate (Xu and Perler 1996). Nevertheless, 27 inteins are known with a penultimate amino acid residue other than His (Perler 2002), at least three of which undergo efficient splicing in *Escherichia coli*. In these cases, the replacement of the penultimate residue with His either slightly reduces splicing efficiency, enhances splicing, or has no significant effect (Mills and Paulus 2001; Nichols and Evans 2004). On the other hand, the *Chlamydomonas eugametos* (*Ceu*) ClpA intein, which has a Gly residue in place of the penultimate His, is unable to splice in *E. coli* unless the Gly residue is replaced by His (Wang and Liu 1997).

Examination of the current intein crystal structures sheds little light on the role of the penultimate His residue in asparagine cyclization. The bond distances reported in the crystal structure of the *Mxe* GyrA intein are consistent with protonation of the scissile bond by His^{197}, which would facilitate its cleavage concomitant with the cyclization of Asn^{198} (Klabunde et al. 1998). On the other hand, the crystal structure of the *Ssp* DnaB intein and adjacent extein fragments is consistent with an H-bond between the Asn carbonyl oxygen of the scissile bond and the imidazole ring of the adjacent His residue (Ding et al. 2003). Asparagine cyclization and cleavage of the downstream scissile bond can still occur when the other steps in protein splicing are blocked by mutations, such as replacement of either the upstream or downstream nucleophilic residue (Hirata and Anraku 1992; Cooper et al. 1993; Chong et al. 1996, 1998a; Xu and Perler 1996; Mathys et al. 1999; Southworth et al. 1999). However, in some cases, the N/O or N/S shift must occur prior to Asn cyclization (Xu and Perler 1996).

2.4 Finishing Reaction

O/N or S/N Acyl Shift. After cyclization of the C-terminal Asn residue and the attendant cleavage of the downstream scissile bond, protein splicing is complete in the sense that the intein is excised and the exteins are linked. However, an ester bond links the extein segments and the excised intein contains an unnatural C-terminal residue, aminosuccinimide. Especially critical for the completion of the splicing process is the O/N or S/N shift of the peptide ester intermediate to yield a spliced product composed entirely of peptide bonds. This is essentially the reverse of the first step of protein splicing, the N/O or N/S acyl shift, which now operates in the thermodynamically favored direction.

Uncatalyzed O/N and S/N shifts have been studied in model peptides. The rates of O/N and S/N acyl shifts of peptide esters are extremely rapid and increase strikingly with pH (Shao and Paulus 1997). Thioesters involving the side chain of Cys residues rearrange about 1000 times more rapidly than oxygen esters. The extremely rapid rates and the virtual irreversibility of O/N and S/N acyl shifts at neutral pH make these acyl shifts ideal finishing reactions that rapidly drive protein splicing to completion under physiological conditions without the need for a catalyst. On the other hand, recent crystallographic studies on a slowly splicing variant of the *Sce* VMA intein, which undergoes protein splicing in the crystal lattice, show the spliced extein to be positioned relative to the excised intein so that its N-terminal amino group can stabilize the oxythiazolidine anion intermediate in the S/N acyl shift (Mizutani et al. 2004). Although facilitation of the acyl shift may not be a critical factor in accelerating the protein-splicing process, its occurrence in the relatively hydrophobic environment of the protein-splicing active center may serve to reduce side reactions such as the hydrolysis of the relatively unstable thioester bond.

Succinimide Hydrolysis. The rate of hydrolysis of C-terminal aminosuccinimides was measured using synthetic tetrapeptides corresponding to the C-terminus of the *Psp*-GBD Pol intein (Shao et al. 1995). The rate of succinimide hydrolysis at 37 °C is strikingly pH-dependent, with a $t_{1/2}$ of 350 h at pH 5.5, which declines to 17 h at pH 7.4. At 90 °C, the $t_{1/2}$ at pH 7.4 is 4 min. These rates are considerably slower than the rates of hydrolysis of N-substituted cyclic imides produced at internal positions of polypeptide chains by nucleophilic attack of the peptide bond N on the carbonyl C of the Asn β-amide (summarized in Shao et al. 1995). The relatively slow rate of hydrolysis of C-terminal aminosuccinimide implies that, at least in mesophilic organisms, a substantial fraction of the excised inteins carry C-terminal aminosuccinimide residues.

2.5 Association of Split Inteins

Protein splicing is ordinarily an intramolecular process and therefore does not involve a reactant association step. However, some inteins can be artificially split, expressed as separate proteins, and reconstituted to yield a functional protein-splicing complex (Mills et al. 1998; Southworth et al. 1998; Yamazaki et al. 1998). The reconstitution of artificially split inteins requires a denaturation/renaturation procedure, probably owing to misfolding that occurs when the intein fragments are expressed separately. It is therefore difficult to measure the parameters of the reassociation reaction, but the affinity of the intein fragments must be quite high, as indicated by nearly quantitative reassociation of the split *Mtu* RecA intein in the 10 μM range with only a two-fold molar excess of the C-terminal fragment (Lew et al. 1999).

However, the discovery of naturally split inteins in cyanobacteria (Wu et al. 1998), in which they presumably serve to reconstitute the separately expressed fragments of the DnaE protein, makes intein fragment association an essential first step in the splicing of DnaE. The association of the N- and C-terminal segments of the *Ssp* DnaE intein occurs with high efficiency without the need for prior denaturation (Evans et al. 2000). The association of the N-terminal segment (N-extein-E_N) with the C-terminal intein segment (E_C) was studied using a competition assay, which suggested that the dissociation of N-extein-E_N from E_C is a very slow and rate-limiting process, consistent with a very low E_N–E_C dissociation constant. In addition, even with a large molar excess of N-extein-E_N over E_C, no turnover of the intein as an active catalyst is observed (Nichols and Evans 2004). This suggests that the product E_N has a much higher affinity than N-extein-E_N for E_C and that its tight binding to E_C blocks further reaction. The binding of N-extein-E_N to E_C may be diminished by the strain energy required to destabilize the scissile bond into a conformation resembling the oxythiazolidine transition state. Since the scissile bond is absent in E_N, strain energy will not need to be diverted from binding energy, and the binding energy should therefore be more favorable for E_N than for N-extein-E_N.

3 Non-canonical Inteins and Their Mechanisms

Significant insight into the canonical mechanism of protein splicing has been afforded by the study of inteins whose activities have been modulated by mutation. In the sections that follow, we will describe the splicing of inteins or intein-like domains that lack conserved intein elements yet still promote protein splicing or splicing side reactions, and how the study of non-canonical inteins broadens our understanding of how inteins promote the individual steps of splicing.

3.1 Substitution of the N-Terminal Nucleophile

Fourteen non-canonical inteins are known that lack an N-terminal nucleophile, of which 13 have N-terminal Ala (Perler 2002). Four of these inteins are known to promote protein splicing (Southworth et al. 2000; Yamamoto et al. 2001; Southworth and Perler 2002), suggesting that they have evolved to splice by bypassing the initial N/S acyl shift (Fig. 2) through a direct attack by the downstream nucleophile on the upstream scissile peptide bond (Fig. 4A). In two such inteins, mutation of the N-terminal Ala to Gly blocks protein splicing, suggesting that the local conformation of the active site is important for coordinating the direct attack of the downstream nucleophile on the peptide bond (Southworth et al. 2000; Yamamoto et al. 2001). In the *Mja* KlbA intein, mutation of the N-terminal Ala to Cys or Ser yields significant amounts of spliced product, and a double mutant (Ala^1Cys and $Cys^{+1}Ala$) promotes an N/S acyl rearrangement (Southworth et al. 2000; Yamamoto et al. 2001), implying that the Cys^1Ala substitution is relatively recent. Mutation of the conserved Thr and His residues of motif N3 greatly reduces the extent of splicing, implying a role for these residues in the attack of the C-terminal nucleophile on the upstream scissile peptide bond.

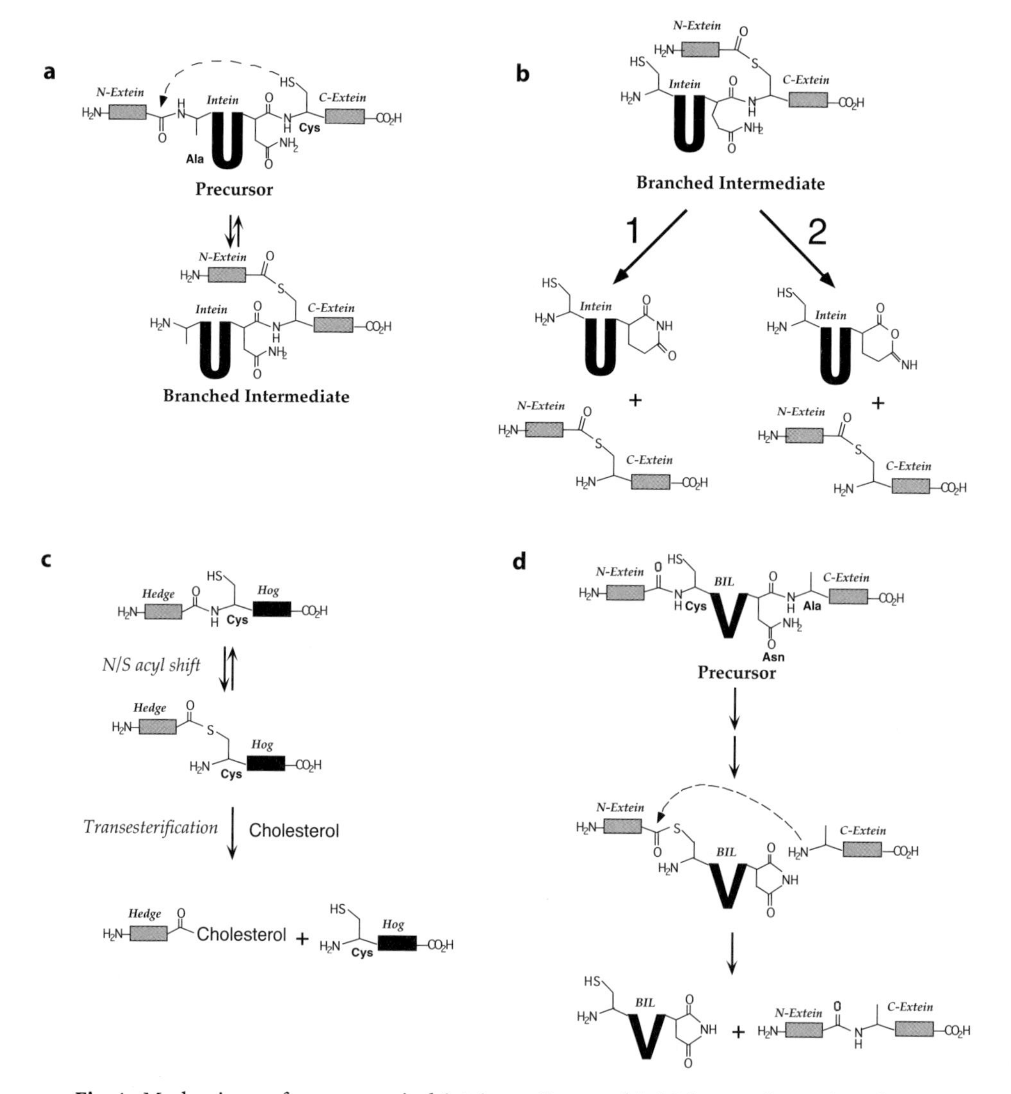

Fig. 4. Mechanisms of non-canonical inteins. **a** Proposed initial step of protein splicing for inteins that lack an N-terminal nucleophile. **b** Possible mechanisms of step 3 in protein splicing for inteins with a C-terminal glutamine residue. Pathway 1 depicts Gln cyclization to yield glutarimide; pathway 2 depicts Gln cyclization to glutaranhydride. **c** Mechanism of Hedgehog protein autoprocessing. The diagram shows the scissile bond linking the N-terminal signaling domain (Hedge) and the C-terminal catalytic domain (Hog). **d** Proposed protein-splicing mechanism involving an A-type BIL domain flanked by Ala at the C-terminus

3.2 Substitution of the C-Terminal Asparagine

Three inteins lacking the conserved C-terminal Asn residue have been described, and all three promote protein splicing (Amitai et al. 2004; Mills et al. 2004). The *Chilo* iridescent virus (*CIV*) RNR intein and the *Pyrococcus abyssi (Pab)* PolII intein each have a C-terminal Gln. Replacement of Gln by Asn slightly improves the rate and extent of splicing in the *Pab* PolII intein (Mills et al. 2004), whereas the splicing yield is reduced sevenfold with the same substitution in the *CIV* RNR intein (Amitai et al. 2004). No evidence for cyclized C-terminal residues in the excised inteins was found with either intein or with their mutants in which the C-terminal Gln was replaced with Asn, suggesting either that the imides are not stable under the experimental conditions, or that the third step of splicing might proceed via a mechanism other than glutarimide formation coupled to peptide bond cleavage (Fig. 4B, pathway 1). An alternate mechanism without a stable cyclized intein intermediate could use the side-chain carbonyl oxygen of the C-terminal Gln as the nucleophile to couple peptide bond cleavage to formation of a C-terminal glutaranhydride (Fig. 4B, pathway 2). There is no experimental evidence to support such a mechanism, but it is attractive since the carbonyl oxygen of the side chain would be a better nucleophile than the amide nitrogen, and the resulting cyclized anhydride would be less stable and therefore more difficult to isolate than the imide.

The *Carboxydothermus hydrogenoformans (Chy)* RNR intein has a C-terminal Asp and is capable of splicing. Splicing mediated by this intein could occur via cyclization of the C-terminal Asp to a succinic anhydride, coupled to peptide bond cleavage (Amitai et al. 2004). C-terminal cleavage is promoted by *Chy* RNR inteins with diverse C-terminal substitutions, including Asn, Gln, Glu and Ala, but is suppressed by a mutation, Cys^{1}Ala, in which the N-terminal nucleophile is replaced by Ala (Amitai et al. 2004). C-terminal cleavage of the Asp^{345}Ala mutant suggests that a mechanism for C-terminal cleavage other than side-chain cyclization might exist, perhaps involving the N-terminal Cys of the intein (Amitai et al. 2004). However, it is also possible that the C-terminal scissile bond is activated for cleavage by an oxygen nucleophile such as the side-chain carboxyl oxygen of Asp, and could be cleaved by water in the Asp^{345}Ala mutant. This is ordinarily a minor side reaction which becomes more prominent in the Asp^{345}Ala mutant in which the normal protein-splicing pathway is inoperable.

3.3 Hedgehog Autoprocessing Domains

Unlike inteins, Hedgehog proteins have not been found in unicellular organisms but occur in metazoans, in which they function as signaling molecules in embryonic development (see Dassa and Pietrokovski, this Vol.). Like inteins, they undergo autoprocessing. Their autoprocessing domains share common sequence motifs with inteins (Pietrokovski 1994) and have a similar structural folding pattern, called the Hint domain (Hall et al. 1997). The first step of Hedgehog autoprocessing is an N/S acyl shift of a highly conserved Gly–Cys peptide bond at the junction of the N-terminal signaling domain and the C-terminal autoprocessing domain (Fig. 4C; Porter et al. 1996b). Evidence for thioester formation comes from in vitro cleavage by DTT or transesterification by peptides with N-terminal Cys (Porter et al. 1996a). Because the crystal structure of the *Drosophila* Hedgehog protein does not include the scissile bond (Hall et al. 1997), it provides no information as to whether catalytic strain plays a role in the N/S acyl shift. However, the highly conserved Thr^{326} and His^{329}, which align with the conserved TXXH motif in intein motif N3, play an essential role in the rearrangement (Hall et al. 1997). The second step in Hedgehog autoprocessing is transesterification of the thioester intermediate by the 3-hydroxyl of cholesterol, resulting in peptide bond cleavage and conjugation of the N-terminal domain with cholesterol (Fig. 4C; Porter et al. 1996b).

3.4 Bacterial Intein-Like Domains

Two new types of Hint domains in a wide range of bacterial species have recently been described (Amitai et al. 2003). These bacterial intein-like (BIL) domains differ from inteins by having additional conserved sequence blocks and being inserted into variable regions of their host proteins, and they differ from Hedgehog autoprocessing domains by their lack of conserved Hedgehog sequence blocks (Amitai et al. 2003; Dassa and Pietrokovski, this Vol.). The BIL domains can be subdivided into two categories: A-type BIL domains contain the conserved intein motifs but often lack a C-terminal nucleophile, whereas B-type BIL domains lack the intein motifs C1 and C2 but usually have a nucleophilic residue (Cys, Ser or Thr) at the penultimate C-terminal position.

An A-type BIL domain from *Clostridium thermocellum* (*Cth*), which is flanked at the C-terminus by Ala, promotes efficient protein splicing (Dassa et al. 2004). This led to the proposal that this BIL domain splices by an alternative mechanism in which the N/S acyl shift leading to the linear thioester and asparagine cyclization leading to C-terminal cleavage occur independently and that the free amino terminus of the C-extein then attacks the linear thioester, thereby effecting extein ligation (Fig. 4D; Dassa et al. 2004). On the other hand, an A-type BIL domain from *Magnetospirillum magnetotacticum (Mma)*, with a C-terminus flanked by Tyr, promotes N- and C-terminal cleavage but requires replacement of its C-terminal Tyr by Cys in order to undergo protein splicing (Southworth et al. 2004).

B-type BIL domains have not been observed to promote splicing reactions, but some can undergo N- or C-terminal cleavage reactions (Dassa et al. 2004). One might speculate that some BIL domains have evolved to promote such cleavage reactions, which resemble the side reactions observed when inteins fail to properly coordinate the steps of splicing owing to disruption by mutation or insertion into foreign extein contexts (see below).

4 Protein Splicing as a System

A remarkable feature of the protein-splicing process is that its three steps occur in a highly coordinated manner without side products (Kane et al. 1990), although each step can proceed independently in inteins modified by mutation (Xu and Perler 1996). These three mechanistically diverse reactions are promoted or "catalyzed" by inteins using a protein-splicing domain that consists of about 150 amino acids. In this section, we will investigate how inteins coordinate the protein-splicing process so that its steps occur in a specific order and yield spliced proteins with high efficiency.

4.1 Side Reactions

Except for the *Sce* VMA intein (Hirata et al. 1990; Kane et al. 1990; Hirata and Anraku 1992; Cooper et al. 1993; Kawasaki et al. 1997; Nogami et al. 1997), no inteins have been studied in their natural host in the context of their natural host protein. Rather, most studies of the mechanism of protein splicing utilized inteins flanked by foreign exteins, usually affinity domains chosen for ease of purification and detection. In addition, these inteins were expressed in *E. coli*, although no inteins occur naturally in this organism. Therefore, it is not surprising that protein splicing is often inefficient and accompanied by side products that result from the improper coordination of the steps in the protein-splicing pathway. For instance, premature asparagine cyclization can lead to C-terminal cleavage uncoupled from splicing, and delay or prevention of transesterification or asparagine cyclization can lead to the accumulation of the ester intermediates, which are susceptible to N-terminal cleavage by *trans*-acting nucleophiles such as hydroxide ion, amines or thiols (Fig. 2). In this section, we examine how the integrity of the protein-splicing process is affected by mutations that affect specific steps in protein splicing or intein structure and by the replacement of the natural exteins.

Role of Intein Residues. The nucleophilic residues at both splice junctions of the *Psp*-GBD Pol intein are Ser, and protein splicing therefore uses an N/O acyl shift and oxygen ester intermediates (Xu and Perler 1996). Blockage of any step by replacing the amino acid that participates in that reaction prevents splicing and diverts the protein-splicing pathway into alternative directions, as indicated in Fig. 2. For instance, accumulation of the C-terminal cleavage product, probably through Asn cyclization, results from the replacement of the upstream nucleophile (Ser^1) with Ala. The branched intermediate and the product of N-terminal cleavage accumulate when Asn at the intein C-terminus is replaced with Asp. The products of N- and C-terminal cleavage, but not the branched ester, accumulate upon replacement of the downstream nucleophile (Ser^{+1}) with Ala (Xu and Perler 1996). Similar results were obtained in studies of the *Sce* VMA intein, which is flanked by Cys residues and therefore uses an N/S acyl shift and thioester intermediates rather than oxygen esters for protein splicing (Cooper et al. 1993). These results are consistent with the notion that all steps in the protein-splicing pathway can operate independently.

The alteration of amino acid residues that do not participate directly in the protein-splicing reactions but function catalytically or structurally can also promote the occurrence of side reactions. For instance, substitution of the penultimate His residue (His^{536}) in the *Psp*-GBD Pol intein completely blocks protein splicing but has no effect on the N/O acyl shift, promoting the accumulation of the branched intermediate (Xu and Perler 1996). Replacement of the conserved His residue in motif N2 (see Fig. 1) of the *Sce* VMA intein promotes C-terminal cleavage by inhibiting the N/S acyl shift or the transesterification steps (Kawasaki et al. 1997). Substitution of the conserved Thr residue in motif C2 of the *Ssp* DnaB intein promotes N-terminal cleavage by attenuating asparagine cyclization (Ding et al. 2003).

Role of Exteins. The observation that the amino acid sequences of inteins inserted at allelic sites in a particular host protein often have diverged more in different organisms than the extein sequences implies that inteins have evolved subsequent to their insertion to optimize protein splicing in the context of the host protein and the host environment. One would expect this optimization to involve specific interactions of the intein with its host protein that improve the efficiency of protein splicing under physiological conditions and minimize the occurrence of side reactions. For instance, upon expression in *E. coli* of a fusion protein consisting of the *Mxe* GyrA intein together with 64–65 GyrA residues at its N- and C-termini, protein splicing occurs in vivo with high efficiency (Telenti et al. 1997). However, when the native GyrA residues at the intein C-terminus are replaced by a foreign polypeptide, protein splicing becomes temperature-sensitive. Moreover, when the native GyrA residues at the intein N-terminus are replaced by a foreign sequence, protein splicing is completely ablated at both 16 and 37 °C, C-terminal cleavage occurs at both temperatures, and quantitative N-terminal cleavage can be induced in vitro when the precursor is incubated with DTT. These observations suggest that interactions between the intein and adjacent extein residues can play an important role in coordinating the reactions to optimize protein splicing.

Possible structural interactions between intein and extein residues were probed by replacing the conserved hydrophobic amino acids in motif C1 at the C-terminus of the *Sce* VMA intein, leading to loss of VMA function, and using an in vivo selection system to isolate intragenic suppressor mutants with restored VMA function (Nogami et al. 1997). A predominant class of suppressors that restores function to the inactive intein involves the three hydrophobic N-extein residues preceding the upstream scissile bond, suggesting a direct interaction between these residues and the intein. The crystal structure of the *Sce* VMA intein with adjacent extein segments shows that there are no direct side-chain interactions between these residues but that they are part of a hydrophobic cluster that also includes certain amino acid side chains in motif C2, which constrains their conformation (Mizutani et al. 2002).

4.2 What Coordinates the Steps in the Protein-Splicing Pathway?

The key to the integration of the protein-splicing reactions into a efficiently coupled process might rest in one or more of the following areas: (1) the relative rates of the protein splicing steps; (2) possible mechanistic linkage of successive reactions; (3) stepwise conformational changes associated with protein splicing; and (4) modulation of the protein-splicing reactions by specific amino acid residues.

Reaction Rates. In a detailed study of the naturally split *Ssp* DnaE intein (Martin et al. 2001), in which protein splicing can be initiated by mixing the N- and C-terminal intein fragments, protein splicing proceeds with a $t_{1/2}$ of about 3 h. DTT-induced N-terminal cleavage is considerably faster than protein splicing, with a $t_{1/2}$ of about 10 min. Because prior incubation of the intein segments in the absence of DTT does not influence the rate of N-terminal cleavage, intein association probably occurs at an even faster rate. The rate of transesterification could not be measured independently, since only small amounts of putative "branched ester" intermediate accumulate in the course of protein splicing (Nichols et al. 2003). On the other hand, C-terminal cleavage coupled to asparagine cyclization proceeds with a $t_{1/2}$ of about 2 h. One can conclude that, for this intein, the N/S acyl rearrangement step of protein splicing is by far more rapid than asparagine cyclization, which assures that the unspliced precursor does not ordinarily undergo C-terminal cleavage as shown in Fig. 2 unless the intein is perturbed by mutation. It is not known whether similar kinetic relationships apply to other inteins. The observation that a major side reaction affecting protein splicing in artificial in vivo systems is premature C-terminal cleavage would suggest otherwise. On the other hand, in some protein-splicing systems reconstituted in *E. coli*, such as the *Mtu* RecA intein, N-terminal cleavage is also significant (Derbyshire et al. 1997; Shingledecker et al. 1998), perhaps owing to the accumulation of ester intermediates as a result of an inappropriately slow C-terminal cleavage reaction.

Mechanistic Linkage. There is some evidence for the interdependence of successive steps in protein splicing. The Ser1Ala mutation in the *Psp*-GBD Pol intein attenuates the rate of Asn cyclization 25-fold (Xu and Perler 1996), suggesting that Asn cyclization depends on a prior N/O acyl shift. A similar interdependence is observed in the *Ssp* DnaE intein (Martin et al. 2001; Nichols and Evans 2004) and in a mutant form of the *Sce* VMA intein (Chong et al. 1998b). In these cases, C-terminal cleavage requires a free intein N-terminus, which can be generated by thiolysis of the linear thioester intermediate. A mechanism in which the nucleophilic residue at the free intein N-terminus participates in asparagine cyclization has been proposed (Paulus 1998) and invoked to explain the properties of non-canonical inteins with a C-terminal Asp residue (Amitai et al. 2004), but there is no direct evidence for its occurrence. In a number of inteins, such as the *Ssp* DnaB intein, C-terminal cleavage is clearly independent of the N/S acyl shift (Mathys et al. 1999; Ding et al. 2003). The observed interdependence of these reactions in certain inteins may not be the result of direct mechanistic coupling, but a reflection of some other regulatory process, such as a sequential conformational change.

Sequential Conformational Change. As discussed in Section 2.2, two intein crystal structures place the nucleophilic side chains at the two scissile bonds 8–10 Å apart (Poland et al. 2000; Ding et al. 2003), such that transesterification cannot occur without a conformational change. However, the observations that protein splicing can occur in the crystal lattice of the *Sce* VMA intein (Mizutani et al. 2002, 2004) and that the structures of the unspliced precursor and the spliced products have an root-mean-square (r.m.s.) difference in the splicing region of only 0.84 Å argue strongly against significant conformational changes in the course of protein splicing.

Specific Regulatory Interactions. Because efficient protein splicing benefits the survival and growth of the microbial host, there probably was strong selective pressure to optimize intein performance, and regulatory sites may have evolved to modulate intein efficiency without playing a specific role in the protein-splicing process. Such sites can be discovered by screening for mutated inteins with altered splicing properties. For example, the selection for mutants of the *Mtu* RecA intein with enhanced C-terminal cleavage activity found an intein with an $Asp^{422}Gly$ mutation in motif C2 (Wood et al. 1999). This suggests that the function of the conserved residue Asp^{422} is not to enhance any of the splicing reactions but specifically to attenuate C-terminal cleavage. It is also possible that residues suggested to have a catalytic role in protein splicing may primarily be regulatory, such as the penultimate intein His residue (see Sect. 2.3).

5 Conclusions

As the number of known inteins and intein-like elements begins to exceed 200, our attention is attracted not so much to their common features, but to their diversity. Although all inteins have a common evolutionary origin, each has evolved further in the context of its particular host protein to optimize its splicing in the physiological context of the microbial host. This has provided as many opportunities for divergence as there are inteins, including the significant mechanistic innovations that we see in the non-canonical inteins. However, even in terms of the details of the canonical protein-splicing mechanism, one has to be cautious not to generalize from one intein to another. For example, in Section 2.1, we cite the disagreement about whether the upstream scissile bond is destabilized by dihedral angle distortion. It is possible that every intein may have evolved its unique way in the context of its exteins to destabilize the precursor structure and promote the N/S or N/O acyl rearrangement, which is not necessarily confined to the destabilization of the upstream scissile bond (e.g. Paulus 2001). Similarly, the mechanisms for integrating the various steps in the protein-splicing pathway into an efficiently functioning system may be unique for almost every intein. Finally, it is possible that some inteins have acquired functions beyond salvaging the product of a gene that has been invaded by selfish DNA. Because protein splicing represents a form of post-translational processing that converts an inactive gene product into a functional one, it could play a role in the control of gene expression, as has been postulated for the *Synechocystis dnaE* gene (Wu et al. 1998). Until now, protein splicing has been studied primarily in artificial contexts for the convenience of the investigator, and *E. coli* is certainly a much more convenient laboratory organism than *M. tuberculosis* or hyperthermophilic archaea. One of the challenges of intein research in the next decade will be to study protein splicing in its physiological context. It is quite possible that such studies may reveal potential regulatory roles of protein splicing as well as mechanisms for its regulation. An important impetus for this type of research should be the fact that, in multicellular organisms, inteins have acquired an important regulatory role in the form of the Hedgehog proteins, which control embryonic development.

References

Amitai G, Belenkiy O, Dassa B, Shainskaya A, Pietrokovski S (2003) Distribution and function of new bacterial intein-like protein domains. Mol Microbiol 47:61–73

Amitai G, Dassa B, Pietrokovski S (2004) Protein splicing of inteins with atypical glutamine and aspartate C-terminal residues. J Biol Chem 279:3121–3131

Chen L, Benner J, Perler FB (2000) Protein splicing in the absence of an intein penultimate histidine, J Biol Chem 275:20431–20435

Chong S, Xu M-Q (1997) Protein splicing of the *Saccharomyces cerevisiae* VMA intein without the endonuclease motifs. J Biol Chem 272:15587–15590

Chong S, Shao Y, Paulus H, Benner J, Perler FB, Xu M-Q (1996) Protein splicing involving the *Saccharomyces cerevisiae* VMA intein: the steps in the splicing pathway, side reactions leading to protein cleavage, and establishment of an in vitro splicing system. J Biol Chem 271:22159–22168

Chong S, Mersha FB, Comb DG, Scott ME, Landry D, Vence LM, Perler FB, Benner J, Kucera R, Hirvonen CA, Pelletier JJ, Paulus H, Xu M-Q (1997) Single-column purification of free recombinant proteins using a self-cleavable affinity tag derived from a protein splicing element. Gene 192:271–281

Chong S, Montello GE, Zhang A, Cantor EJ, Liao W, Xu M-Q, Benner J (1998a) Utilizing the C-terminal cleavage activity of a protein splicing element to purify recombinant proteins in a single chromatographic step. Nucleic Acids Res 26:5109–5115

Chong S, Williams KS, Wotkovicz C, Xu M-Q (1998b) Modulation of protein splicing of the *Saccharomyces cerevisiae* vacuolar membrane ATPase intein. J Biol Chem 273:10567–10577

Clarke S (1987) Propensity for spontaneous succinimide formation from aspartyl and asparaginyl residues in cellular proteins. Int J Peptide Protein Res 30:808–821

Cooper AA, Chen Y-J, Lindorfer MA, Stevens TH (1993) Protein splicing of the yeast *TFP1* intervening protein sequence: a model for self-excision. EMBO J 12:2575–2583

Dalgaard JZ, Moser MJ, Hughey R, Mian IS (1997) Statistical modeling, phylogenetic analysis and structure prediction of a protein splicing domain common to inteins and hedgehog proteins. J Comput Biol 4:193–214

Dassa B, Haviv H, Amitai G, Pietrokovski S (2004) Protein splicing and auto-cleavage of bacterial intein-like domains lacking a C′-flanking nucleophilic residue. J Biol Chem 279:32001–32007

Derbyshire V, Wood DW, Wu W, Dansereau JT, Dalgaard LZ, Belfort M (1997) Genetic definition of a protein-splicing domain: functional mini-inteins support structure predictions and a model for intein evolution. Proc Natl Acad Sci USA 94:11466–11471

Ding Y, Xu M-Q, Ghosh I, Chen X, Ferrandon S, Lesage G, Rao Z (2003) Crystal structure of a mini-intein reveals a conserved catalytic module involved in side chain cyclization of asparagine during protein splicing. J Biol Chem 278:39133–39142

Duan X, Gimble FS, Quiocho FA (1997) Crystal structure of PI-SceI, a homing endonuclease with protein splicing activity. Cell 89:555–564

Evans TC, Xu M-Q (2002) Mechanistic and kinetic considerations of protein splicing. Chem Rev 102:4869–4883

Evans TC, Martin D, Kolly R, Panne D, Sun L, Ghosh I, Chen L, Benner J, Liu X-Q, Xu M-Q (2000) Protein *trans*-splicing and cyclization by a naturally split intein from the *dnaE* gene of *Synechocystis* species PCC6803. J Biol Chem 275:9091–9094
Geiger T, Clarke S (1987) Deamidation, isomerization, and racemization at asparaginyl and aspartyl residues in peptides. Succinimide-linked reactions that contribute to protein degradation. J Biol Chem 262:785–794
Hall TM. T, Porter JA, Young KE, Koonin EV, Beachy PE, Leahy DJ (1997) Crystal structure of a hedgehog protein autoprocessing domain: homology between hedgehog and self-splicing proteins. Cell 91:85–97
Hirata R, Anraku Y (1992) Mutations at the putative junction sites of the yeast VMA1 protein, the catalytic subunit of the vacuolar H^+-ATPase, inhibit its processing by protein splicing. Biochem Biophys Res Commun 188:40–47
Hirata R, Ohsumi Y, Nakano A, Kawasaki H, Suzuki K, Anraku Y(1990) Molecular structure of a gene, *VMA1*, encoding the catalytic subunit of H^+-translocating adenosine triphosphatase from vacuolar membranes of *Saccharomyces cerevisiae*. J Biol Chem 265:6726–6733
Iwai K, Ando T (1967) N-O acyl rearrangement. Methods Enzymol 11:262–282
Jencks WP, Cordes S, Carriulo J (1960) The free energy of thiol ester hydrolysis. J Biol Chem 235:3608–3614
Kane PM, Yamashiro CT, Wolczyk DF, Neff N, Goebl M, Stevens TH (1990) Protein splicing converts the yeast *TFP1* gene product to the 69-kD subunit of the vacuolar H^+-adenosine triphosphatase. Science 250:651–657
Kawasaki M, Nogami S, Satow Y, Ohya Y, Anraku Y (1997) Identification of three core regions essential for protein splicing of the yeast Vma1 protozyme. A random mutagenesis study of the entire Vma1-derived endonuclease sequence. J Biol Chem 272:15668–15674
Klabunde T, Sharma S, Telenti A, Jacobs WR, Sacchettini JC (1998) Crystal structure of GyrA intein from *Mycobacterium xenopi* reveals structural basis of protein splicing. Nat Struct Biol 5:31–36
Lew BM, Mills KV, Paulus H (1999) Characteristics of protein splicing in trans mediated by a semisynthetic intein. Biopolymers (Peptide Sci) 51:355–362
Martin DD, Xu M-Q, Evans TC (2001) Characterization of a naturally occurring trans-splicing intein from *Synechocystis* sp. PCC6803. Biochemistry 40:1393–1402
Mathys S, Evans TC, Chute IC, Wu H, Chong S, Benner J, Liu X-Q, Xu M-Q (1999) Characterization of a self-splicing mini-intein and its conversion into autocatalytic N- and C-terminal cleavage elements: facile production of protein building blocks for protein ligation. Gene 231:1–13
Mills KV, Paulus H (2001) Reversible inhibition of protein splicing by zinc ion. J Biol Chem 241:10832–10838
Mills KV, Lew BM, Jiang S-Q, Paulus H (1998) Protein splicing in *trans* by purified N- and C-terminal fragments of the *Mycobacterium tuberculosis* RecA intein. Proc Natl Acad Sci USA 95:3453–3458
Mills KV, Manning JS, Garcia AM, Wuerdeman LA (2004) Protein splicing of a *Pyrococcus abyssi* intein with a C-terminal glutamine. J Biol Chem 279:20685–20691
Mizutani R, Nogami S, Kawasaki M, Ohya Y, Anraku Y, Satow Y (2002) Protein splicing reaction via a thiazolidine intermediate; crystal structure of the VMA1-derived endonuclease bearing the N- and C-terminal propeptides. J Mol Biol 316:919–929
Mizutani R, Anraku Y, Satow Y (2004) Protein splicing of yeast VMA1-derived endonuclease via thiazolidine intermediates. J Synchrotron Rad 11:109–112
Muir TW (2003) Semisynthesis of proteins by expressed protein ligation. Annu Rev Biochem 72:249–289

Nichols NM, Evans TC (2004) Mutational analysis of protein splicing, cleavage, and self-association reactions mediated by the naturally split Ssp DnaE intein. Biochemistry 43:10265–10276

Nichols NM, Benner JS, Martin DD, Evans TC Jr (2003) Zinc ion effects on individual *Ssp* DnaE intein splicing steps: regulating pathway progression. Biochemistry 42:5301–5311

Nogami S, Satow Y, Ohya Y, Anraku Y (1997) Probing novel elements for protein splicing in the yeast VmaI protozyme: a study of replacement mutagenesis and intragenic suppression. Genetics 147:73–85

Noren CJ, Wang J, Perler FB (2000) Dissecting the chemistry of protein splicing and its applications. Angew Chem Int Ed 39:451–466

Paulus H (1998) The chemical basis of protein splicing. Chem Soc Rev 27:375–386

Paulus H (2000) Protein splicing and related forms of protein autoprocessing. Annu Rev Biochem 69:447–496

Paulus H (2001) Inteins as enzymes. Bioorg Chem 29:119–129

Perler FB (2002) InBase, the intein database. Nucleic Acids Res 30:383–384

Perler FB, Xu M-Q, Paulus H (1997) Protein splicing and autoproteolysis mechanisms. Curr Opin Chem Biol 1:292–299

Pietrokovski S (1994) Conserved sequence features of inteins (protein introns) and their use in identifying new inteins and related proteins. Protein Sci 3:2340–2350

Pietrokovski S(1998) Modular organization of inteins and C-terminal autocatalytic domains. Protein Sci 7:64–71

Poland BW, Xu M-Q, Quiocho FA (2000) Structural insights into the protein splicing mechanism of PI-SceI. J Biol Chem 275:16408–16413

Porter JA, Ekker SC, Park WJ, Kessler DP, Young KE, Chen CH, Ma Y, Woods AS, Cotter RJ, Koonin EV, Beachy PA (1996a) Hedgehog patterning activity: role of lipophilic modification mediated by the carboxy-terminal autoprocessing domain. Cell 86:21–34

Porter JA, Young KE, Beachy PA (1996b) Cholesterol modification of hedgehog signaling proteins in animal development. Science 274:255–259

Ramachandran GM, Mitra AK (1976) An explanation for the rare occurrence of cis peptide units in proteins and polypeptides. J Mol Biol 107:85–92

Romanelli A, Shekhtman A, Cowburn D, Muir TW (2004) Semisynthesis of a segmental isotopically labeled protein splicing precursor: NMR evidence for an unusual peptide bond at the N-extein-intein junction. Proc Natl Acad Sci USA 101:6397–6402

Shao Y, Paulus H (1997) Protein splicing: estimation of the rate of O-N and S-N acyl rearrangements, the last step of the splicing process. J Peptide Res 50:193–198

Shao Y, Xu M-Q, Paulus H (1995) Protein splicing: characterization of the aminosuccimide residue at the carboxyl terminus of the excised intervening sequence. Biochemistry 34:10844–10850

Shao Y, Xu M-Q, Paulus H (1996) Protein splicing: evidence for an N-O acyl rearrangement as the initial step in the splicing process. Biochemistry 35:3810–3815

Shingledecker K, Jiang S-Q, Paulus H (1998) Molecular dissection of the *Mycobacterium tuberculosis* RecA intein: design of a minimal intein and of a trans-splicing system involving two intein fragments. Gene 207:187–195

Shingledecker K, Jiang S-Q, Paulus H (2000) Reactivity of the cysteine residues in the protein splicing active center of the *Mycobacterium tuberculosis* RecA intein. Arch Biochem Biophys 375:138–144

Southworth MW, Adam E, Panne D, Byer R, Kautz R, Perler FB (1998) Control of protein splicing by intein fragment reassembly. EMBO J 17:918–926

Southworth MW, Amaya K, Evans TC, Xu M-Q, Perler FB (1999) Purification of proteins fused to either the amino or carboxy terminus of the *Mycobacterium xenopii* gyrase A intein. Biotechniques 27:110–120
Southworth MW, Benner J, Perler FB (2000) An alternative protein splicing mechanism for inteins lacking an N-terminal nucleophile. EMBO J 19:1–8
Southworth MW, Perler FB (2002) Protein splicing of the *Deinococcus radiodurans* strain R1 Snf2 intein. J Bacteriol 184:6387–6388
Southworth MW, Yin J, Perler FB (2004) Rescue of protein splicing activity from a *Magnetospirillum magnetotacticum* intein-like element. Biochem Soc Trans 32:250–254
Telenti A, Southworth M, Alcaide F, Daugelat S, Jacobs WR, Perler FB (1997) The *Mycobacterium xenopi* GyrA protein splicing element: characterization of a minimal intein. J Bacteriol 179:6378–6382
Voorter CEM, de Haard-Hoekman WA, van den Oetelaar PJM, Bloemendal H, de Jong WW (1988) Spontaneous peptide bond cleavage in aging α-crystallin through a succinimide intermediate. J Biol Chem 263:19020–19023
Wang S, Liu X-Q (1997) Identification of an unusual intein in chloroplast ClpP protease of *Chlamydomonas eugametos*. J Biol Chem 272:11869–11873
Wood DW, Wu W, Belfort G, Derbyshire V, Belfort M (1999) A genetic system yields self-cleaving inteins for bioseparations. Nat Biotechnol 17:889–892
Wu H, Hu Z, Liu X-Q (1998) Protein *trans*-splicing by a split intein encoded in a split DnaE gene of *Synechocystis* sp. PCC6803. Proc Natl Acad Sci USA 95:9226–9231
Xu M-Q, Perler FB (1996) The mechanism of protein splicing and its modulation by mutation. EMBO J 15:5146–5153
Xu M-Q, Southworth MW, Mersha FB, Hornstra LJ, Perler FB (1993) In vitro protein splicing of purified precursor and the identification of a branched intermediate. Cell 75:1371–1377
Xu M-Q, Comb DG, Paulus H, Noren CJ, Shao Y, Perler FB (1994) Protein splicing: an analysis of the branched intermediate and its resolution by succinimide formation. EMBO J 13:5517–5522
Yamamoto K, Low B, Rutherford SA, Rajagopalan M, Madiraju MV (2001) The *Mycobacterium avium-intracellulare* complex dnaB locus and protein intein splicing. Biochem Biophys Res Commun 280:896–903
Yamazaki T, Otomo T, Oda N, Kyogoku Y, Uegaki K, Ito N, Ishino Y, Nakamura H (1998) Segmental isotope labeling for protein NMR using peptide splicing. J Am Chem Soc 120:5591–5592

The Structure and Function of Intein-Associated Homing Endonucleases

CARMEN M. MOURE, FLORANTE A. QUIOCHO

1 Introduction

Homing endonucleases are a large group of proteins that are characterized by their ability to recognize long (14–40 base pairs, bp) asymmetric or pseudo-palindromic double-stranded DNA sequences as cleavage sites (Chevalier and Stoddard 2001). By cleaving DNA, they assist in the homing process of their encoding genes into other genes. Homing endonucleases include both intron-encoded and intein- (for intervening protein) associated members that self-splice at the mRNA and protein level, respectively (Belfort and Roberts 1997). Those inteins that contain associated endonuclease domains are especially interesting due to their bifunctionality (Gogarten et al. 2002; Liu 2000) and their exceptionally long DNA recognition abilities. With two exceptions, all of the approximately 160 inteins that have been characterized to date (Perler 2002) are associated with the LAGLIDADG homing endonuclease subfamily, which also includes numerous intron-encoded endonucleases.

This chapter is devoted mainly to the structural studies of intein-associated LAGLIDADG homing endonucleases that have revealed the domain organization and architecture of this type of protein, providing insights into the combination of protein splicing and site-specific DNA recognition and cleavage across a single peptide chain. The knowledge gained in these structure–function studies has been successfully exploited in biotechnology to devise

C.M. Moure (e-mail: mmoure@bcm.tmc.edu)
Department of Biochemistry and Molecular Biology, Baylor College of Medicine, Houston, Texas 77030, USA

F.A. Quiocho (e-mail: faq@bcm.tmc.edu)
Department of Biochemistry and Molecular Biology, Howard Hughes Medical Institute, Baylor College of Medicine, Houston, Texas 77030, USA

Nucleic Acids and Molecular Biology, Vol. 16
Marlene Belfort et al. (Eds.)
Homing Endonucleases and Inteins

systems for protein purification (Xu et al. 2000; David et al. 2004) and to study protein–protein interactions (Ozawa et al. 2000) using the splicing capability of inteins, and in gene targeting studies (Rouet et al. 1994; Jasin 1996; Egli et al. 2004), which use their rare-cutting endonuclease properties. These applications are discussed in other chapters in this volume.

PI-SceI was the first intein and LAGLIDADG homing endonuclease whose structure was elucidated (Duan et al. 1997). PI-SceI, also known as VDE (for VMA1-derived endonuclease), is a 53 kDa (454 amino acids) intein that homes within the *VMA1* gene (encoding the vacuolar H^+-ATPase) of *Saccharomyces cerevisae* (Gimble and Thorner 1992, 1993). Its crystal structures in the apo (PDB entries 1VDE, 1DFA; Duan et al. 1997; Hu et al. 2000) and DNA-bound forms (PDB entries 1LWS, 1LWT; Moure et al. 2002) have been determined. In addition, the structure of the intein endonuclease PI-PfuI (PDB entry 1DQ3), also a 53-kDa protein that is inserted in a ribonucleotide reductase in the archaea *Pyrococcus furiosus*, has been determined in its apo form (Ichiyanagi et al. 2000). Interestingly, PI-PfuI shows binding specificity for Holliday junction DNA, an intermediate in the double-strand break repair pathway that allows the homing process of intein or intron genes (Komori et al. 1999).

This chapter presents the domain organization of inteins and related intron-encoded homing endonucleases. It follows with a discussion of the structure of PI-SceI bound to its DNA substrate. We compare the mechanism of DNA recognition by PI-SceI to other LAGLIDADG homing endonucleases for which the DNA complex has been determined, specifically I-CreI (PDB entry 1G9Y; Jurica et al. 1998; Chevalier et al. 2001) and I-SceI (PDB entry 1R7M; Moure et al. 2003; Chevalier et al., this Vol.). Flexibility of the DNA-binding interactions and divergence in catalysis emerge as features in these structure-function studies of these proteins. The structure–function studies that relate to the splicing mechanism of inteins are discussed extensively by Mills and Paulus (this Vol.) on mechanisms of protein splicing.

2 Architecture of Inteins: A Two-Domain Organization

Despite their divergent primary sequences, the structures of PI-SceI and PI-PfuI (Fig. 1) reveal a common architecture that consists of two domains with independent functions. In PI-SceI, the splicing domain (residues 1–180 and 411–454) is formed by segments from the amino- or N-terminal and carboxy- or C-terminal ends and contains residues involved in protein splicing. The endonuclease domain (residues 181–410) harbors the catalytic sites involved in double-stranded DNA cleavage. This domain is also present in intron-encoded homing endonucleases. The functional independence of the two domains was demonstrated by deleting the endonuclease domain in PI-SceI and PI-

MtuI, the recA intein of *Mycobacterium tuberculosis*, which did not cause significant reduction in protein-splicing activity in vivo, although kinetics of the reaction were not determined (Chong and Xu 1997; Derbyshire et al. 1997). Virtually all inteins that are observed to be associated with homing endonucleases belong to the LAGLIDADG family (except for Ssp GyrB and Npu gyrB of cyanobacteria, which are associated with an HNH endonuclease domain). These inteins should share a common architecture with PI-SceI and PI-PfuI.

The free-standing HO endonuclease, which is responsible for the mating-type switch in *Saccharomyces cerevisiae*, has a strong sequence similarity (50%) to PI-SceI. A model of this protein has been built using the PI-SceI structure as a template (Bakhrat et al. 2004). The HO endonuclease contains

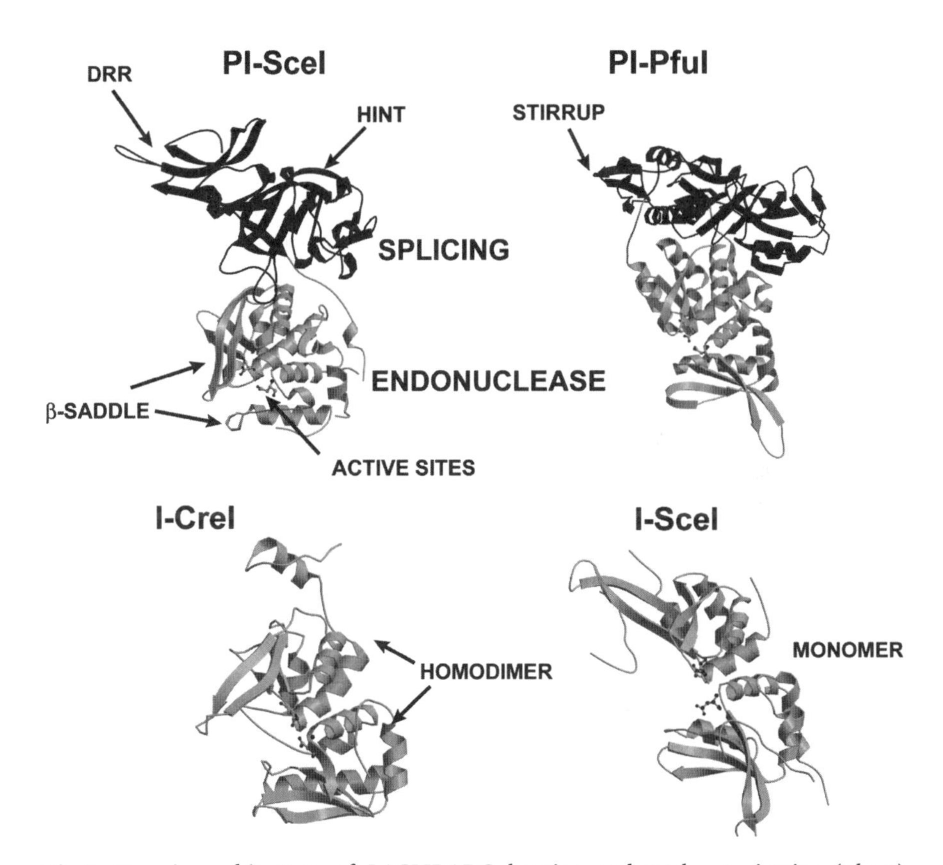

Fig. 1. Domain architecture of LAGLIDADG homing endonucleases, inteins (*above*) and intron-encoded homing endonucleases (*below*). The structures depicted represent the unbound state except for I-SceI for which only the DNA-bound structure has been determined. The domains and the β-saddle DNA-binding motif are labeled in PI-SceI. Two active sites are located at the end of the LADLIDADG helices

all the motifs that typify inteins but does not auto-splice, presumably because of mutations in essential residues in the intein domain active site (experiments to recover splicing activity by reverting these mutations have not been reported). In addition, it contains a 53 residue Zn-finger motif in its C-terminal end that is essential for endonuclease activity.

The two-domain architecture of intein-associated homing endonucleases and the functional independence of their domains suggest that they originated by the invasion of a homing endonuclease gene into a gene that encoded for a self-splicing mini-intein. In this association, both components would benefit from each other (Duan et al. 1997). The intein provides a convenient means of excising itself from the host protein without causing deleterious effects while the homing endonuclease function allows for their mobility and persistence. To this end, an artificial bifunctional intein was made by inserting a homing endonuclease, the intron-encoded I-CreI, into the GyrA mini-intein of *Mycobacterium xenopi* (Fitzsimons Hall et al. 2002).

2.1 The Splicing Domain

The splicing domain resembles a horseshoe and consists almost exclusively of β-strands that have a two-fold pseudo-symmetric arrangement (Fig. 1). This suggests that self-splicing inteins originated from the gene duplication and fusion of two genes that encoded proteins with cleavage activity (Hall et al. 1997). The splicing domain was later shown to have the same fold as the Hedgehog C-terminal autoprocessing domain from *Drosophila melanogaster* (PDB entry 1AT0; Hall et al. 1997), which indicated a common ancestor of both inteins and Hedgehog signaling domains. This fold is now termed the Hint module (for Hedgehog and intein; see Dassa and Pietrokovski, this Vol.). The structures of the mini-intein Mxe GyrA (PDB entry 1AM2; Klabunde et al. 1998) of *Mycobacterium xenopi* that lacks an endonuclease domain and the splicing domain of the Ssp dnaB intein of *Synechocystis* (PDB entry 1MI8; Ding et al. 2003) also contain a Hint fold. Recently, new types of Hint domains which harbor protein-splicing activity but are not inteins have been found in bacteria (Amitai et al. 2003). The conserved residues that are involved in protein splicing in PI-SceI (the N-terminal Cys1, His 79, and C-terminal His 453 and Asn 454) are located within the hydrophobic core of the Hint domain in close proximity to each other. The corresponding residues in PI-PfuI, Mxe GyrA mini-intein, and Ssp DnaB intein are superimposable on PI-SceI, indicating a conservation in the geometry of the splicing catalytic site. The protein-splicing domains in PI-SceI and PI-PfuI contain unique subdomain insertions. For example, in PI-SceI, the DNA recognition region (DRR, residues 90–133) is inserted in the N-terminal region of the splicing domain. This subdomain is composed of three antiparallel β-strands and a single α-helix and contains residues that are critical for DNA binding (He et al. 1998). Moreover, the loop that resides at the interface of the splicing and endonuclease domains is also unique to PI-SceI and is also involved in DNA binding (Wende et al. 2000). In contrast, PI-PfuI contains a stirrup subdomain (residues 339–414) at the interface between the endonuclease and the C-terminal region of the splicing domain. This subdomain contains a three-stranded β-sheet with positively charged residues lining its surface, which suggests a likely DNA-binding activity (Ichiyanagi et al. 2000).

2.2 Endonuclease Domain

The endonuclease domains of PI-SceI and PI-PfuI have an α/β fold with an internal pseudo-two-fold axis (Duan et al. 1997) that is described in detail by Chevalier et al. in the chapter on free-standing LAGLIDADG endonucleases. The domain has been proposed to emerge from gene duplication and fusion as well. The domain is characterized by the presence of two conserved LAGLIDADG sequence motifs that are located along the two helices that define the interface between the N-terminal and C-terminal subdomains. It is similar to the structures of intron-encoded homing endonucleases such as the homodimeric I-CreI (PDB entry 1AF5; Heath et al. 1997) and I-MsoI (PDB entry 1M5X; Chevalier et al. 2003), in which each monomer provides a LAGLIDADG motif, and the monomeric I-DmoI (PDB entry 1B24; Silva et al. 1999) and I-SceI (Moure et al. 2003), which contain two motifs. The structures of I-CreI and I-SceI are depicted in Fig. 1.

3 DNA Binding to the Splicing and Endonuclease Domains of PI-SceI

The X-ray structure of the intein PI-SceI in complex with a 36-bp duplex DNA that contains its 31-bp recognition sequence (Fig. 2) is the only structure determined thus far for an intein-associated homing endonuclease in complex with its DNA target (Moure et al. 2002). The interactions of other inteins with their substrates could show similar features, especially those that have a substantial degree of sequence homology, such as the HO endonuclease and other VMA1 inteins which recognize similar sequences. In addition, the cocrystal structures of several intron-encoded homing endonucleases have also been determined: the homodimeric I-CreI bound to its 22-bp DNA substrate (Fig. 2) and its product reaction (Jurica et al. 1998; Chevalier et al. 2001), I-MsoI (the isoschizomer of I-CreI) bound to a 22-bp substrate (Chevalier et al.

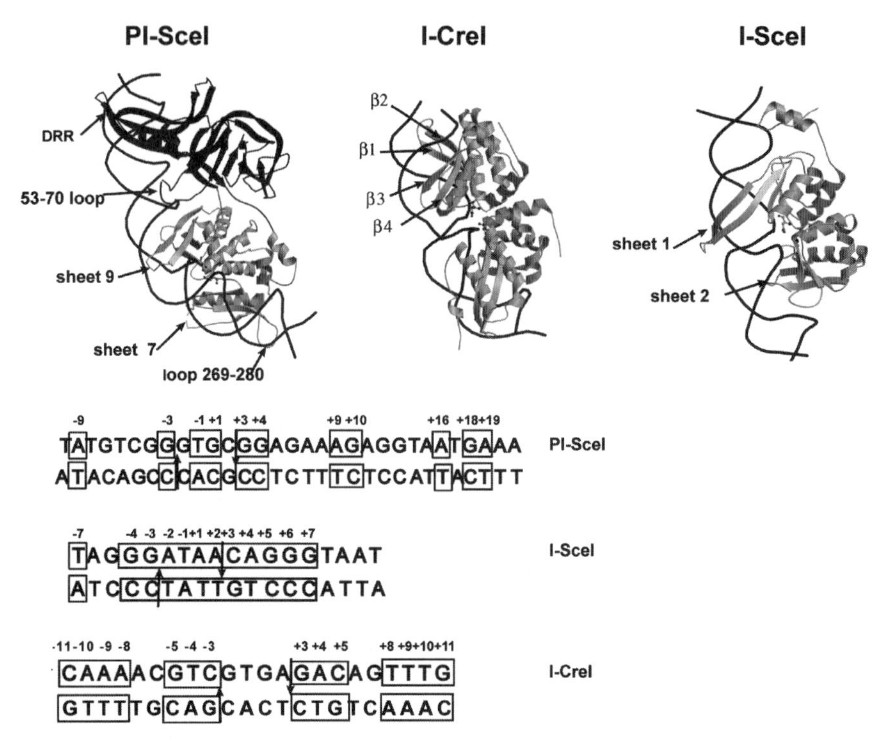

Fig. 2. DNA binding to intein and homing endonucleases. Above DNA complex structures of PI-SceI, I-CreI, and I-SceI. PI-SceI and I-SceI show an asymmetric DNA-binding mode. In I-CreI, DNA binding is symmetric. The bending of DNA bound to PI-SceI is more pronounced than in I-CreI and especially I-SceI. Below DNA recognition sequences for PI-SceI, I-SceI and I-CreI. Essential bases are *boxed*

2003), the monomeric I-SceI bound to a 24-bp substrate (Fig. 2; Moure et al. 2003), and I-AniI, which also functions as a maturase, bound to a 31-bp substrate (PDB entry 1P8K; Bolduc et al. 2003). In this section, we discuss the salient features of DNA recognition by this type of enzyme, including DNA bending and flexibility of the protein–DNA interactions. The comparison of divalent metal binding to the active sites in the high-resolution structures of the symmetric I-CreI and asymmetric I-SceI provides a framework in which we address the catalytic mechanism of homing endonucleases.

3.1 Protein–DNA Contacts Across the Splicing and Endonuclease Domains

The PI-SceI/DNA cocrystal structure (Fig. 2) shows the DNA bound along a large groove that is about 100 Å long that extends from the splicing domain to the endonuclease domain and covers about three and a half turns of the double helix. This is in agreement with biochemical evidence showing that binding of DNA to both domains is necessary for endonuclease activity (Grindl et al. 1998; He et al. 1998). The splicing domain contacts about 17 bp from +6 to +21 while the endonuclease domain contacts 15 bp from −10 to +5 (the region of the DNA target flanking its cleavage sites). In the splicing domain, the DRR and the 53-70 loop interact with the DNA in the major and minor groove, respectively. The DRR and the 53-70 loop in the splicing domain and the 269-284 loop in the endonuclease domain are specific to PI-SceI, other VMA1 inteins, and the HO endonuclease, suggesting that inteins may have developed specificity by acquiring different DNA-binding modules.

Biochemical studies have shown that the splicing domain of PI-SceI is capable of binding DNA independently (Grindl et al. 1998). In fact, the interactions by residues of the DRR region, most notably Arg 94 and His 170, are critical for DNA binding and their mutation abolishes DNA binding completely (He et al. 1998). In contrast, mutations of residues in the endonuclease domain significantly affect only cleavage activity (Gimble and Wang 1996), which supports a coevolution of the two domains. By acquiring the DRR region and the 53-70 loop, the splicing domain of PI-SceI evolved DNA-binding activity and simultaneously the endonuclease domain lost the ability to bind DNA independently. It has not been demonstrated that the splicing domain of other inteins bind DNA. In PI-PfuI, which also binds a long DNA substrate of 30 bp, the positively charged stirrup subdomain has been suggested to bind DNA (Ichiyanagi et al. 2000). Intron-encoded homing endonucleases, which lack a splicing domain, provide binding and catalysis in a single domain, resulting in an overall less extensive binding surface (Fig. 2). However, the length of the β-saddle in PI-SceI is about 40 Å, whereas this surface in I-CreI is about 70 Å long, supporting the idea that PI-SceI lost its extended saddle architecture when it became associated with a protein-splicing domain that acquired DNA-binding activity (Gimble et al. 1998).

3.2 Protein Conformational Changes

Homing endonucleases do not undergo large global conformational changes upon DNA binding, but several regions at the protein–DNA interface become ordered or experience conformational changes (Jurica et al. 1998; Moure et al. 2002). The superimposed unbound and DNA-bound structures of PI-SceI and I-CreI reveal that regions disordered in the unbound structures become ordered only in the presence of DNA. In the DRR of PI-SceI, the 93–102 loop becomes organized, extending the length of the strands and enabling them to interact with the DNA in the major groove. In the endonuclease domain, the 269–284 loop is also disordered in the unbound structures while in the bound structure the loop is ordered and internally stabilized by two hydrogen bonds. The long β-hairpin loop (369–375) that forms part of the saddle undergoes a hinge-flap motion relative to the ligand-free structure, to interact with the major groove of the DNA. The 53–70 loop that straddles between the two domains moves also in a hinge-flap motion to interact with the DNA in the minor groove. This loop is responsible for domain intercommunication, coupling the binding by the DRR region in the splicing domain to DNA distortion and catalysis in the endonuclease domain (Noel et al. 2004). Disulfide bonds inserted between this β-hairpin loop and the 53–70 loop constrain their flexibility and have been used to switch the enzyme between inactive and active DNA-binding conformations to create an endonuclease that can be turned on and off (Posey and Gimble 2002).

In I-CreI, the most prominent change is that the loop that connects the β1 and β2 strands of the saddle adopts a twisted conformation, allowing it to maintain interactions over a long distance of 9 bp. The 138–153 loop, located at the C-terminal end of the protein, also becomes ordered upon DNA binding. Because of the large movements of the regions involved in DNA binding, as well as distortions of the bound DNA (described below), modeling of a canonical DNA to an unbound protein structure might not be an effective way to obtain an accurate picture of the protein–DNA interactions.

3.3 DNA Bending

The bound DNA substrates also experience conformational changes with respect to a canonical B-DNA structure. These conformational changes mirror those of the protein and maximize protein–DNA interactions. The most significant feature of the DNA bound to homing endonucleases is its compression in the minor groove region, which allows for the simultaneous insertion of the top and bottom scissile phosphates into the active sites (see Fig. 3; Jurica et al. 1998; Moure et al. 2002, 2003). The narrowing of the minor groove is the result of the combination of deviations from the canonical B-DNA conformation in the form of negative rolls in the active site regions. Since different endonucleases compress the minor groove in different degrees there is no correlation between the overall degree of bending of the DNA and the dimensions of the minor groove at the active sites. For instance, PI-SceI bends the

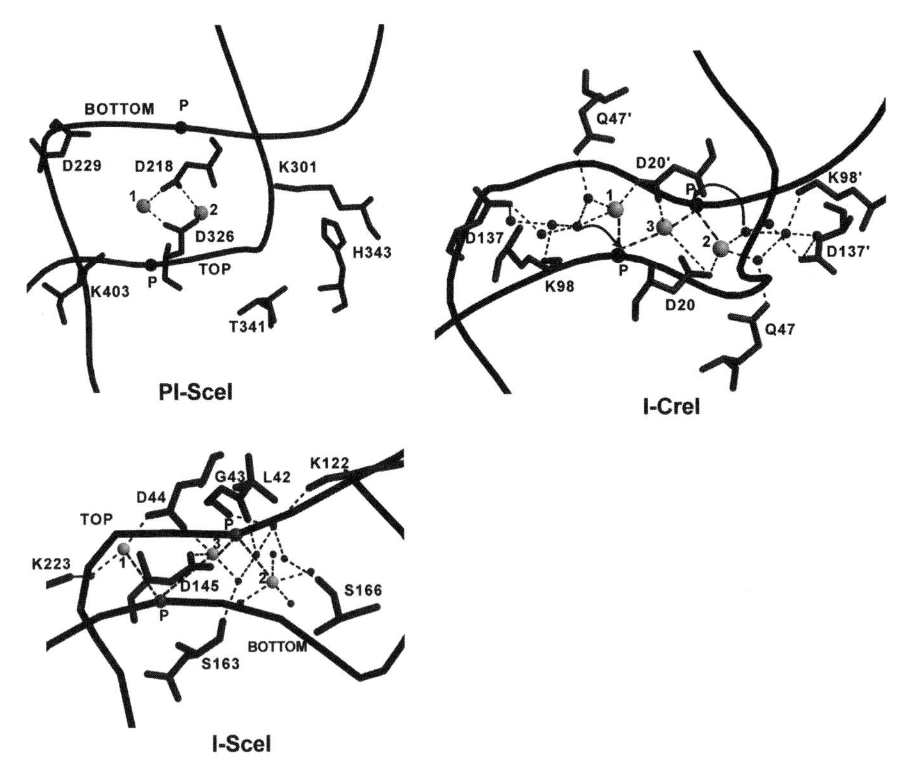

Fig. 3. Active site regions of PI-SceI, I-CreI and I-SceI showing the arrangements of metal ions at the active sites looking down through the minor groove of the DNA. Metal ions are depicted as *light colored spheres* and are numbered; water molecules are represented as *small dark spheres*. The scissile phosphates on the DNA are also labeled. In I-CreI, the nucleophilic waters are identified by *arrows* pointing at the scissile phosphates

DNA much more severely (60°) than I-SceI which shows no appreciable substrate bending. However, the minor groove is more compressed in I-SceI than in PI-SceI (see Fig. 3), resulting in a distance of only 5.5 Å between the scissile phosphates in I-SceI as opposed to 9.8 Å in PI-SceI. The less compressed minor groove in PI-SceI is the consequence of the presence of positive rolls in the immediate downstream region from the active sites that widen the minor groove above standard values (Moure et al. 2002). Another important feature is that the minor groove compression can be symmetric, such as in the homodimeric I-CreI, or asymmetric such as in I-SceI, where the cleavage site of the bottom strand is more buried than that of the top strand (Moure et al. 2003). This feature correlates with differences in the cleavage mechanism of these enzymes (see section 4.2).

3.4 DNA Recognition

Homing endonucleases are remarkably specific enzymes. However, they can tolerate base-pair substitutions at certain positions within their recognition region (Fig. 2), allowing for the degeneracy of their homing sites without decreasing specificity. For instance, in the 31-bp recognition region of PI-SceI, only 11 bases are essential for DNA cleavage activity as revealed by mutation studies on the DNA (Gimble and Wang 1996). This level of recognition is sufficient to ensure a single cut in the yeast genome (Bremer et al. 1992). I-CreI recognizes a pseudo-palindromic 22-bp substrate (Fig. 2) for which inverted repeats of the left or right half-sites are cleaved with similar activities. Estimations of substrate specificity indicate that I-CreI will recognize only one out of a billion random sequences of 22 bp (Argast et al. 1998).

Although free-standing homing endonucleases or the endonuclease domains of inteins are relatively small (200–250 amino acids), they can recognize long DNA substrates. The two β-sheets that form the saddle structure provide an extensive binding surface that can interact over long DNA distances. The β1 and β2 strands in I-CreI (Jurica et al. 1998) and sheet 9 in PI-SceI (Moure et al. 2002) interact with nine consecutive base pairs. The β-strands provide hydrogen-bond donors and acceptors that interact with bases edges in a sequence-specific manner and are critical to preserve the bending of the DNA into the actives sites. For instance, in PI-SceI, distortions of the bound DNA in both the splicing and endonuclease domains occur at locations where base-specific interactions are made (Moure et al. 2002). The different lengths of the β-strands coupled to the low sequence homology of their primary sequences add flexibility to the protein–DNA interface. The undersaturation of the DNA-binding surface contributes to the flexibility of DNA interactions and might explain the recognition of degenerate substrates. In structural studies carried out on I-CreI and I-MsoI, which recognize almost identical sequences, it was found that the divergence of the binding surface was greater than expected from the alignment of the primary sequences and that different sets of amino acids were involved in recognizing almost identical DNA substrates (Chevalier et al. 2003). The same principle may also apply to other homing endonucleases that recognize homologous substrates.

4 DNA Cleavage

4.1 Active Sites

Homing endonucleases cleave their double-stranded substrate DNAs at positions +2/+3 and −2/−3 in the presence of Mg^{2+} and other divalent ions to generate cohesive four-base extension overhangs. DNA cleavage activity has not been demonstrated for all isolated inteins that harbor associated endonuclease domains. Some intein-endonucleases fail to cleave DNA due to accumulated mutations that affect binding or catalysis. The catalytic center of homing endonucleases consists of two overlapping active sites, each defined by a conserved aspartate, that are located at the end of the LAGLIDADG helices (Christ et al. 1999). The aspartates coordinate the divalent metals that act as cofactors in DNA cleavage (Schöttler et al. 2000). Other divalent metals can substitute for Mg^{2+} but not Ca^{2+} which acts as competitive inhibitor of the hydrolytic reaction. Mn^{2+} relaxes the specificity of some inteins, such as PI-SceI (Gimble and Thorner 1992), and rescues the activity of mutants with reduced activities.

When superimposed, the active sites of homing endonucleases show no homology at the primary level except for the catalytic aspartates which bind the essential Mg^{2+} ions and one or two lysines which are also related by a symmetric or pseudo-symmetric two-fold axis (Fig. 3). The lysines have been proposed to act as general bases in the catalytic mechanism. In fact, in I-CreI, Lys 98 and Lys 98′ have been found to be critical for catalysis (Seligman et al. 1997), whereas in PI-SceI only one of them, Lys 301, seems to be essential for cleavage activity (Gimble et al. 1998). Mutations of nearby residues such as Asp 229 and its pseudo-symmetry related His 343 in PI-SceI (Gimble and Wang 1996) and Gln 47 and Gln 47′ in I-CreI (Seligman et al. 1997) also decrease cleavage activity. These residues are not conserved among the active sites of homing endonucleases and inteins whose structures have been solved (Fig. 3), indicating the high divergence of this family. Their participation in hydrogen-bonding water networks, which are likely to vary depending on the protein, might explain their function in catalysis.

4.2 Divalent Metal Binding at the Active Sites

The DNA-bound structure of the homodimeric I-CreI was the first to reveal the presence of three metals at the active sites that are symmetrically distributed (Fig. 3) and are coordinated to the aspartates and scissile phosphates (Chevalier et al. 2001). One of the metals (number 3) is shared between the two aspartates, suggesting two overlapping active sites with a two-metal mechanism, similar to the two-metal mechanism of some restriction endonucleases (Galburt and Stoddard 2002). A water molecule bound to each of the unshared metals acts as nucleophilic water since it is absent in the structure of I-CreI bound to its product. The identity of the general base that abstracts a proton from the nucleophilic water is not known. Instead, the water network that connects the nucleophilic water to nearby residues which are important for cleavage and to the leaving group was proposed to act as a nucleophile. The monomeric I-SceI (Moure et al. 2003) also has three metal ions at the active sites, two unshared and one shared between the aspartates, but, in contrast to I-CreI, they are arranged in an asymmetric manner (Fig. 3). The asymmetric arrangement of the metal ions is due largely to the asymmetric bending of the top and bottom DNA strands resulting in the bottom strand being closer to the active sites. Consequently, the unshared metal ion that coordinates the scissile phosphate of the top strand is buried deep into the minor groove of the DNA and is not coordinated to the either of the aspartates. The PI-SceI/DNA/Ca^{2+} ternary structure was obtained at 3.5 Å resolution (Moure et al. 2002). Despite the low-resolution structure, difference electron density maps clearly showed the distinctive presence of two peaks in the active sites that correspond to two divalent metals that are 4.2 Å apart. The structure of the homing endonuclease and maturase I-AniI also shows two metals at the active sites (Bolduc et al. 2003).

The differences in the number of metal ions and their locations as well as the presence of dissimilar DNA conformations (Fig. 3) and water networks at the active sites suggest divergence in DNA-cleavage mechanisms. For instance, I-CreI, which cannot discriminate between the top and bottom strands of its DNA substrate, cleaves both strands at the same rate which is reflected in the symmetry of the active sites. In contrast, I-SceI, which preferentially cleaves the bottom strand, shows an asymmetric arrangement of metals ions in which the bottom strand cleavage site seems ready to be cleaved first (Moure et al. 2003). The metal bound to the top strand cleavage site is coordinated to DNA bases via water molecule interactions, indicating the possibility of substrate-assisted catalysis. For PI-SceI, the two metals found at the active sites are colinear with the scissile phosphate of the top strand, suggesting that the top strand might be cut first. Although no nicking activity has been detected for PI-SceI, mutational and metal mapping studies (Christ et al. 1999) are consistent with a preference for top strand cleavage. Biochemical experiments show that some inteins (e.g. PI-PabII) cut the DNA in a concerted manner (Saves et al. 2002), while others (e.g. PI-PfuI, PI-MtuI, PI-PabI, and PI-TfuI) cut the DNA in a sequential mode (Ichiyanagi et al. 2000; Guhan and Muniyappa 2002; Saves et al. 2002; Thion et al. 2002). In the apo structure of PI-PfuI, the absence of the residue equivalent to PI-SceI's Lys 301 results in two non-equivalent catalytic sites, which the authors suggest as the source of different rates of cleavage for the top and bottoms strands (Ichiyanagi et al. 2000). It is also possible that all homing endonucleases use a similar mechanism in which both strands are cut simultaneously but with different efficiencies by the two active sites.

References

Amitai G, Belenkiy O, Dassa B, Shainskaya A, Pietrokovski S (2003) Distribution and function of new bacterial intein-like protein domains. Mol Microbiol 47:61–73

Argast GM, Stephens KM, Emond MJ, Monnat RJ Jr (1998) I-PpoI and I-CreI homing site sequence degeneracy determined by random mutagenesis and sequential in vitro enrichment. J Mol Biol 280:345–353

Bakhrat A, Jurica MS, Stoddard BL, Raveh D (2004) Homology modeling and mutational analysis of Ho endonuclease of yeast. Genetics 166:721–728

Belfort M, Roberts RJ (1997) Homing endonucleases: keeping the house in order. Nucleic Acids Res 25:3379–3388

Bolduc JM, Spiegel PC, Chatterjee P, et al (2003) Structural and biochemical analyses of DNA and RNA binding by a bifunctional homing endonuclease and group I intron splicing factor. Genes Dev 17:2875–2888

Bremer MCD, Gimble FS, Thorner J, Smith CL (1992) VDE endonuclease cleaves *Saccharomyces cerevisiae* genomic DNA at a single site: physical mapping of the *VMA1* gene. Nucleic Acids Res 20:5484

Chevalier B, Stoddard BL (2001) Homing endonucleases: structural and functional insight into the catalysts of intron/intein mobility. Nucleic Acids Res 29:3757–3774

Chevalier B, Monnat RJ Jr, Stoddard BL (2001) The homing endonuclease I-CreI uses three metals, one of which is shared between the two active sites. Nat Struct Biol 8:312–316

Chevalier B, Turmel M, Lemieux C, Monnat RJ Jr, Stoddard BL (2003) Flexible DNA target site recognition by divergent homing endonuclease isoschizomers I-CreI and I-MsoI. J Mol Biol 329:253–269

Chong S, Xu MQ (1997) Protein splicing of the *Saccharomyces cerevisiae* VMA intein without the endonuclease motifs. J Biol Chem 272:15587–15590

Christ F, Schoettler S, Wende W, Steuer S, Pingoud A, Pingoud V (1999) The monomeric homing endonuclease PI-SceI has two catalytic centres for cleavage of the two strands of its DNA substrate. EMBO J 18:6908–6916

David R, Richter MP, Beck-Sickinger AG (2004) Expressed protein ligation. Method and applications. Eur J Biochem 271:663–677

Derbyshire V, Wood DW, Wu W, Dansereau JT, Dalgaard JZ, Belfort M (1997) Genetic definition of a protein-splicing domain: functional mini-inteins support structure predictions and a model for intein evolution. Proc Natl Acad Sci USA 94:11466–11471

Ding Y, Xu MQ, Ghosh I, Chen X, Ferrandon S, Lesage G, Rao Z (2003) Crystal structure of a mini-intein reveals a conserved catalytic module involved in side chain cyclization of asparagine during protein splicing. J Biol Chem 278:39133–39142

Duan X, Gimble FS, Quiocho FA (1997) Crystal structure of PI-SceI, a homing endonuclease with protein splicing activity. Cell 89:555–564

Egli D, Hafen E, Schaffner W (2004) An efficient method to generate chromosomal rearrangements by targeted DNA double-strand breaks in *Drosophila melanogaster*. Genome Res 14:1282–1393

Fitzsimons Hall M, Noren CJ, Perler FB, Schildkraut I (2002) Creation of an artificial bifunctional intein by grafting a homing endonuclease into a mini-intein. J Mol Biol 323:173–179

Galburt EA, Stoddard BL (2002) Catalytic mechanisms of restriction and homing endonucleases. Biochemistry 41:13851–13860

Gimble FS, Thorner J (1992) Homing of a DNA endonuclease gene by meiotic gene conversion in *Saccharomyces cerevisiae*. Nature 357:301–306

Gimble FS, Thorner J (1993) Purification and characterization of VDE, a site-specific endonuclease from the yeast *Saccharomyces cerevisiae*. J Biol Chem 268:21844–21853

Gimble FS, Wang J (1996) Substrate recognition and induced DNA distortion by the PI-SceI endonuclease, an enzyme generated by protein splicing. J Mol Biol 263:163–180
Gimble FS, Hu D, Duan X, Quiocho FA (1998) Identification of Lys403 in the PI-SceI homing endonuclease as part of a symmetric catalytic center. J Biol Chem 273:30524–30529
Gogarten JP, Senejani AG, Zhaxybayeva O, Olendzenski L, Hilario E (2002) Inteins: structure, function, and evolution. Annu Rev Microbiol 56:263–287
Grindl W, Wende W, Pingoud V, Pingoud A (1998) The protein splicing domain of the homing endonuclease PI-SceI is responsible for specific DNA binding. Nucleic Acids Res 26:1857–1862
Guhan N, Muniyappa K (2002) *Mycobacterium tuberculosis* RecA intein possesses a novel ATP-dependent site-specific double-stranded DNA endonuclease activity. J Biol Chem 277:16257–16264
Hall TMT, Porter JA, Young KE, Koonin EV, Beachy PA, Leahy DJ (1997) Crystal structure of a hedgehog autoprocessing domain: homology between hedgehog and self-splicing proteins. Cell 91:85–97
He Z, Crist M, Yen H, Duan X, Quiocho FA, Gimble FS (1998) Amino acid residues in both the protein splicing and endonuclease domains of the PI-SceI intein mediate DNA binding. J Biol Chem 273:4607–4615
Heath PJ, Stephens KM, Monnat RJ Jr, Stoddard BL (1997) The structure of I-CreI, a group I intron-encoded homing endonuclease. Nat Struct Biol 4:468–476
Hu D, Crist M, Duan X, Quiocho FA, Gimble FS (2000) Probing the structure of the PI-SceI-DNA complex by affinity cleavage and affinity photocross-linking. J Biol Chem 275:2705–2712
Ichiyanagi K, Ishino Y, Ariyoshi M, Komori K, Morikawa K (2000) Crystal structure of an archaeal intein-encoded homing endonuclease PI-PfuI. J Mol Biol 300:889–901
Jasin M (1996) Genetic manipulation of genomes with rare-cutting endonucleases. Trends Genet 12:224–228
Jurica MS, Monnat RJ Jr, Stoddard BL (1998) DNA recognition and cleavage by the LAGLI-DADG homing endonuclease I-CreI. Mol Cell 2:469–476
Klabunde T, Sharma S, Telenti A, Jacobs WR, Sacchettini JC (1998) Crystal structure of gyrA intein from *Mycobacterium xenopi* reveals structural basis of protein splicing. Nat Struct Biol 5:31–36
Komori K, Fujita N, Ichiyanagi K, Shinagawa H, Morikawa K, Ishino Y (1999) PI-PfuI and PI-PfuII, intein-coded homing endonucleases from *Pyrococcus furiosus*. I. Purification and identification of the homing-type endonuclease activities. Nucleic Acids Res 27:4167–4174
Liu XQ (2000) Protein-slicing intein: genetic mobility, origin, and evolution. Annu Rev Genet 34:61–76
Moure CM, Gimble FS, Quiocho FA (2002) Crystal structure of the intein homing endonuclease PI-SceI bound to its recognition sequence. Nat Struct Biol 9:764–770
Moure CM, Gimble FS, Quiocho FA (2003) The crystal structure of the gene targeting homing endonuclease I-SceI reveals the origins of its target site specificity. J Mol Biol 334:685–695
Noel AJ, Wende W, Pingoud A (2004) DNA recognition by the homing endonuclease PI-SceI involves a divalent metal ion cofactor-induced conformational change. J Biol Chem 279:6794–6804
Ozawa T, Nogami S, Sato M, Ohya Y, Umezawa Y (2000) A fluorescent indicator for detecting protein-protein interactions in vivo based on protein splicing. Anal Chem 72:5151–5157
Perler FB (2002) InBase: the intein database. Nucleic Acids Res 30:383–384

Posey KL, Gimble FS (2002) Insertion of a reversible redox switch into a rare-cutting DNA endonuclease. Biochemistry 41:2184–2190

Rouet P, Smih F, Jasin M (1994) Introduction of double-strand breaks into the genome of mouse cells by expression of a rare-cutting endonuclease. Mol Cell Biol 14:8096–8106

Saves I, Morlot C, Thion L, Rolland JL, Dietrich J, Masson JM (2002) Investigating the endonuclease activity of four *Pyrococcus abyssi* inteins. Nucleic Acids Res 30:4158–4165

Schöttler S, Wende W, Pingoud V, Pingoud A (2000) Identification of Asp218 and Asp326 as the principal Mg^{2+} binding ligands of the homing endonuclease PI-SceI. Biochemistry 39:15895–15900

Seligman LM, Stephens KM, Savage JH, Monnat RJ Jr (1997) Genetic analysis of the *Chlamydomonas reinhardtii* I-CreI mobile intron homing system in *Escherichia coli*. Genetics 147:1653–1664

Silva GH, Dalgaard JZ, Belfort M, van Roey P (1999) Crystal structure of the thermostable archaeal intron-encoded endonuclease I-DmoI. J Mol Biol 286:1123–1136

Thion L, Laurine E, Erard M, Burlet-Schiltz O, Monsarrat B, Masson JM, Saves I (2002) The two-step cleavage activity of PI-TfuI intein endonuclease demonstrated by matrix-assisted laser desorption ionization time-of-flight mass spectrometry. J Biol Chem 277:45442–45450

Wende W, Schottler S, Grindl W, Christ F, Steuer S, Noel AJ, Pingoud V, Pingoud A (2000) Analysis of binding and cleavage of DNA by the gene conversion PI-SceI endonuclease using site-directed mutagenesis. Mol Biol (Mosk) 34:1054–1064

Xu MQ, Paulus H, Chong S (2000) Fusions to self-splicing inteins for protein purification. Methods Enzymol 326:376–418

Harnessing Inteins for Protein Purification and Characterization

Shaorong Chong, Ming-Qun Xu

1 Introduction

Inteins are internal protein segments embedded in various host proteins and are excised by a post-translational process called protein splicing (Perler et al. 1994). During protein splicing, the intein catalyzes the precise cleavage of peptide bonds at its termini and ligation of its flanking protein sequences (termed exteins) without the need for any exogenous factors (Paulus 2000). Characterization of key catalytic residues in the protein-splicing pathway has led to a number of novel strategies for protein research. This chapter is dedicated to a brief review of various intein-related applications in protein purification, and protein/peptide ligation. In particular, we introduce some new approaches in protein manipulation and characterization such as protein *trans*-cleavage, kinase assays, and peptide arrays.

2 Intein Fusion Systems for Protein Purification

Protein purification can be simplified by fusion of the protein of interest to an affinity tag, such as poly-His tag, MBP tag, etc. (LaVallie and McCoy 1995). However, removal of the affinity tag by protease treatment and subsequent purification steps are often costly and time-consuming. The discovery of inteins and the elucidation of protein-splicing mechanisms provided a novel approach to affinity protein purification (Xu et al. 2000). The ability of an intein to catalyze peptide bond cleavage at its termini is the basis for intein-mediated affinity protein purification (Fig. 1). An intein fusion system consists of

S. Chong, M.-Q. Xu (e-mail: xum@neb.com)
32 Tozer Road, New England Biolabs, Inc., Beverly, Massachusetts 01915 USA

Nucleic Acids and Molecular Biology, Vol. 16
Marlene Belfort et al. (Eds.)
Homing Endonucleases and Inteins

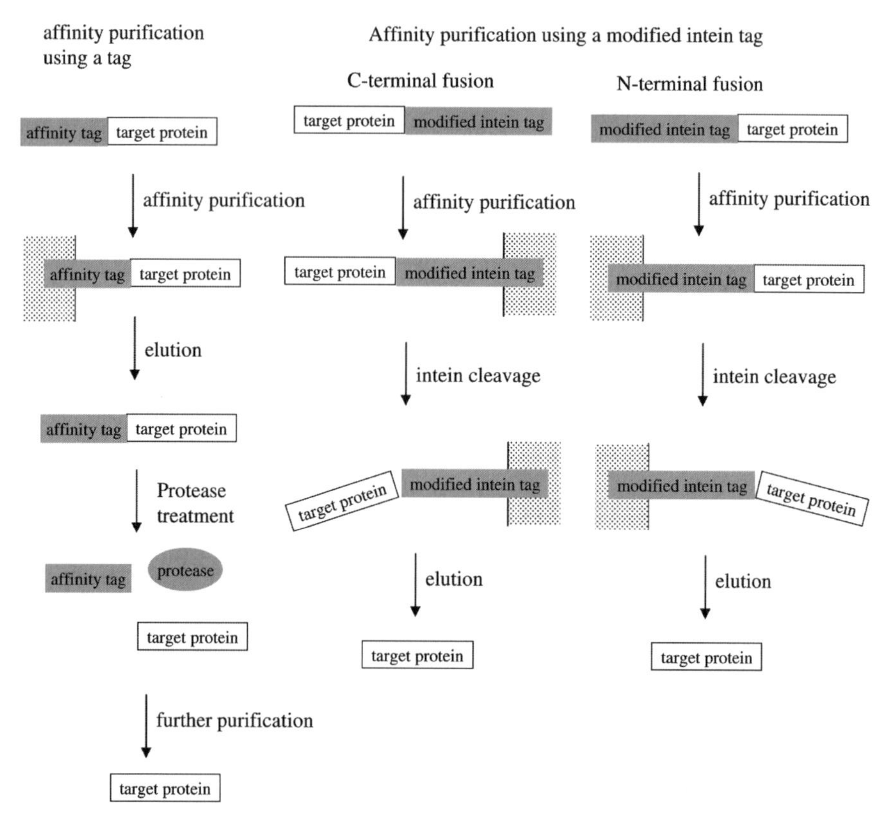

Fig. 1. A schematic comparison between conventional affinity purification and intein-mediated affinity purification

a modified intein fused to the C-terminus (C-terminal fusion) or N-terminus (N-terminal fusion) of a target protein. The intein is linked to an affinity tag and is modified to catalyze peptide bond cleavage at one or both of its termini. The affinity tag immobilizes the intein fusion precursor on an affinity column and a subsequent intein-catalyzed cleavage step releases the target protein from the column-bound affinity tag. Therefore, protein purification using an intein fusion system can lead to elution of a single target protein species after just one chromatographic step (Fig. 2).

Several intein fusion systems have been developed to ensure that the intein-mediated protein purification is applicable to as many target proteins as possible (Wood et al. 1999; Xu et al. 2000). These systems differ by the modified intein tags (the modified intein+affinity tag) and by whether the intein tags are at the N-terminus or at the C-terminus. Two other important differences

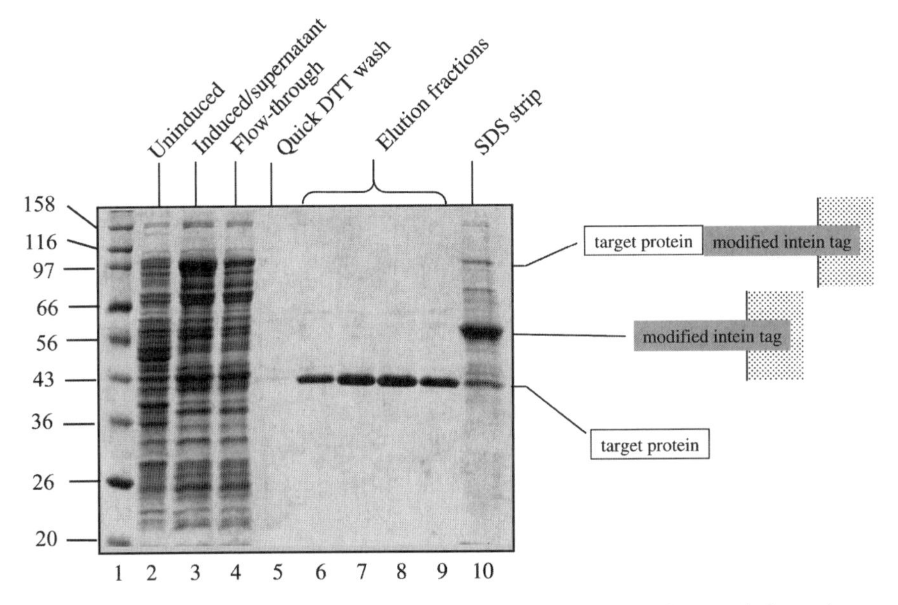

Fig. 2. Single column purification of the recombinant protein HhaI methyltransferase from *E. coli* expressed from a C-terminal intein fusion vector. Samples taken from different steps during the expression and purification procedures were separated by SDS-PAGE and the gel was stained with Coomassie blue. *Lane 1* Protein molecular weight standards (kDa, NEB); *lane 2* uninduced cell extract; *lane 3* induced cell extract; *lane 4* flow through from the load. After loading, the column is washed with column buffer until the protein content of the eluate reaches a minimum. *Lane 5* Flow through from the quick DTT flush; *lanes 6–9* the first four fractions of the elution after overnight incubation at 4 °C in the presence of DTT; *lane 10 a* fraction from the SDS elution of the chitin resin

are the residues at the fusion sites that are favorable for intein cleavage and the conditions under which intein cleavage is induced.

2.1 C-Terminal Fusion System

In a C-terminal fusion system, the C-terminal residue of a target protein is fused to the first N-terminal residue (usually a cysteine) of a modified intein. The intein is mutated to catalyze only an N/S acyl rearrangement at the N-terminal cysteine, which leads to a labile thioester bond linkage between the intein and the target protein (Fig. 3). The modified intein is in turn fused at its C-terminus to a small chitin-binding domain (CBD) from *Bacillus circulans* as an affinity tag, making it possible to purify the fusion precursor in vitro.

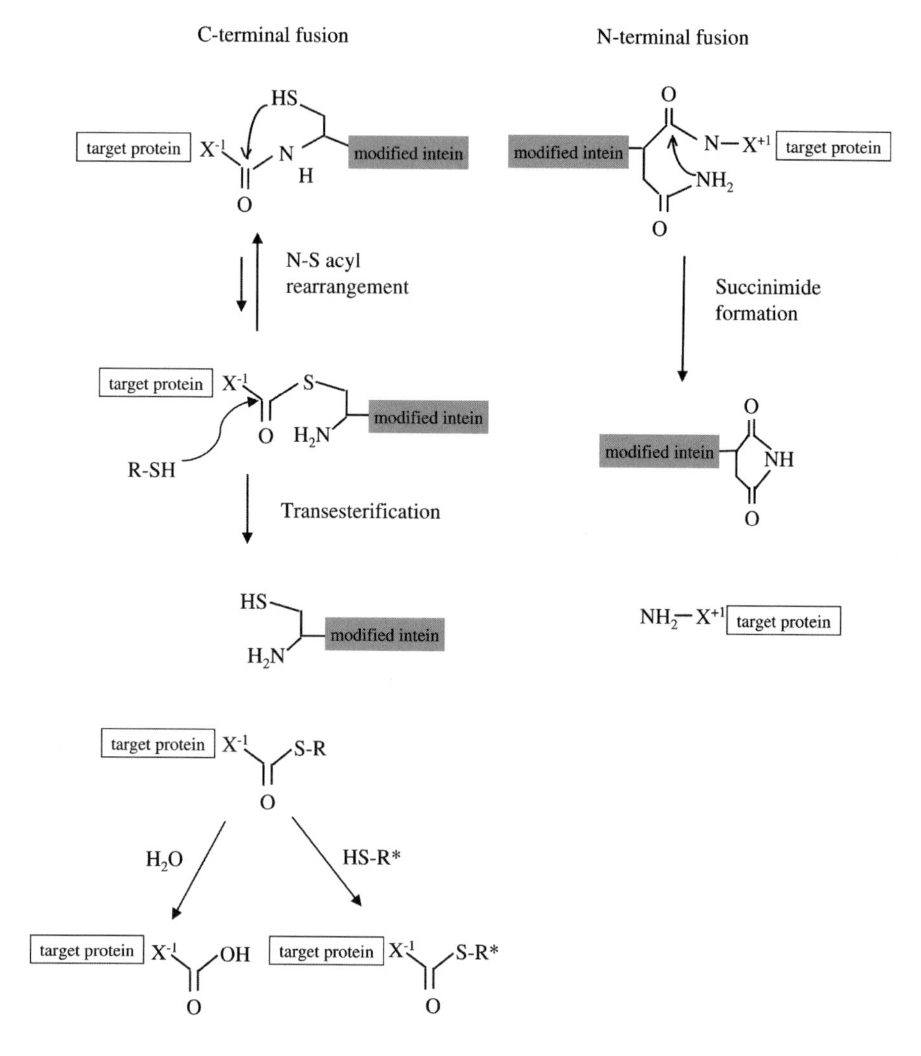

Fig. 3. Chemical mechanism of intein-catalyzed peptide bond cleavage at the fusion junctions between a target protein and a modified intein. In a C-terminal fusion (*left*), the first N-terminal residue, a Cys, of the modified intein is linked to the last C-terminal residue of the N-extein, termed X^{-1} residue, e.g., the last residue of a target protein. The N/S acyl rearrangement generates a thioester linkage which can be cleaved by an exogenous thiol compound, R-SH, resulting a reactive thioester bond at the C-terminus of the target protein. The C-terminal thioester can be hydrolyzed by H_2O to generate the native carboxyl end or attacked by a compound containing a reactive thiol group (SH-R*). This latter reaction is the basis of protein ligation and labeling applications. In an N-terminal fusion (*right*), the last C-terminal residue, an Asn, of the modified intein is linked to the first N-terminal residue of the C-extein, termed X^{+1} residue, e.g. the first residue of a target protein. Under certain conditions, the modified intein catalyzes succinimide formation to release the target protein from the intein moiety

Addition of a thiol compound cleaves the thioester bond at the intein N-terminus by transesterification. This thiol-induced cleavage when performed in vitro on an affinity matrix (e.g., chitin beads) can result in selective elution of only the target protein from the affinity column (Fig. 2).

The 454-residue Sce VMA intein was the first intein engineered for protein purification (Chong et al. 1997). As a C-terminal fusion system, the last residue, Asn 454, of the Sce VMA intein was replaced with an Ala, yielding a modified intein lacking splicing and C-terminal cleavage activities but capable of thioester bond formation at its N-terminus. Addition of thiol compounds, such as dithiothreitol (DTT), β-mercaptoethanol or free cysteine, efficiently cleaved the thioester bond linkage, thereby separating the target protein from the intein tag. Mini-inteins, with smaller molecular weights due to their lack of internal endonuclease domains, were later used for protein purification. These mini-inteins include the 198-residue Mxe GyrA intein and the 134-residue Mth RIR1 intein (Evans et al. 1999b; Southworth et al. 1999). Both inteins were modified by changing the last Asn to Ala to block the splicing reaction and allow the thioester formation at the intein N-termini. DTT and 2-mercaptoethanesulfonic acid (MESNA) were both found to induce the intein cleavage efficiently.

In a C-terminal fusion system, protein synthesis and folding of the upstream target protein determine, to a large extent, the expression and solubility of the whole fusion protein. Some target proteins were found to express well as upstream fusion domains, whereas others expressed better when fused downstream of the modified intein (Chong et al. 1998). To facilitate successful expression of a variety of target proteins, inteins were also modified to generate the N-terminal fusion system.

2.2 N-Terminal Fusion System

In an N-terminal fusion system, the N-terminal residue of a target protein is fused to the last C-terminal residue (usually an Asn) of a modified intein. The cleavage of the peptide bond between the intein and the target protein is achieved by the cyclization of the intein C-terminal Asn (Fig. 3). Based on the proposed protein-splicing mechanism (Paulus 2000), the nucleophilic attack of the Asn β-amide N on the carbonyl C of the peptide bond leads to cyclization of the Asn and concomitant peptide bond cleavage. Spontaneous cyclization of Asn residues coupled to peptide cleavage occurs at a low rate in proteins under physiological conditions (Clarke 1987). A unique feature of inteins is to allow efficient peptide bond cleavage at their C-terminal Asn either independently or in the context of protein splicing. In an N-terminal fusion system for protein purification, the inteins are modified to minimize the C-terminal cleavage in vivo, and, when the fusion protein is immobilized on an affinity matrix, the C-terminal cleavage is induced by addition of a thiol compound or a change in pH and/or temperature.

The Sce VMA intein was also modified to allow inducible C-terminal cleavage (Chong et al. 1998). The first C-extein residue Cys 455 was mutated to Ala to block the splicing reaction. Since the conserved first and last intein residues Cys 1 and Asn 454 were not mutated, the intein retained the cleavage activities at both its termini. The penultimate His 453 of the intein was then replaced by Gln with the purpose of attenuating in vivo cleavage activity. As expected, the Sce VMA intein containing this double mutation exhibited very limited in vivo cleavage at either terminus. Remarkably, the modified intein in the purified fusion protein catalyzed efficient in vitro cleavage at both termini in the presence of thiols such as DTT or free cysteine. The cleavage reaction was more efficient at 23 °C than at 4 °C and was inhibited at pH 6.0 and below. It seemed that the thiols induced N-terminal cleavage prior to C-terminal cleavage. Exactly how C-terminal cleavage was induced by an upstream cleavage event remained unclear. Nevertheless, the modified Sce VMA intein has been successfully used as an N-terminal fusion system to purify proteins, including proteins whose purification was unsuccessful in a C-terminal fusion system (Chong et al. 1998).

The N-terminal fusion system also includes several modified mini-inteins: Ssp DnaB, Mtu RecA and Mth RIR1 mini-inteins (Wu et al. 1998b; Evans et al. 1999b; Wood et al. 1999). Both Ssp DnaB (154-residue) and Mtu RecA mini-inteins (168-residue) were derived from full-length inteins by deletion of the endonuclease domains, whereas the Mth RIR1 intein is a natural 134-residue mini-intein. To achieve C-terminal cleavage, the first N-terminal Cys of all three mini-inteins was replaced by Ala. In the case of Ssp DnaB and Mth RIR1 mini-inteins, this single substitution was sufficient to allow the C-terminal cleavage to occur in a pH- and temperature-dependent manner. The cleavage reaction was most favored at pH 6.0–7.0, inhibited at pH 8.5, and was accelerated with increasing temperatures (4–25 °C). For the Mtu RecA mini-intein, additional mutations were made by a selection system to achieve efficient C-terminal cleavage (Wood et al. 1999). The cleavage activity of the modified Mtu RecA mini-intein was remarkably pH- and temperature-sensitive. The cleavage rates increased sharply as the pH was reduced (from 8.5 to 6.0) and the temperature was elevated.

2.3 Choosing an Appropriate Intein Fusion System

One reason for the development of different intein fusion systems (Fig. 4) is to accommodate the expression, folding and purification requirement of a variety of recombinant proteins. The C-terminal intein fusion system has the advantage of generating a target protein with an unmodified N-terminus and, for protein labeling and ligation applications, a reactive thioester bond at its

C-terminal intein fusion constructs

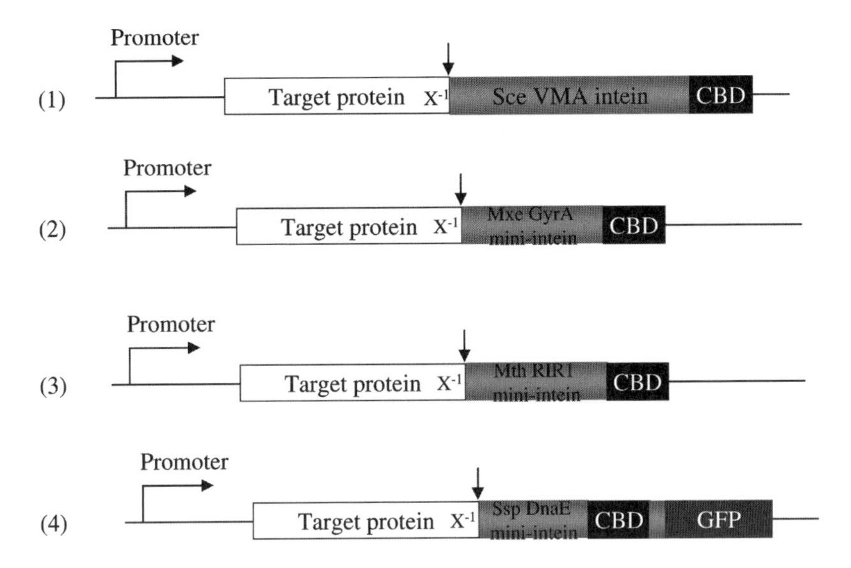

N-terminal intein fusion constructs

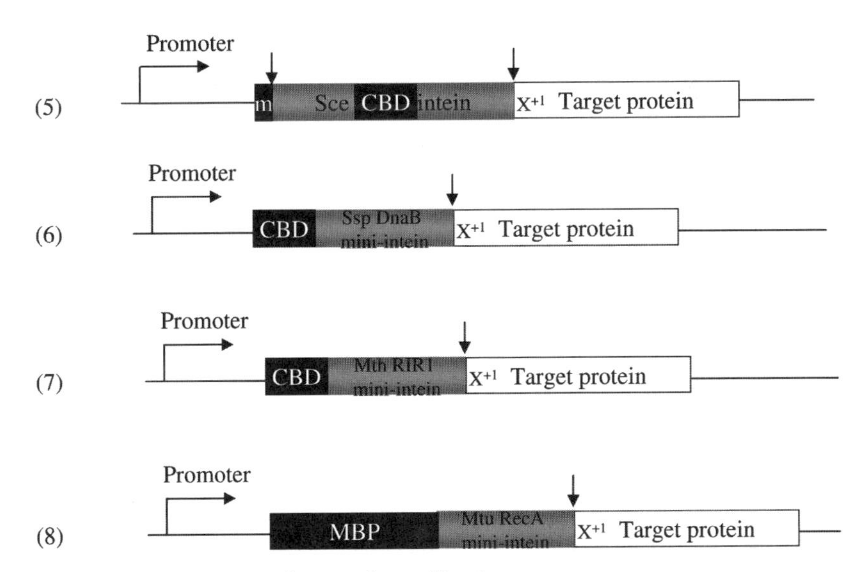

Fig. 4. Intein fusion constructs for protein purification

C-terminus. For many target proteins, use of a smaller intein, e.g., the Mxe GyrA mini-intein, results in a higher expression level and yield. Nevertheless, the Sce VMA intein fusion system has been shown to work well in eukaryotes such as insect cells (Pradhan et al. 1999).

The N-terminal intein fusion system has the advantage of generating a target protein with an N-terminal residue other than Met and, in the case of mini-inteins, inducing intein cleavage without using a thiol compound. In addition, the N-terminal intein fusion allows the translation initiation to be optimized. For instance, the Sce VMA intein fusion system [Fig. 4, construct (5)] encodes the first ten residues from MBP at its N-terminus to ensure a good translational start (Chong et al. 1998). The Mtu RecA mini-intein system [Fig. 4, construct (8)] encodes the entire MBP as an N-terminal domain to enhance protein expression and solubility. The only N-terminal fusion system that uses a thiol compound to induce C-terminal cleavage is that of the modified Sce VMA intein. However, this system cannot generate a target protein with an N-terminal Cys, as Cys at the fusion site (X^{+1} position) results in protein splicing rather than the C-terminal cleavage. Other N-terminal fusion systems contain mini-inteins (Fig. 4), which allow induction of the intein C-terminal cleavage by a shift in pH and temperature. Furthermore, these systems are capable of producing target proteins with an N-terminal Cys, useful for protein ligation and other applications.

Fusion of a target protein upstream or downstream of a modified intein tag would change the dynamics of protein synthesis and folding. This may explain the observation that the N- and C-terminal intein fusions often gave quite different expression and purification results for a given target protein (Chong et al. 1998). There are no sufficient data to support one fusion system over the other for all target proteins. Successful expression and purification of a target protein depend not only on its location in the fusion protein, but also on the type of the intein tag. In addition, the residue at the fusion junction between the target protein and the modified intein-tag sometimes plays a critical role.

2.4 Choosing an Appropriate Residue at the Fusion Junction

The active site of an intein is located at its termini. Fusion of a target protein to either terminus of an intein places the N- or C-terminal residue of the target protein adjacent to the catalytic residues of the intein. In an intein fusion protein, the target protein residues adjacent to the intein N-terminal and C-terminal residues are named minus 1 residue (X^{-1}) and plus 1 residue (X^{+1}), respectively (Figs. 3 and 4). The X^{-1} and X residues exhibit more effect on the intein cleavage efficiency than any other residues further away from the intein cleavage sites. The most favorable and unfavorable residues for intein cleavage in different fusion systems are listed in Table 1. The complete lists including all 20 amino acids for certain intein fusion systems can be found in the original references (see Table 1). It should be noted that these data were generated by examining a particular test protein. Therefore, the conclusion may not be completely applicable to other target proteins. Nevertheless, the effect of some amino acid residues at the fusion sites has been proven consistent in multiple target proteins. For instance, Asp as the X^{-1} residue invariably results in almost complete in vivo cleavage in a C-terminal fusion system, whereas Pro as the X^{-1} or X^{+1} residue leads to inhibition of the intein cleavage in the C-terminal or N-terminal fusion system, respectively. For a target protein with a C-terminal Asp or Pro, the N-terminal fusion system rather than the C-terminal fusion system should be used in order to obtain a recombinant protein with a native sequence. If addition of one or more extra residues at the target protein termini is not expected to have an adverse effect on its structure and activity, both N-terminal and C-terminal fusion systems can be used. Addition of one or two residues favorable for intein cleavage (Table 1) can often enhance significantly both the expression level of the fusion protein and the efficiency of intein cleavage.

2.5 Conditions for Intein Cleavage

A unique advantage of the intein fusion system is the separation of a target protein from its affinity tag by inducing intein cleavage under mild conditions. Different intein fusion systems require different cleavage conditions (Table 1). Thiol reagents such as DTT and (less expensive) β-mercaptoethanol can induce efficient cleavage at around pH 8.0. For recombinant proteins that are sensitive to reducing reagents such as DTT, free cysteine can be used, which can attach to the C-terminus of the target protein through a peptide bond after cleavage. This modification of the target protein C-terminus occurs only when the C-terminal fusion system is used. The target protein is not modified by free cysteine when the Sce VMA intein is used in the N-terminal fusion system. After inducing intein cleavage, DTT also modifies the target protein by forming a thioester bond with the C-terminal residue. At pH 8.0, the DTT tag is not stable in solution and is normally hydrolyzed. For applications such as protein labeling and ligation, a stable and reactive thioester bond at the C-terminus of a target protein is desirable. It was found that MESNA could also form a reactive thioester that allows for more efficient ligation reaction than when DTT was used (Evans and Xu 1999). Since MESNA induces more efficient cleavage by the modified Mxe GyrA mini-intein than the Sce VMA intein, the Mxe GryA mini-intein and MESNA are recommended for purification of recombinant proteins to be used for protein labeling and ligation.

Table 1. Appropriate residues at the intein fusion sites and the conditions to induce intein cleavage

Intein fusion constructs	Modified intein tags	Residues preferred at fusion sites	Residues to be avoided at fusion sites	Conditions for intein cleavage
C-terminal fusion	Sce VMA intein CBD	Gly, Ala	Arg, Asp, Asn, Cys, Pro	Addition of DTT, free cysteine, or MESNA, pH 8–8.5 at 4–25 °C
	Mxe GyrA mini-intein CBD	Met, Tyr, Phe	Asp, Pro	
	Mth RIR1 mini-intein CBD	Gly, Ala	Pro	
	Ssp DnaE mini-intein (CBD)-GFP	Gly, Ala	Asp, Val, Pro	
N-terminal fusion	Sce VMA intein (CBD)	Ala, Met, Gly	Ser, Cys, Pro	Addition of DTT, free cysteine, pH 8–8.5 at 4–25 °C
	CBD-Ssp DnaB mini-intein	Cys, Cys-Arg	Pro, Lys, Arg, Ile, Leu, Asn, Gln	pH 6.0 at 4–25 °C
	CBD-Mth RIR1 mini-intein	Cys	Not determined	
	MBP-Mtu RecA mini-intein	Met, Cys	Not determined	

Note: In an intein fusion protein, the target protein residues adjacent to the intein N- and C-terminal residues are named minus 1 residue (X^{-1}) and plus 1 residue (X^{+1}), respectively (Figs. 3 and 4). In the C- and N-terminal fusion systems, residues preferred at fusion sites listed in this table are the minus 1 or plus 1 residues, respectively. The complete lists including all 20 amino acid residues at fusion sites for certain intein fusion systems can be found in the original references (Chong et al. 1997, 1998; Southworth et al. 1999; Zhang et al. 2001)

2.6 Intein Fusion Systems for High-Throughput and Large-Scale Applications

Though intein fusion systems have simplified affinity purification, intein fusion proteins have to be expressed in a soluble, correctly folded form in order for intein-mediated purification to be effective. Fusion proteins overexpressed in *E. coli* sometimes misfold resulting in very low expression or inclusion bodies. It would be advantageous if the level of soluble expression of an intein fusion protein is known before the cell lysis and electrophoretic analyses steps. This is especially desirable if a large number of recombinant proteins are to be expressed and purified. The intein fusion system potentially useful for high-throughput protein purification is a C-terminal fusion system consisting of a modified Ssp DnaE mini-intein whose C-terminus is fused to green fluorescent protein (GFP) as an expression and solubility reporter (Zhang et al. 2001; Fig. 4). A CBD domain is inserted between the N- and C-terminal domains of the Ssp DnaE mini-intein, which is modified to allow thiol-inducible cleavage at its N-terminus. A variety of target proteins were expressed and purified in this intein-GFP fusion system (Zhang et al. 2001). The data suggested a positive linear correlation between GFP fluorescence of induced cultures and the final target protein yields after intein-mediated purification. The intein-GFP system can be used to screen for culture conditions, strain/vector variants, mutations, etc., that improve soluble expression of a particular fusion protein. Once an optimized condition is found, the target protein can be directly purified on a single column. Alternatively, the system can be used to screen for soluble expression of a large number of recombinant proteins for rapid purification in high-throughput experiments.

Since intein fusion systems purify proteins on an affinity matrix and remove the affinity tag in the same chromatographic step, it has the potential to significantly reduce recovery costs for the industrial production of recombinant proteins. Studies have shown that intein fusion proteins remain stable in vivo in *E. coli* during high-cell-density fermentation and can catalyze efficient peptide bond cleavage (Sharma et al. 2003). It is therefore feasible to use intein fusion systems for industrial-scale production of recombinant proteins.

3 Protein *trans*-Splicing and Cleavage Systems

The N- and C-terminal segments of an artificially or naturally split intein, each segment being fused to a foreign protein sequence, are able to assemble and mediate protein *trans*-splicing, yielding a native peptide bond between the two foreign protein sequences (Lew et al. 1998; Mills et al. 1998; Southworth et al. 1998; Wu et al. 1998a,b; Yamazaki et al. 1998; Evans et al. 2000). In particular, studies of a naturally occurring split intein from the *dna*E gene of *Synechocystis* sp. PCC6803 (Ssp DnaE intein; Wu et al. 1998a) have led to its use in protein ligation, cyclization and purification. It has been shown that the N-terminal 123-residue segment and the C-terminal 36-residue segment of the Ssp DnaE intein associate with high affinity and mediate proficient ligation of two target protein sequences through a native peptide bond in vivo or under mild conditions in vitro (Evans et al. 2000; Martin et al. 2001). The *trans*-splicing approach has been employed to split a target gene in two halves for expressing the target gene into two primary translational products. The functional target gene product can be subsequently reconstituted in vivo and in vitro. One successful example is the reconstitution of herbicide resistance from a split gene encoding 5-enolpyruvylshkimate-3-phosphate synthase (EPSPS) in transgenic plants (Chin et al. 2003). Another utilization of the *trans*-splicing approach is in vivo cyclization of proteins and peptides, which aims to produce presumably more stable circular polypeptides in cells. The split intein-mediated circular ligation of peptide and proteins (SICLOPPS) was achieved by sandwiching a target protein or peptide between the C- and N-terminal intein halves such that protein splicing during protein expression ligates the two termini of the target protein/peptide, yielding circular proteins or peptides in vivo (Scott et al. 1999; Evans et al. 2000).

The *trans*-splicing activity can be modulated to a *trans*-cleavage activity by employing a similar mutational strategy as that of the *cis*-splicing inteins (Chong et al 1997; Nichols et al. 2003). For instance, changing the last Asn residue of the C-terminal fragment of the split Ssp DnaE mini-intein to Ala converts a *trans*-splicing system into a *trans*-cleavage system (Fig. 5). A target protein fused to the N-terminal fragment can be expressed as a stable product. Addition of the intein C-terminal fragment containing the Asn-to-Ala mutation led to the reconstitution of the intein cleavage activity and, under inducing conditions, the release of the target protein. Use of the *trans*-cleavage fusion system may avoid the problem of in vivo premature cleavage of the intein-tag occasionally encountered by other intein fusion systems.

→

Fig. 5. Protein *trans*-cleavage utilizing the split Ssp DnaE intein. **a** diagram illustrating protein *trans*-cleavage. The protein of interest is fused to the intein N-terminal 123 amino acid residues, *In(N)*. The intein C-terminal fragment of 36 amino acid residues, *In(C)*, carries a substitution of the intein's last residue, Asn159 with an Ala, which blocks splicing activity and cleavage at the C-terminal splice site. The fusion of a small chitin-binding domain *(CBD)* to each intein segment facilitates the isolation of the expressed intein

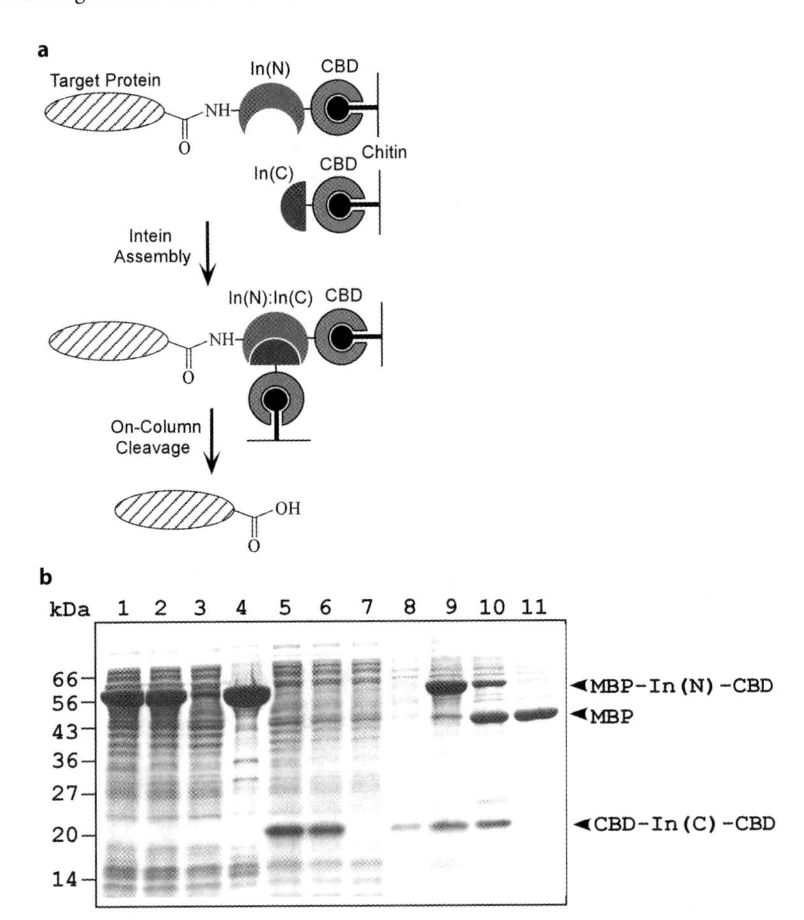

fusion proteins by binding to chitin resin. Protein *trans*-cleavage is initiated by mixing the two chitin-bound protein fractions, resulting in the reconstitution of a cleavage-proficient intein under native conditions. The target protein is released upon overnight incubation at 4 °C in column buffer containing 20 mM Tris-HCl (pH 7.0), 0.5 M NaCl and 30 mM DTT. The *trans*-cleavage product can be used for IPL by the replacement of DTT with 2-mercaptoethanesulfonic acid (MESNA). **b** Example of protein *trans*-cleavage resulting in single chitin column purification of maltose-binding protein (MBP). MBP-In(N)-CBD fusion protein and CBD-In(C)-CBD fusion protein were expressed in *E. coli* from pMEB14 and pBEB, respectively, and each was isolated by chitin resin. The cleavage reaction was initiated by mixing the chitin-bound protein fractions in the presence of 30 mM DTT at 4 °C. Samples taken from different purification steps were analyzed by 12% SDS-PAGE stained with Coomassie Blue. *Lane 1* Induced cell extract of MBP-In(N)-CBD (MEB); *lane 2* clarified cell extract of MEB; *lane 3* MEB extract after passage through a chitin column; *lane 4* chitin-bound MEB after washing with a column buffer (10 mM Tris-HCl, pH 8.5, 0.5 M NaCl); *lane 5* induced cell extract of CBD-In(C)-CBD (BEB); *lane 6* clarified cell extract of BEB; *lane 7* BEB extract after passage through a chitin column; *lane 8* chitin-bound BEB after washing with column buffer; *lane 9* a fraction of mixed MEB and BEB; *lane 10* a fraction of the mixed chitin resin after the 4 °C overnight incubation in the presence of DTT; *lane 11* the supernatant containing the MBP product after centrifugation of the chitin resin shown in *lane 10*

4 Intein-Mediated Protein Ligation (IPL)

Unlike other affinity fusion systems that mainly serve to simplify protein purification, the intein fusion systems can be modulated to perform numerous other tasks, one of which is intein-mediated protein ligation (IPL; Evans et al. 1998; Evans and Xu 1999) or expressed protein ligation (Muir et al. 1998; Severinov and Muir 1998). The IPL technique has been used for incorporation of non-coded amino acids into a protein sequence, segmental labeling of proteins for NMR analysis, addition of fluorescent probes to create biosensors, and synthesis of cytotoxic proteins (Cotton and Muir 1999, 2000; Otomo et al. 1999; Xu et al. 1999). These applications utilize essentially the same purification protocols. By altering cleavage conditions, the intein fusion systems can generate a protein with a reactive C-terminal thioester or an N-terminal cysteine. The facile generation of thioester-tagged recombinant proteins has greatly expanded the utility of native chemical ligation (Dawson et al. 1994; Tam et al. 1995). The latter method was developed for fusion of two synthetic peptides, one possessing a C-terminal thioester and the other possessing an N-terminal cysteine. The ability of IPL to generate large recombinant protein substrates under mild conditions circumvents the problem associated with synthetic peptide intermediates which are usually limited to less than 100 amino acids. Further advance of the IPL method has been achieved by employing the intein C-terminal cleavage system to generate proteins with an N-terminal cysteine (Evans et al. 1999b; Mathys et al. 1999). The utility of various intein fusion systems has contributed to an alternate approach in backbone cyclization of large recombinant proteins as well as peptides. Cyclic proteins/ peptides can be generated by creating an N-terminal cysteine residue on the same protein possessing a C-terminal thioester (Evans et al. 1999a; Iwai and Pluckthun 1999; Xu et al. 1999). The following sections describe some of our new applications of IPL technology.

4.1 Intein-Mediated Peptide Array (IPA)

Peptides are widely used to generate antibodies, define antibody epitopes and determine substrate specificities of protein modification enzymes such as protein kinases and phosphatases. Binding of peptides to commonly used membranes such as nitrocellulose is ineffective and variable, resulting in low sensitivity and inconsistency. A novel method for making arrays of synthetic peptides has been developed by ligating the peptide substrates to a carrier protein which is generated by the intein fusion system and has a high binding affinity for nitrocellulose (Sun et al. 2004). For example, when the synthetic peptide antigens were ligated to a thioester-tagged methylase from

Haemophilus haemolyticus (M.Hha, 39 kDa) or the paramyosinΔSal fragment (27 kDa) from *Dirofilaria immitis*, the use of the ligation products resulted in improved retention of peptides and an increase in sensitivity of up to 10^4-fold in immunoassay and epitope scanning experiments (Fig. 6). Because the carrier protein contains one peptide reactive site and dominates the binding of ligated products, the amount of peptides arrayed onto the membranes can be

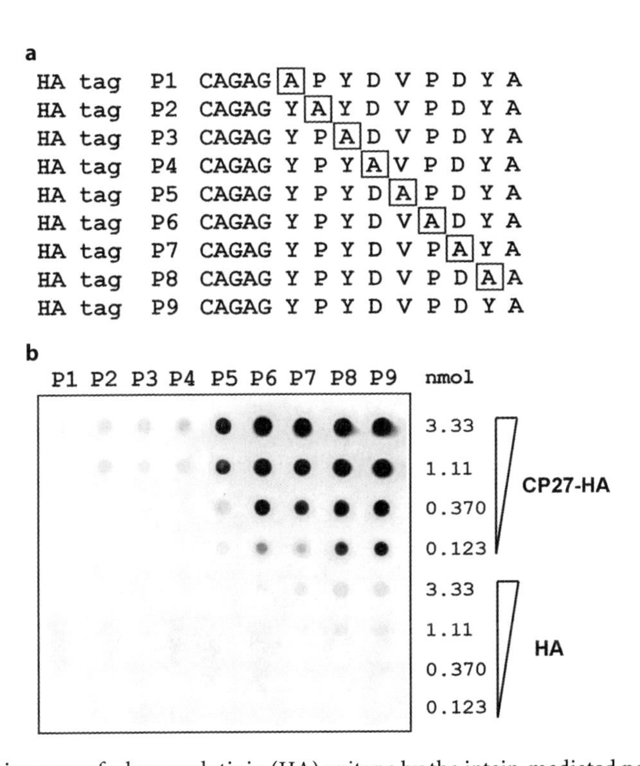

Fig. 6. Alanine scan of a hemagglutinin (HA) epitope by the intein-mediated peptide array (IPA) method (Sun et al. 2004). **a** An alanine-scanning HA peptide library was synthesized with an N-terminal cysteine. P9 contains the wild-type sequence corresponding to residues 98 to 106 of HA protein (YPYDVPDYA). Each of the other eight peptides carries a single substitution with an alanine residue. Each peptide (0.5 mM) was ligated to thioester-tagged paramyosin (25 μM) at 4 °C overnight. **b** The ligated products (*rows 1–4*) and the unligated peptides (*rows 5–8*) were threefold serially diluted in phosphate-buffered saline (PBS) and arrayed on a 0.45-μm nitrocellulose membrane. Immunoblotting was performed with anti-HA monoclonal antibody (Zymed Laboratories) and detection was with the Phototope-HRP Western blot detection system (Cell Signaling Technology). The amount of total peptide present in each sample is indicated on the right side. The data indicated that the residues mutated in P1, P2, P3, and P4 are essential for antibody recognition and the unligated peptides did not generate a detectable signal

effectively normalized. This IPL generated peptide array (IPA) approach permits synthetic peptides to be uniformly arrayed onto commonly used membranes (i.e., nitrocellulose, Nylon, and PVDF) thereby making it convenient and economical to produce peptide arrays in a research laboratory.

4.2 Kinase Assays Using Carrier Protein–Peptide Substrates

Another example of the application of IPL is the generation of improved peptide substrates for kinase assays and subsequent Western blot analysis and array (Fig. 7). Synthetic peptide substrates have become a convenient tool to determine kinase specificities and to conduct mutational analysis but they are not suitable for Western blot analysis due to their small size. IPL allows for ef-

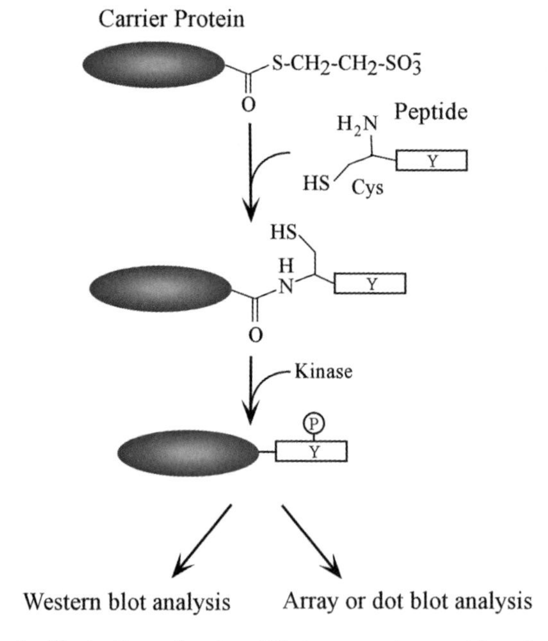

Fig. 7. A schematic illustration of using IPL to generate peptide substrates. A carrier protein possessing a cysteine-reactive C-terminal thioester is released following intein-mediated cleavage by the addition of 2-mercaptoethanesulfonic acid (MESNA). Ligation of this carrier protein to a synthetic peptide containing an N-terminal cysteine as well as a potential phosphorylation site (tyrosine or Y) results in a native peptide bond between the two reacting species. The ligated carrier protein–peptide product serves as substrate for kinase assays and subsequent Western blot analysis with a phospho-tyrosine antibody (Xu et al. 2004). Phosphorylation of the tyrosine residue in the peptide sequence (indicated by a *circled P*) results in a positive signal

ficient ligation of a synthetic peptide substrate with an N-terminal cysteine residue to an intein-generated carrier protein (with a size range of 20–40 kDa; Evans and Xu 1999). A distinct advantage of this procedure is that since each carrier protein molecule ligates only one peptide, the resulting ligation product migrates as a single band on sodium dodecyl sulfate/polyacrylamide gel electrophoresis (SDS-PAGE). The peptide substrates may contain a desired amino acid substitution or chemical modification. This design has led to the mutational analysis of the peptide substrates derived from human cyclin-dependent kinase, Cdc2, that contains a phosphorylation site for human c-Src protein tyrosine kinase (Xu et al. 2004). A peptide possessing a phosphorylation site of interest is synthesized with an amino-terminal cysteine residue. The peptide is ligated to the cysteine-reactive carboxyl terminus of a carrier protein via a peptide bond. Following assays with the Src kinase, protein phosphorylation is subsequently examined by Western blot analysis with a phospho-specific antibody that recognizes the phosphorylated tyrosine epitopes.

4.3 Purification of Peptide-Specific Antibody

Affinity purification of a peptide-specific antibody relies on conjugation of peptide antigens to agarose or other matrices. This procedure is often time-consuming and generates chemical waste. By taking advantage of the IPL technique and the high affinity of the chitin-binding domain (CBD) for chitin, a method has been developed to produce a peptide affinity column (Sun et al. 2003). A reactive thioester is first generated at the C-terminus of the CBD from the chitinase A1 of *Bacillus circulans* WL-2 by thiol-induced cleavage of the peptide bond between the CBD and a modified intein. Peptide epitopes possessing an N-terminal cysteine were subsequently ligated to the chitin-bound CBD tag. It has been shown that the resulting peptide columns are highly specific and efficient in affinity purification of antibodies from animal sera.

5 Remarks and Conclusions

The precise cleavage and creation of specific peptide bonds by self-splicing inteins have provided enormous opportunities for scientists to develop novel strategies in protein research. Various intein fusion systems have been invented to utilize modified or unmodified inteins to fulfill different goals. Now, naturally occurring inteins, some possessing distinct properties, can be tailored to accomplish a wide range of tasks including protein expression and isolation, protein semisynthesis and labeling, backbone cyclization, antigen and antibody characterization, enzymatic assays, and peptide array.

Acknowledgements. We sincerely thank Dr. Thomas C. Evans, Jr., for his critical reading of the manuscript and Dr. Donald G. Comb for his support. We also thank Drs. Inca Ghosh and Luo Sun for providing their unpublished data presented in Fig. 6.

References

Chin HG, Kim GD, Marin I, Mersha F, Evans TC Jr, Chen L, Xu M-Q, Pradhan S (2003) Protein *trans*-splicing in transgenic plant chloroplast: reconstruction of herbicide resistance from split genes. Proc Natl Acad Sci USA 100:4510–4515

Chong S, Mersha FB, Comb DG, Scott ME, Landry D, Vence LM, Perler FB, Benner J, Kucera RB, Hirvonen CA, Pelletier JJ, Paulus H, Xu M-Q (1997) Single-column purification of free recombinant proteins using a self-cleavable affinity tag derived from a protein splicing element. Gene 192:271–281

Chong S, Montello GE, Zhang A, Cantor EJ, Liao W, Xu M-Q, Benner J (1998) Utilizing the C-terminal cleavage activity of a protein splicing element to purify recombinant proteins in a single chromatographic step. Nucleic Acids Res 26:5109–5115

Clarke S (1987) Propensity for spontaneous succinimide formation from aspartyl and asparaginyl residues in cellular proteins. Int J Peptide Protein Res 30:808–821

Cotton GJ, Muir TW (1999) Peptide ligation and its application to protein engineering. Chem Biol 6:R247–256

Cotton GJ, Muir TW (2000) Generation of a dual-labeled fluorescence biosensor for Crk-II phosphorylation using solid-phase expressed protein ligation. Chem Biol 7:253–261

Dawson PE, Muir TW, Clark-Lewis I, Kent SB (1994) Synthesis of proteins by native chemical ligation. Science 266:776–779

Evans TC Jr, Benner J, Xu M-Q (1998) Semisynthesis of cytotoxic proteins using a modified protein splicing element. Protein Sci 7:2256–2264

Evans TC Jr, Benner J, Xu M-Q (1999a) The cyclization and polymerization of bacterially expressed proteins using modified self-splicing inteins. J Biol Chem 274:18359–18363

Evans TC Jr, Benner J, Xu M-Q (1999b) The in vitro ligation of bacterially expressed proteins using an intein from *Methanobacterium thermoautotrophicum.* J Biol Chem 274:3923–3926

Evans TC Jr, Martin D, Kolly R, Panne D, Sun L, Ghosh I, Chen L, Benner J, Liu XQ, Xu M-Q (2000) Protein *trans*-splicing and cyclization by a naturally split intein from the *dnaE* gene of *Synechocystis* species PCC6803. J Biol Chem 275:9091–9094

Evans TC Jr, Xu M-Q (1999) Intein-mediated protein ligation: harnessing nature's escape artists. Biopolymers 51:333–342

Ghosh I, Sun L, Evans TC Jr, Xu M-Q. An improved method for utilization of peptide substrates for antibody characterization and enzymatic assays. J Immunol Methods 293:85–95

Iwai H and Pluckthun A (1999) Circular beta-lactamase: stability enhancement by cyclizing the backbone. FEBS Lett 459:166–172

LaVallie ER, McCoy JM (1995) Gene fusion expression systems in *Escherichia coli.* Curr Opin Biotechnol 6:501–506

Lew BM, Mills KV, Paulus H (1998) Protein splicing in vitro with a semisynthetic two-component minimal intein. J Biol Chem 273:15887–15890

Martin DD, Xu M-Q, Evans TC (2001) Characterization of a naturally occurring *trans*-splicing intein from *Synechocystis* sp. PCC6803. Biochemistry 40:1393–1402

Mathys S, Evans TC, Chute IC, Wu H, Chong S, Benner J, Liu XQ, Xu M-Q (1999) Characterization of a self-splicing mini-intein and its conversion into autocatalytic N- and C-terminal cleavage elements: facile production of protein building blocks for protein ligation. Gene 231:1–13

Mills KV, Lew BM, Jiang S, Paulus H (1998) Protein splicing in trans by purified N- and C-terminal fragments of the *Mycobacterium tuberculosis* RecA intein. Proc Natl Acad Sci USA 95:3543–3548

Muir TW, Sondhi D, Cole PA (1998) Expressed protein ligation: a general method for protein engineering. Proc Natl Acad Sci USA 95:6705–6710

Nichols NM, Benner JS, Martin DD, Evans TC Jr (2003) Zinc ion effects on individual Ssp DnaE intein splicing steps: regulating pathway progression. Biochemistry 42:5301–5311

Otomo T, Ito N, Kyogoku Y, Yamazaki T (1999) NMR observation of selected segments in a larger protein: central-segment isotope labeling through intein-mediated ligation. Biochemistry 38:16040–16044

Paulus H (2000) Protein splicing and related forms of protein autoprocessing. Annu Rev Biochem 69:447–496

Perler FB, Davis EO, Dean GE, Gimble FS, Jack WE, Neff N, Noren CJ, Thorner J, Belfort M (1994) Protein splicing elements: inteins and exteins – a definition of terms and recommended nomenclature. Nucleic Acids Res 22:1125–1127

Pradhan S, Bacolla A, Wells RD, Roberts RJ (1999) Recombinant human DNA (cytosine-5) methyltransferase. I. Expression, purification, and comparison of de novo and maintenance methylation. J Biol Chem 274:33002–33010

Scott CP, Abel-Santos E, Wall M, Wahnon DC, Benkovic SJ (1999) Production of cyclic peptides and proteins in vivo. Proc Natl Acad Sci USA 96:13638–13643

Severinov K, Muir TW (1998) Expressed protein ligation, a novel method for studying protein-protein interactions in transcription. J Biol Chem 273:16205–16209

Sharma S, Zhang A, Wang H, Harcum SW, Chong S (2003) Study of protein splicing and intein-mediated peptide bond cleavage under high-cell-density conditions. Biotechnol Prog 19:1085–1090

Southworth MW, Adam E, Panne D, Byer R, Kautz R, Perler FB (1998) Control of protein splicing by intein fragment reassembly. EMBO J 17:918-926

Southworth MW, Amaya K, Evans TC, Xu M-Q, Perler FB (1999) Purification of proteins fused to either the amino or carboxy terminus of the *Mycobacterium xenopi* gyrase A intein. Biotechniques 27:110–120

Sun L, Ghosh I, Xu M-Q (2003) Generation of an affinity column for antibody purification by intein-mediated protein ligation. J Immunol Methods 282:45–52

Sun L, Rush J, Ghosh I, Maunus JR, Xu M-Q (2004) Producing peptide arrays for epitope mapping by intein-mediated protein ligation. Biotechniques 37:430–443

Tam JP, Lu YA, Liu CF, Shao J (1995) Peptide synthesis using unprotected peptides through orthogonal coupling methods. Proc Natl Acad Sci USA 92:12485–12489

Wood DW, Wu W, Belfort G, Derbyshire V, Belfort M (1999) A genetic system yields self-cleaving inteins for bioseparations. Nat Biotechnol 17:889–892

Wu H, Hu Z, Liu XQ (1998a) Protein trans-splicing by a split intein encoded in a split *DnaE* gene of *Synechocystis* sp. PCC6803. Proc Natl Acad Sci USA 95:9226–9231

Wu H, Xu M-Q, Liu XQ (1998b) Protein trans-splicing and functional mini-inteins of a cyanobacterial dnaB intein. Biochim Biophys Acta 1387:422–432

Xu J, Sun L, Ghosh I, Xu M-Q (2004) Western blot analysis of Src kinase assays using peptide substrates ligated to a carrier protein. Biotechniques 36:976–81

Xu M-Q, Paulus H, Chong S (2000) Fusions to self-splicing inteins for protein purification. Methods Enzymol 326:376–418

Xu R, Ayers B, Cowburn D, Muir TW (1999) Chemical ligation of folded recombinant proteins: segmental isotopic labeling of domains for NMR studies. Proc Natl Acad Sci USA 96:388–393

Yamazaki T, Otomo T, Oda N, Kyogoku Y, Uegaki K, Ito N, Ishino Y, Nakamura H (1998) Segmental isotope labeling for protein NMR using peptide splicing. J Am Chem Soc 120:5591–5592

Zhang A, Gonzalez SM, Cantor EJ, Chong S (2001) Construction of a mini-intein fusion system to allow both direct monitoring of soluble protein expression and rapid purification of target proteins. Gene 275:241–252

Production of Cyclic Proteins and Peptides

ALI TAVASSOLI, TODD A. NAUMANN, STEPHEN J. BENKOVIC

1 Introduction

The development of biomimetic routes to cyclic proteins and peptides is an area of great current interest. Cyclization confers conformational constraints that are thought to result in higher biological potency and stability of the resulting protein or peptide. This strategy has been utilized by nature, with cyclosporin A and gramicidin S being two well-studied examples of naturally occurring cyclic peptides (Kahan 1982; Polin and Egorov 2003).

1.1 Naturally Occurring Cyclic Peptides

Marine sponges have proven to be a rich source of cyclic peptides that are wide ranging in structure and biological activity (Milanowski et al. 2004; Schmidt et al. 2004), many of which contain non-proteinogenic amino acids and polyketide-derived moieties. Fungi isolated from various marine organisms have also yielded several biologically active cyclic peptides (Abarzua and Jakubowski 1995; Bringmann et al. 2004).

Naturally occurring cyclic peptides can be divided in two classes. Cyclic non-ribosomal peptides (such as the immunosuppressant cyclosporin A) are synthesized enzymatically by the ordered condensation of monomer building blocks to produce a linear peptide that is cyclized by a thioester domain at the C-terminus at the end of the biological process (Kohli and Walsh 2003).

The structures of a different second class of naturally occurring cyclic peptides have been reported recently (Trabi and Craik 2002; Craik et al. 2003). Present in a variety of organisms, they are distinct from non-ribosomal cyclic

A. Tavassoli, T.A. Naumann, S.J. Benkovic
Department of Chemistry, Pennsylvania State University, University Park, Pennsylvania 16802, USA

Nucleic Acids and Molecular Biology, Vol. 16
Marlene Belfort et al. (Eds.)
Homing Endonucleases and Inteins

peptides in that they are produced via transcription and translation of genes and have a well-defined three-dimensional structure. They differ from linear peptides only in their post-translational modification to join the N- and C-termini. In bacteria the currently known cyclic peptides of this class range from 21 to 78 amino acids and vary considerably in primary sequence and structure; no homology is evident between the amino acids involved in the *de novo* peptide bond or the flanking regions (Martinez-Bueno et al. 1994; Solbiati et al. 1999). Their biosynthesis involves diverse auxiliary proteins and a common mechanism of cyclization is thought unlikely (Craik et al. 2003).

In plants, an interesting group of cyclic peptides known as the cyclotides (Craik et al. 1999) exist that contain six conserved cysteine residues that form a series of disulfide bonds in their protein core (Fig. 1; Rosengren et al. 2003). The multiple internal disulfide bonds confer interesting structural characteristics upon the cyclotides resulting in twisted, knotted, and ringed cyclic peptides (Rosengren et al. 2003). The resulting structural rigidity is thought to be responsible for their resistance to proteolysis and stability; cyclotides have also been found to have a range of bioactivities, ranging from anti-HIV (Daly et al. 1999) to potent insecticides (Jennings et al. 2001).

The advantage of cyclic over linear peptides is demonstrated by the naturally occurring antibacterial mammalian peptide rhesus theta defensin (RTD-1), biosynthesized by ligation of two linear peptides. RTD-1 in its native cyclic form has been shown to have three times the antibacterial activity of the linear synthetic analog (Tang et al. 1999). This difference has been attributed to the extra stability provided by the cyclic form in vivo and not structural changes upon cyclization.

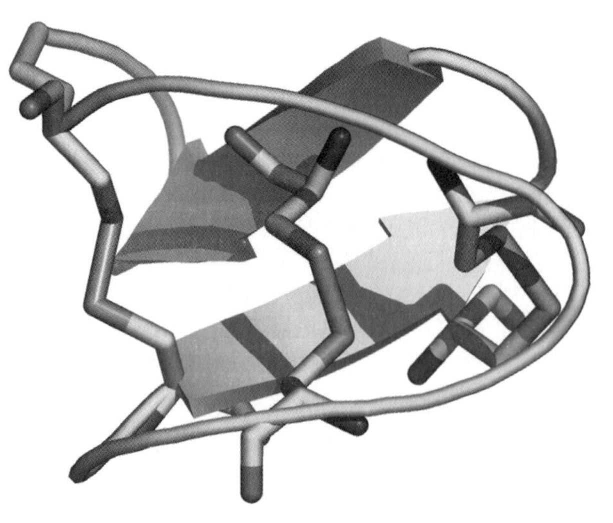

Fig. 1. The structure of Kalata B1. The three internal disulfide bonds are highlighted

1.2 Intein-Mediated Cyclization

Utilization of intein-based methods has resulted in the use of bacterial protein expression for the synthesis of cyclic versions of linear proteins. Cyclization has been shown to offer two main advantages, increased in vivo stability and biological activity, as a result of resistance to degradation by proteases and decreasing conformational flexibility over the linear protein counterpart (Hruby 1982; Hruby et al. 1990). Here, we describe recent developments in intein-based methodology for the biosynthesis of cyclic peptides and cyclization of linear proteins. Although there are several examples of recently developed methods and many studies into the use of cyclic peptides as pharmaceutical agents, it should be noted that the field is still in relative infancy.

Protein splicing is a self-catalyzed post-translational process in which an intervening internal sequence (intein) excises itself out of a precursor polypeptide resulting in the concomitant linkage of the flanking sequences (exteins) by a native peptide bond. Over 100 inteins have been identified in bacteria and unicellular eukaryotic organisms (Perler 2000); whilst they vary in size (134–600 amino acids), a set of highly conserved residues exist at the splicing junction (Pietrokovski 1994, 1998; Perler et al. 1997). Inteins generally begin with serine or cysteine and end in asparagine; the C-terminal of the intein is always followed by a serine, threonine or cysteine. There is little homology in the extein sequence thus the native exteins can be replaced by a foreign sequence without a dramatic adverse effect on splicing and cleavage. Naturally occurring inteins can be subdivided into three groups: those containing a homing endonuclease between the splicing domains, those lacking a homing endonuclease region, and those in which the splicing domain is split, the so-called *trans*-inteins.

2 Intein-Mediated Ligation

A technique known as intein-mediated ligation (Iwai and Pluckthun 1999) allows for the in vitro cyclization of proteins. In this method the protein to be cyclized is fused at its C-terminus to a disabled *Saccharomyces cerevisiae* VMA intein (Hirata et al. 1990; Kane et al. 1990), which is itself fused at its C-terminus to a chitin-binding domain (CBD; Chong et al. 1997). The intein is disabled by mutation of the C-terminal asparagine residue to alanine, thus disabling the splicing function at the C-terminus. The N-terminus of the protein of interest is altered if necessary, so that the second residue of the protein is cysteine. This relies on the in vivo activity of *E. coli* methionyl-aminopeptidase (Hirel et al. 1989) to remove the terminal methionine resulting in an N-terminal cysteine. After expression the fusion protein is then captured on chitin affinity resin. The partially functioning intein catalyzes an N/S acyl shift, which replaces the peptide bond between the protein and the intein with a thioester bond. Addition of an external nucleophile such as MESNA (2-mercaptoethanesulfonic acid) attacks the thioester, resulting in cleavage of the fusion protein (Xu and Evans 2001).

A linear protein contains an N-terminal cysteine and a C-terminal thioester, which can react to form a peptide bond in either an inter- or intramolecular fashion yielding a polymeric or a cyclized protein, respectively. A drawback of this methodology is that the intramolecular cyclization has to compete with intermolecular processes such as polymerization and hydrolysis of the thioester group thus reducing the yield and complicating purification of the cyclic protein (Fig. 2).

An example of the application of this technique is the cyclization of TEM-1 lactamase (BLA; Iwai and Pluckthun 1999). The proximity of the N- and C-

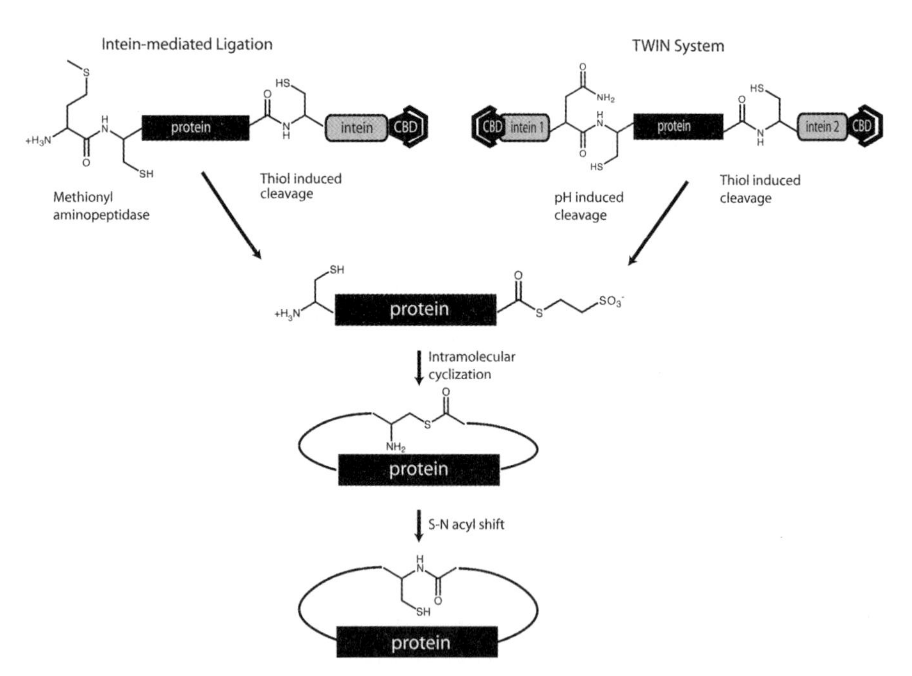

Fig. 2. Intein-mediated ligation and TWIN. Intein-mediated ligation relies on the removal of the initial methionine by MAP which results in an N-terminal cysteine. In the TWIN system the N-terminal nucleophile is unmasked by the processing of the C-terminal portion of intein 1. In both cases the N-terminal intein undergoes an N/S acyl shift followed by thiol-induced cleavage by an external nucleophile. The N-terminal cysteine can react intramolecularly with the C-terminal thioester; the product undergoes a further S/N acyl shift forming the cyclic protein

termini of BLA and the addition of a 15 amino acid linker allow its cyclization. The circular form of BLA was found to be more stable to irreversible aggregation upon heating than the linear form. The linear protein had fully precipitated after heating at 50 °C for 1 h, whereas 50% of the cyclic form stayed in the soluble fraction; the midpoint of the transition (T_m) increased by 5 °C upon cyclization (45 °C for linear, 50 °C for cyclic). This difference was utilized in the purification of the cyclic form away from linear and polymerized forms of BLA. As the free energies of related proteins are often well correlated with the aggregation temperature, the stabilizing effect of cyclization is consistent with the decreased conformational entropy of the unfolded state, as predicted by polymer theory.

3 TWIN

Another method which uses a similar in vitro approach for cyclization of peptides and proteins is the two inteins system (TWIN) developed by New England Biolabs (Evans et al. 1999). TWIN uses two self-cleaving inteins, one with N-terminal cleavage activity (*Mycobacterium xenopi* GyrA intein), and the other with C-terminal cleavage activity (Ssp DnaB intein), to generate a protein possessing both an N-terminal cysteine and a C-terminal thioester. The fusion of CBDs to the inteins allows affinity purification of the precursor protein by chitin resin. The Ssp DnaB intein undergoes cleavage at its C-terminus by cyclization of Asn154 (pH induced) whereas the Mxe GyrA intein undergoes thiol-induced (addition of MESNA) cleavage of the thioester bond (formed as a result of N/S acyl shift at cysteine). Upon release of the target protein the free thiol at the N-terminus reacts with the thioester of the C-terminus intramolecularly, leading to cyclization, or intermolecularly, resulting in polymerization (Fig. 2).

The cleavage efficiency was found to be somewhat determined by the amino acids adjacent to the scissile peptide (Evans et al. 1999). Significant in vivo cleavage of >50% was observed with the Cys-Gly-Ala sequence adjacent to the C-terminus of the SspI DnaB mini-intein; this rate was reduced to <10% with Cys-Arg-Ala. Thus cleavage can be modulated through modification of the extein sequence without changing the intein.

The TWIN system has been used in the cyclization of several proteins and peptides (Evans et al. 1999). Thioredoxin (135 amino acids) was cyclized using a 12 amino acid (9 on the N-terminus, 3 on the C-terminus) linker and >80% of the isolated protein was found to be cyclic. Interestingly, increasing the length of the linker to 26 amino acids (3 on the N-teminus, 23 on the C-terminus) decreased the rate of cyclization to 50%. In another example, maltose-binding protein (395 amino acids) also showed a 50% cyclization rate when a 26 amino acid linker was used.

Several cyclic peptides were also isolated including RGD (10 amino acids), an inhibitor of platelet aggregation (Yamamoto et al. 1995), and CDR-H3/C2 (16 amino acids; Levi et al. 1993), which inhibits HIV-1 replication. Previously, using chemical synthesis, these peptides were cyclized with cysteines at the N- and C-termini then oxidized to form a disulfide bond. The biological synthesis is postulated to be superior as the peptides are cyclized with a native peptide bond between the N- and C-termini, which is thought to endow resistance to reducing environments (such as the intracellular environment) and exoproteases.

Neither intein-mediated ligation nor TWIN allows for the total biasing of the final step toward cyclization or polymerization; although the rate of polymerization is concentration-dependent (increasing concentration increases the likelihood of an intermolecular reaction), other factors such as the protein structure, flexibility of the linkers and the termini, and the propensity to self-associate also determine the direction of the final step.

4 *trans*- and *cis*-Splicing

Protein *trans*-splicing involves the coming together of two halves of an intein sequence through their high affinity for each other followed by ligation of the two extein sequences. The only example of a naturally occurring split-intein was identified in the *dnaE* gene encoding the catalytic subunit of DNA polymerase III of *Synechocystis* sp. PCC6803 (Wu et al. 1998). The N-terminal half of DnaE (followed by a 123 amino acid N-intein sequence) is encoded by an open reading frame more than 745 kilobases away from that which encodes the C-terminal half of the protein (preceded by a 36 amino acid C-intein sequence).

The promiscuity of the Ssp DnaE split-intein (with respect to the extein sequence) was demonstrated by the in vivo production of a functional foreign protein. Herbicide-resistant acetolactate synthase II was reconstituted from two unlinked gene fragments fused to the Ssp DnaE split-intein. The function of the reconstituted protein was confirmed by the herbicide resistance that it conferred to the host *E. coli* cell (Sun et al. 2001). It should be noted that prior to the discovery of the Ssp DnaE intein, artificially split inteins were used to perform protein *trans*-splicing (Shingledecker et al. 1998; Southworth et al. 1998).

That split-inteins mediate efficient in vitro and in vivo splicing of foreign proteins has led to the development of several useful technologies, amongst them new methods for the cyclization of proteins and peptides.

5 Cyclization with Artificially Split Inteins

A truncated *Mycobacterium tuberculosis* RecA intein (Shingledecker et al. 1998) was used for the in vivo cyclization of the soluble (IIA^{GLC}) and membrane-bound ($IICB^{GLC}$) subunits of the glucose transporter in *E. coli*. The two subunits sequentially transfer phosphoryl groups from the phosphoryl carrier protein Hpr to glucose. Upon cyclization the activity of the purified soluble (IIA^{GLC}) and membrane-bound ($IICB^{GLC}$) proteins was found to have increased 100 and 25%, respectively (Siebold and Erni 2002).

The artificially split *Pyrococcus furiosus* PI-PfuI intein, which had been previously used for *trans*-ligation in vitro (Yamazaki et al. 1998; Otomo et al. 1999; Ichiyanagi et al. 2000), served in the in vivo synthesis of cyclic green fluorescent protein (GFP; Crameri et al. 1996). Although the cyclization of GFP using a six amino acid linker had been previously reported (Baird et al. 1999; Topell et al. 1999), this constituted the first report of an artificially split intein functioning in vivo (Iwai et al. 2001). The presence of cyclic protein in the cell was demonstrated by Western blotting, and the pure cyclic form was obtained from *E. coli* cell extracts with no contamination from the linear protein. This contrasts with *cis*-splicing using Ssp DnaE, which has been reported to produce linear by-products in vivo (Evans et al. 2000). This is thought to be due to the difference in N/S acyl migration rates between Ssp DnaE and PI-PfuI inteins. The cyclic GFP was shown to have similar structural properties to linear GFP, but it was found to unfold at half the rate of the linear form upon chemical denaturation and be more resilient to unfolding at high temperatures as expected.

The Ssp DnaB split-intein was synthesized *de novo* from overlapping oligonucleotides with optimal codon usage to ensure good expression in *E. coli* (Williams et al. 2002). Once suitably rearranged, this synthetic split intein was used for the in vivo cyclization of the NH_2 terminal domain of the *E. coli* replicative helicase DnaB, selected because the NMR structural data available indicated the proximity of the N- and C-termini in the folded state. A nine-residue linker was used to join the termini and the cyclization was found to proceed efficiently, with little accumulation of the products of incomplete splicing. The solution structure of cyclic DnaB was not found to differ significantly from that of the linear protein by NMR. As with other cyclized proteins, an increase in thermostability was observed, with the cyclic protein unfolding reversibly ~14 °C higher than the linear form. The thermodynamic stabilization of the structure upon cyclization was measured to be about 2 kcal/mol, which was in line with estimates from the reduced conformation entropy in the unfolded form.

6 SICLOPPS

By rearranging the order of the elements in the Ssp DnaE *trans*-intein, an active *cis*-intein (C-terminal intein:target peptide:N-terminal intein) was obtained that upon splicing resulted in cyclization of the target protein (Scott et al. 1999; Evans et al. 2000). Named split intein-mediated circular ligation of peptide and proteins (SICLOPPS; Scott et al. 1999; Abel-Santos et al. 2003), this method allows for the genetic encoding and in vivo production of cyclic peptides and proteins. After translation of the C-intein:peptide:N-intein fusion, in vivo formation of a peptide bond between the first and last amino acids of the target sequence is catalyzed by the reconstituted intein, thereby creating a cyclic peptide or protein (Fig. 3).

SICLOPPS was used to cyclize the *E. coli* enzyme dihydrofolate reductase (DHFR; Scott et al. 1999). DHFR had been previously circularly permuted (Buchwalder et al. 1992; Uversky et al. 1996; Iwakura and Nakamura 1998) and cyclized via a disulfide bond (Iwakura and Honda 1996). A three amino acid linker (or longer) was demonstrated to be required between the amino and carboxy termini in order to retain wild-type enzyme activity. The cyclic

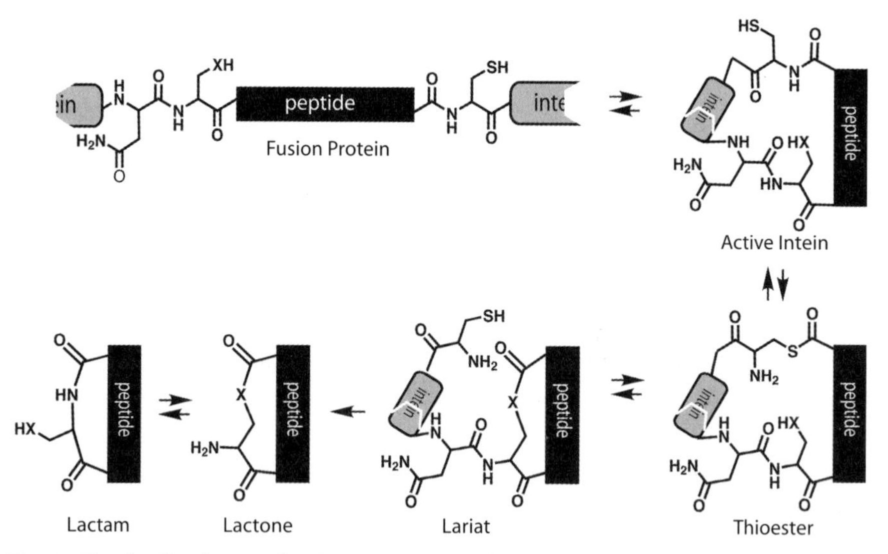

Fig. 3. Circular ligation mechanism. An expressed fusion protein folds to form an active protein intein. The enzyme catalyzes an N/S acyl shift at the target N-terminal intein junction to produce a thioester, which undergoes transesterification with a side-chain nucleophile at the C-terminal intein junction to form a lariat intermediate. Asparagine side-chain cyclization liberates the cyclic product as a lactone, and an X/N acyl shift generates the thermodynamically favored lactam product in vivo

DHFR produced by SICLOPPS was shown to be resistant to proteolysis and had steady-state kinetic parameters; and substrate, cofactor, and methotrexate dissociation constants that were indistinguishable from those of the wild-type enzyme at 25 °C. Activity assays conducted after preincubation of wild-type and cyclic DHFR at 65 °C showed an improvement in the thermostability of the cyclic enzyme.

SICLOPPS was also used for the intracellular production of the tyrosinase inhibitor peptide pseudostellarin F (cyclo-SGGYLPPL; Morita et al. 1994). The serine residue in pseudostellarin F served as the nucleophile for the transesterification portion of the protein ligation mechanism, instead of the cysteine that the wild-type Ssp DnaE C-terminal intein uses. NMR and mass spectroscopy verified the in vivo production of the peptide. Pseudostellarin F produced by SICLOPPS was shown to inhibit recombinant *Streptomyces antibioticus* tyrosinase in vivo demonstrating this integration of intracellular cyclization and functional screening (Scott et al. 1999).

The versatility of the SICLOPPS methodology was demonstrated by its use in the production of combinatorial libraries of small molecules (Scott et al. 2001). Compound libraries are an important tool in the search for inhibitors of ever increasing biological targets and therefore in the battle against disease. Traditionally, organic chemistry has been called upon for the construction of such libraries, thus limiting their size (10^3–10^5 members) and ease of construction (Gallop et al. 1994; Gordon et al. 1994; Czarnik 1997); genetic encoding in contrast allows the elaboration of very large libraries (10^6–10^9 members) that can be readily interfaced with biological selection. Libraries of cyclic peptides with five randomly variable amino acids and either one or four fixed residues were prepared (Scott et al. 2001). Randomized codons were introduced into the libraries via oligonucleotide synthesis followed by polymerase chain reaction (PCR) cloning, yielding between 10^7 and 10^8 transformants. Thus in each individual clone the peptide-encoding region between the intein halves contains a series of five distinct codons that determine the variable region of the cyclic peptide. As the cyclic peptide libraries are biosynthesized within the cell, SICLOPPS can be combined with a functional genetic selection and high-throughput screening to yield a powerful method for identifying potential inhibitors of intracellular targets.

The intracellular production of a library of random cyclic peptides in human cells was recently demonstrated. Retroviral vectors were used in the expression of the DnaB intein from Ssp PCC6803 in the human B cell line BJAB, enabling the function of bacterial split inteins in the cytoplasm of mammalian cells (Kinsella et al. 2002). This has been combined with a functional genetic screen for inhibitors of the interleukin-4 signaling pathway in the BJAB cell line (Schreiber and Crabtree 1992). Disruption of this pathway is a potential therapeutic approach for lowering IgE levels (IL-4 stimulates the germline epsilon gene, a sterile transcript required for IgE isotype class switching) to ameliorate diseases such as allergy and asthma (Oettgen and Geha 2001). Using the genetic screen, 13 cyclic peptides consisting of serine and four random amino acids were isolated. The selective inhibition of germline epsilon transcription by the selected cyclic peptides demonstrated the power and utility of combining random cyclic peptide production with genetic screening.

7 Summary

The true strength of the newly discovered biosynthetic methods outlined above lies in their potential combination with other biochemical methodologies, resulting in new powerful techniques. This is a new area in which there is ample room for future growth and further development.

References

Abarzua S, Jakubowski S (1995) Biotechnological investigation for the prevention of biofouling. 1. Biological and biochemical principles for the prevention of biofouling. Mar Ecol Prog Ser 123:301–312

Abel-Santos E, Scott CP, Benkovic SJ (2003) Use of inteins for the in vivo production of stable cyclic peptide libraries in *E. coli*. Methods Mol Biol 205:281–294

Baird GS, Zacharias DA, Tsien RY (1999) Circular permutation and receptor insertion within green fluorescent proteins. Proc Natl Acad Sci USA 96:11241–11246

Bringmann G, Lang G, Steffens S, Schaumann K (2004) Petrosifungins A and B, novel cyclodepsipeptides from a sponge-derived strain of *Penicillium brevicompactum*. J Nat Prod 67:311–315

Buchwalder A, Szadkowski H, Kirschner K (1992) A fully active variant of dihydrofolate reductase with a circularly permuted sequence. Biochemistry 31:1621–1630

Chong S, Mersha FB, Comb DG, Scott ME, Landry D, Vence LM, Perler FB, Benner J, Kucera RB, Hirvonen CA, Pelletier JJ, Paulus H, Xu M-Q (1997) Single-column purification of free recombinant proteins using a self-cleavable affinity tag derived from a protein splicing element. Gene 192:271–281

Craik DJ, Daly NL, Bond T, Waine C (1999) Plant cyclotides: a unique family of cyclic and knotted proteins that defines the cyclic cystine knot structural motif. J Mol Biol 294:1327–1336

Craik DJ, Daly NL, Saska I, Trabi M, Rosengren KJ (2003) Structures of naturally occurring circular proteins from bacteria. J Bacteriol 185:4011–21

Crameri A, Whitehorn EA, Tate E, Stemmer WP (1996) Improved green fluorescent protein by molecular evolution using DNA shuffling. Nat Biotechnol 14:315–319

Czarnik AW (1997) Encoding methods for combinatorial chemistry. Curr Opin Chem Biol 1:60–66

Daly NL, Koltay A, Gustafson KR, Boyd MR, Casas-Finet JR, Craik DJ (1999) Solution structure by NMR of circulin A: a macrocyclic knotted peptide having anti-HIV activity. J Mol Biol 285:333–345

Evans TC Jr, Benner J, Xu M-Q (1999) The cyclization and polymerization of bacterially expressed proteins using modified self-splicing inteins. J Biol Chem 274:18359–18363
Evans TC Jr, Martin D, Kolly R, Panne D, Sun L, Ghosh I, Chen L, Benner J, Liu XQ, Xu M-Q (2000) Protein trans-splicing and cyclization by a naturally split intein from the dnaE gene of *Synechocystis* species PCC6803. J Biol Chem 275:9091–9094
Gallop MA, Barrett RW, Dower WJ, Fodor SP, Gordon EM (1994) Applications of combinatorial technologies to drug discovery. 1. Background and peptide combinatorial libraries. J Med Chem 37:1233–1251
Gordon EM, Barrett RW, Dower WJ, Fodor SP, Gallop MA (1994) Applications of combinatorial technologies to drug discovery. 2. Combinatorial organic synthesis, library screening strategies, and future directions. J Med Chem 37:1385–1401
Hirata R, Ohsumk Y, Nakano A, Kawasaki H, Suzuki K, Anraku Y (1990) Molecular structure of a gene, VMA1, encoding the catalytic subunit of H(+)-translocating adenosine triphosphatase from vacuolar membranes of *Saccharomyces cerevisiae*. J Biol Chem 265:6726–6733
Hirel PH, Schmitter MJ, Dessen P, Fayat G, Blanquet S (1989) Extent of N-terminal methionine excision from *Escherichia coli* proteins is governed by the side-chain length of the penultimate amino acid. Proc Natl Acad Sci USA 86:8247–8251
Hruby VJ (1982) Conformational restrictions of biologically active peptides via amino acid side chain groups. Life Sci 31:189–199
Hruby VJ, al-Obeidi F, Kazmierski W (1990) Emerging approaches in the molecular design of receptor-selective peptide ligands: conformational, topographical and dynamic considerations. Biochem J 268:249–262
Ichiyanagi K, Ishino Y, Ariyoshi M, Komori K, Morikawa K (2000) Crystal structure of an archaeal intein-encoded homing endonuclease PI-PfuI. J Mol Biol 300:889–901
Iwai H, Pluckthun A (1999) Circular beta-lactamase: stability enhancement by cyclizing the backbone. FEBS Lett 459:166–172
Iwai H, Lingel A, Pluckthun A (2001) Cyclic green fluorescent protein produced in vivo using an artificially split PI-PfuI intein from *Pyrococcus furiosus*. J Biol Chem 276:16548–16554
Iwakura M, Honda S (1996) Stability and reversibility of thermal denaturation are greatly improved by limiting terminal flexibility of *Escherichia coli* dihydrofolate reductase. J Biochem (Tokyo) 119:414–420
Iwakura M, Nakamura T (1998) Effects of the length of a glycine linker connecting the N- and C-termini of a circularly permuted dihydrofolate reductase. Protein Eng 11:707–713
Jennings C, West J, Waine C, Craik D, Anderson M (2001) Biosynthesis and insecticidal properties of plant cyclotides: the cyclic knotted proteins from *Oldenlandia affinis*. Proc Natl Acad Sci USA 98:10614–10619
Kahan BD (1982) Cyclosporin a: a new advance in transplantation. Tex Heart Inst J 9:253–266
Kane PM, Yamashiro CT, Wolczyk DF, Neff N, Goebl M, Stevens TH (1990) Protein splicing converts the yeast TFP1 gene product to the 69-kD subunit of the vacuolar H(+)-adenosine triphosphatase. Science 250:651–657
Kinsella TM, Ohashi CT, Harder AG, Yam GC, Li W, Peelle B, Pali ES, Bennett MK, Molineaux SM, Anderson DA, Masuda ES, Payan DG (2002) Retrovirally delivered random cyclic peptide libraries yield inhibitors of interleukin-4 signaling in human B cells. J Biol Chem 277:37512–37518
Kohli RM, Walsh CT (2003) Enzymology of acyl chain macrocyclization in natural product biosynthesis. Chem Commun (Camb) 297–307

Levi M, Sallberg M, Ruden U, Herlyn D, Maruyama H, Wigzell H, Marks J, Wahren B (1993) A complementarity-determining region synthetic peptide acts as a miniantibody and neutralizes human immunodeficiency virus type 1 in vitro. Proc Natl Acad Sci USA 90:4374–4378

Martinez-Bueno M, Maqueda M, Galvez A, Samyn B, van Beeumen J, Coyette J, Valdivia E (1994) Determination of the gene sequence and the molecular structure of the enterococcal peptide antibiotic AS-48. J Bacteriol 176:6334–6339

Milanowski DJ, Rashid MA, Gustafson KR, O'Keefe BR, Nawrocki JP, Pannell LK, Boyd MR (2004) Cyclonellin, a new cyclic octapeptide from the marine sponge *Axinella carteri*. J Nat Prod 67:441–444

Morita H, Kayashita T, Kobata H, Gonda A, Takeya K, Itokawa H (1994) Pseudostellarins D–F, new tyrosinase inhibitory cyclic-peptides from pseudostellaria-heterophylla. Tetrahedron 50:9975–9982

Oettgen HC, Geha RS (2001) IgE regulation and roles in asthma pathogenesis. J Allergy Clin Immunol 107:429–440

Otomo T, Teruya K, Uegaki K, Yamazaki T, Kyogoku Y (1999) Improved segmental isotope labeling of proteins and application to a larger protein. J Biomol NMR 14:105–114

Perler FB (2000) InBase, the intein database. Nucleic Acids Res 28:344–345

Perler FB, Olsen GJ, Adam E (1997) Compilation and analysis of intein sequences. Nucleic Acids Res 25:1087–1093

Pietrokovski S (1994) Conserved sequence features of inteins (protein introns) and their use in identifying new inteins and related proteins. Protein Sci 3:2340–2350

Pietrokovski S (1998) Modular organization of inteins and C-terminal autocatalytic domains. Protein Sci 7:64–71

Polin AN, Egorov NS (2003) Structural and functional characteristics of gramicidin S in connection with its antibiotic activity. Antibiot Khimioter 48:29–32

Rosengren KJ, Daly NL, Plan MR, Waine C, Craik DJ (2003) Twists, knots, and rings in proteins. Structural definition of the cyclotide framework. J Biol Chem 278:8606–8616

Schmidt EW, Raventos-Suarez C, Bifano M, Menendez AT, Fairchild CR, Faulkner DJ (2004) Scleritodermin A, a cytotoxic cyclic peptide from the lithistid sponge *Scleritoderma nodosum*. J Nat Prod 67:475–478

Schreiber SL, Crabtree GR (1992) The mechanism of action of cyclosporin A and FK506. Immunol Today 13:136–142

Scott CP, Abel-Santos E, Wall M, Wahnon DC, Benkovic SJ (1999) Production of cyclic peptides and proteins in vivo. Proc Natl Acad Sci USA 96:13638–13643

Scott CP, Abel-Santos E, Jones AD, Benkovic SJ (2001) Structural requirements for the biosynthesis of backbone cyclic peptide libraries. Chem Biol 8:801–815

Shingledecker K, Jiang SQ, Paulus H (1998) Molecular dissection of the *Mycobacterium tuberculosis* RecA intein: design of a minimal intein and of a trans-splicing system involving two intein fragments. Gene 207:187–195

Siebold C, Erni B (2002) Intein-mediated cyclization of a soluble and a membrane protein in vivo: function and stability. Biophys Chem 96:163–171

Solbiati JO, Ciaccio M, Farias RN, Gonzalez-Pastor JE, Moreno F, Salomon RA (1999) Sequence analysis of the four plasmid genes required to produce the circular peptide antibiotic microcin J25. J Bacteriol 181:2659–2662

Southworth MW, Adam E, Panne D, Byer R, Kautz R, Perler FB (1998) Control of protein splicing by intein fragment reassembly. EMBO J 17:918–926

Sun L, Ghosh I, Paulus H, Xu M-Q (2001) Protein trans-splicing to produce herbicide-resistant acetolactate synthase. Appl Environ Microbiol 67:1025–1029

Tang YQ, Yuan J, Osapay G, Osapay K, Tran D, Miller CJ, Ouellette AJ, Selsted ME (1999) A cyclic antimicrobial peptide produced in primate leukocytes by the ligation of two truncated alpha-defensins. Science 286:498–502

Topell S, Hennecke J, Glockshuber R (1999) Circularly permuted variants of the green fluorescent protein. FEBS Lett 457:283–289

Trabi M, Craik DJ (2002) Circular proteins – no end in sight. Trends Biochem Sci 27:132–138

Uversky VN, Kutyshenko VP, Protasova N, Rogov VV, Vassilenko KS, Gudkov AT (1996) Circularly permuted dihydrofolate reductase possesses all the properties of the molten globule state, but can resume functional tertiary structure by interaction with its ligands. Protein Sci 5:1844–1851

Williams NK, Prosselkov P, Liepinsh E, Line I, Sharipo A, Littler DR, Curmi PM, Otting G, Dixon NE (2002) In vivo protein cyclization promoted by a circularly permuted *Synechocystis* sp. PCC6803 DnaB mini-intein. J Biol Chem 277:7790–7798

Wu H, Hu Z, Liu XQ (1998) Protein trans-splicing by a split intein encoded in a split DnaE gene of *Synechocystis* sp. PCC6803. Proc Natl Acad Sci USA 95:9226–9231

Xu MQ, Evans TC Jr (2001) Intein-mediated ligation and cyclization of expressed proteins. Methods 24:257–277

Yamamoto Y, Almehdi M, Katow H, Sofuku S (1995) Structure and activity of fibronectin-related peptides – the role of amino-acid residues of position-5 and position-6 in rgdspass-containing cyclic decapeptide (Fr-1). Chem Lett 11–12

Yamazaki T, Otomo T, Oda N, Kyogoku Y, Uegaki K, Ito N, Ishino Y, Nakamura H (1998) Segmental isotope labeling for protein NMR using peptide splicing. J Am Chem Soc 120:5591–5592

Inteins for Split-Protein Reconstitutions and Their Applications

Takeaki Ozawa, Yoshio Umezawa

Abstract

Our knowledge of biological systems relies increasingly on the ability of quantifying and imaging intracellular signals and events in living subjects. The development of novel methods and advances in biotechnology have provided many basic tools that allow analyses of the complex biological systems in living cells. Since the discovery of protein splicing in 1990, the elucidation of the splicing mechanism and the identification of key amino acid residues involved in the dissection and ligation of the peptide bonds have facilitated the molecular engineering of inteins for different applications in protein chemistry. These include protein purification, protein ligation and peptide cyclization, construction of split reporter proteins, regulation of protein activity, and introduction of non-natural amino acids. In this chapter, we focus on the construction of split reporter proteins and their applications for detecting protein-protein interactions, identification of organelle-localized proteins, growing safer transgenic plants, and screening antimycobacterial agents.

1 Introduction

Our knowledge of biological systems relies increasingly on the ability to quantify and image intracellular signals and events in living subjects. Development of novel methods and advances in biotechnology have provided many basic tools that allow analyses of the complex biological systems in living cells. Since the discovery of protein splicing in 1990, the elucidation of the splicing mech-

---

T. Ozawa, Y. Umezawa (e-mail: umezawa@chem.s.u-tokyo.ac.jp)
Department of Chemistry, School of Science, The University of Tokyo, Hongo, Bunkyo-ku, Tokyo 113-0033, Japan, and Japan Science and Technology Agency, Tokyo, Japan

Nucleic Acids and Molecular Biology, Vol. 16
Marlene Belfort et al. (Eds.)
Homing Endonucleases and Inteins

anism and the identification of key amino acid residues involved in the dissection and ligation of the peptide bonds have facilitated the molecular engineering of inteins for different applications in protein chemistry. These include protein purification, protein ligation and peptide cyclization, construction of split reporter proteins, regulation of protein activity, and introduction of non-natural amino acids. In this chapter, we focus on the construction of split reporter proteins and their applications for detecting protein–protein interactions, identification of organelle-localized proteins, growing safer transgenic plants, and screening antimycobacterial agents.

2 General Characteristics of Protein Splicing

Protein biosynthesis was initially thought to be a simple process; the genetic information in DNA is directly copied into messenger RNAs, which in turn direct the biosynthesis of proteins. However, an unexpected discovery was made by two groups independently in 1990 (Hirata et al. 1990; Kane et al. 1990). In *Saccharomyces cerevisiae*, the nascent 120-kDa translation product of the *VMA1* gene autocatalytically excises a 50-kDa site-specific endonuclease (*VMA1*-derived endonuclease; VDE, also called as PI-SceI) and splices the two external polypeptides (exteins) to form the 70-kDa catalytic subunit of vaculoar H^+-ATPase. This discovery led to the conclusion that the post-translational removal of the internal polypeptide (intein) occurs by protein splicing (Kawasaki et al. 1996). Since the initial discovery of the VMA1 intein, more than 170 putative inteins have been identified in eubacteria, archea and eukaryotic unicellular organisms (see InBase, the Intein registry website at http://www.neb.com/neb/inteins.html; New England Biolabs). Protein splicing is a multi-step processing event involving excision of an intein segment from a primary translation product with concomitant ligation of the flanking extein sequences. A typical intein segment consists of 400–500 amino acid residues (Fig. 1). It contains six conserved protein-splicing motifs, N1, N2, N3, N4, C1 and C2, as well as a homing endonuclease sequence embedded between motifs N4 and C2 (Pietrokovski 1998). The endonuclease sequence can be deleted from the intein sequence without abolishing protein splicing (Chong and Xu 1997). At the intein-extein junctions, three conserved amino acid residues are directly involved in the protein-splicing reaction (Evans and Xu 2002; David et al. 2004). They include a Ser or Cys at the intein amino terminus (the first amino acid in motif A), an Asn or Gln at the intein carboxy terminus (the last amino acid in motif G), and a Ser, Thr, or Cys at the beginning of the C-extein. While some inteins can be artificially split into two fragments and still retain activity, a functional naturally split intein-coding sequence was found in 1998 by Liu and colleagues (Wu et al. 1998). This intein is associated with the

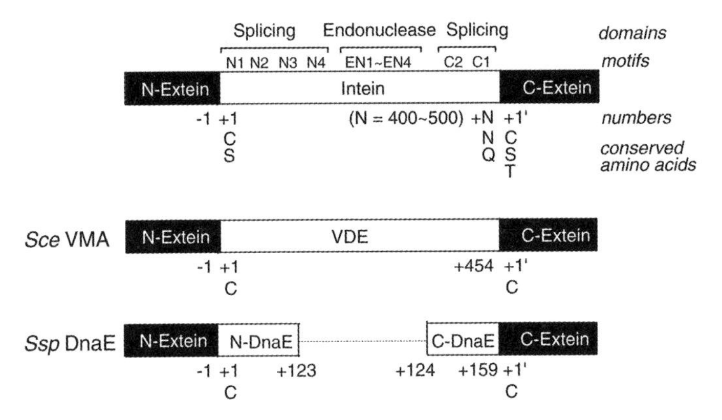

Fig. 1. Intein domains, motifs and conserved amino acids. Domains N1, N2, N3, and N4 contain the N-terminal splicing motifs, while domains C1 and C2 contain the C-terminal splicing motifs. The first amino acid residue upstream of the intein is numbered −1, and the intein is numbered sequentially beginning with the N-terminal amino acid residue. The first C-extein residue is numbered +1′. Conserved amino acid residues at position +1 and +1′ are indicated

split *dnaE* gene in the genome of *Synechocystis* sp. PCC6803 (Wu et al. 1998). The Ssp DnaE intein can mediate a trans-splicing reaction when fused to foreign proteins. Compared to artificially split inteins where the endonuclease domain is eliminated, the DnaE intein catalyzes the protein *trans*-splicing reaction with a higher efficiency (Martin et al. 2001). An important feature of protein splicing is the self-catalyzed excision of the intein and ligation of the flanking exteins without any external enzymes. This feature has been used to devise novel split reporter reconstitutions and their various applications.

3 Reporter Protein Reconstitution

Fluorescent or bioluminescent reporter proteins have proven to be of great use for detecting a variety of cellular activities (Grimm 2004). A typical example is the reporter gene assay with a firefly or *Renilla* luciferase reporter gene. Activation of transcription factors leads to gene expression of the luciferase, which leads to a detectable light signal in the presence of the substrate, d-luciferin or coelenterazine. Green fluorescent protein (GFP) and its variants are also used as a marker of gene expression in living cells and living subjects (Zhang et al. 2002) because GFPs form their own chromophores without external substrates (Nishiuchi et al. 1998). Reporter gene assays have also been applied to detect protein–protein and protein–DNA interactions, by use of yeast or mammalian two-hybrid assays (Fields and Song 1989; Shioda et al. 2000). One of the drawbacks of the reporter gene assay is that transcription factor activation occurs only in the nucleus, and therefore it is impossible to detect protein–protein interactions outside the nucleus (Ozawa and Umezawa 2002).

To overcome this drawback, novel reporters have been developed by our group, in which protein splicing generates a reconstituted reporter protein from two split protein fragments (Ozawa et al. 2000). One example is reconstitution of split GFP. The structure of GFP is composed of eleven β-strands that form a barrel structure with short α-helices forming lids on each end (Ormo et al. 1996; Yang et al. 1996; Brejc et al. 1997). The fluorescence-active center of GFP is located inside the barrel. When the GFP is dissected between amino acid positions 128 and 129, located at the end of the sixth β-strand of GFP, the fluorescence is completely lost. However, insertion of the VDE intein at this position results in the ligation of the amino- and carboxy-terminal fragments of GFP by protein splicing (Fig. 2). The ligated fragments fold correctly to form its fluorophore, and recover its fluorescence in vitro and in vivo. The basic concept of reconstitution by protein splicing is of general use for any reporter protein. Reconstitutions of split firefly luciferase and split *Renilla* luciferase by protein splicing have also been demonstrated (Ozawa et al. 2001a; Paulmurugan et al. 2002).

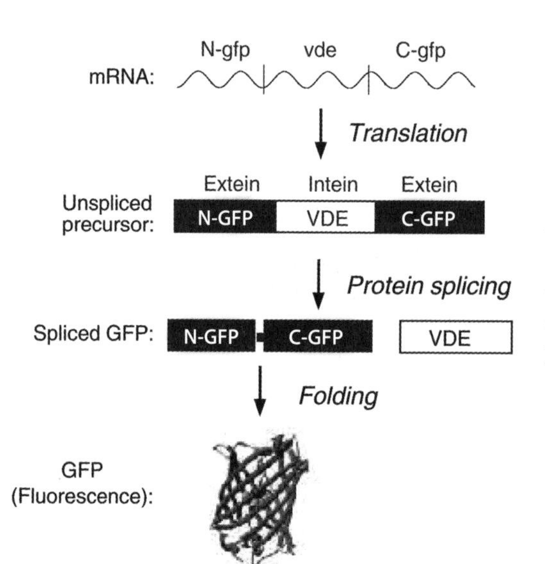

Fig. 2. Reconstitution of split GFP by protein splicing. A single polypeptide, composed of 128 amino acid residues of N-GFP, 454 residues of VDE and 110 residues of C-GFP, undergoes protein splicing and thereby N- and C-terminal fragments of GFP ligate by a peptide bond. The EGFP fragment thus formed folds correctly and its fluorophore is formed

3.1 Detection of Protein–Protein Interactions

In living cells, protein-protein interactions constitute essential regulatory steps that modulate the activity of signaling pathways. Identification of the interactions and characterization of their physiological significance is one of the main goals of current research in different fields of biology. Towards this goal, several technologies have been developed for detecting protein–protein interactions without the need to disrupt living cells. We describe herein a method for detecting protein–protein interactions based on reporter protein reconstitution and some potential applications of the technique.

The basic principle of detecting protein–protein interactions using the VDE intein is shown in Fig. 3 (Ozawa et al. 2000). An N-terminal fragment of VDE (N-VDE; 1–184 amino acids) is fused to an N-terminal fragment of GFP (N-GFP; 1–128 amino acids), and the C-terminal fragment of VDE (C-VDE; 389–454 amino acids) is fused to the rest of GFP (C-GFP; 129–239 amino acids). Each of these fusion proteins is linked to a protein of interest (protein A) and its target (protein B). When an interaction occurs between the two proteins of interest, the N- and C-VDE fragments are brought into close proximity and undergo correct folding, which induces a splicing event. The N-GFP and C-GFP become linked by a peptide bond, and thereby the mature GFP forms its fluorophore with an emission maximum at 510 nm. The extent of the protein–protein interaction can be evaluated by measuring the magnitude of fluorescence intensity generated by the formation of GFP. In a proof of principle, calmodulin (CaM) and its target peptide (M13) were used to show that their interaction in *E. coli* results in the formation of fluorescent GFP (Ozawa et al. 2000).

On the basis of this design, we have demonstrated a bacterial screening and selection system with several advantages (Ozawa et al. 2001b). Several bacterial one- and two-hybrid systems have been proposed, where there is a common principle: when the proteins interact, they trigger transcriptional activation of a reporter gene to produce a signal protein that is accumulated in the bacteria. Unlike the earlier protein interaction assays, the split-GFP system in bacteria involves the reconstitution of GFP, and does not require that a reporter gene is translated into its protein or that an enzyme substrate be present. This will make the method more generally useful in eukaryotic cells and allow the interactions to be screened in the cytosol, intracellular organelles or at the inner-membrane level.

Using the same concept as the GFP reconstitution system, a luciferase reconstitution system has been developed for monitoring protein–protein interactions in mammalian cells (Ozawa et al. 2001a). Firefly luciferase oxidizes its substrate d-luciferin to result in light emission (bioluminescence), which enables background-free and high-sensitivity detection. X-ray analysis

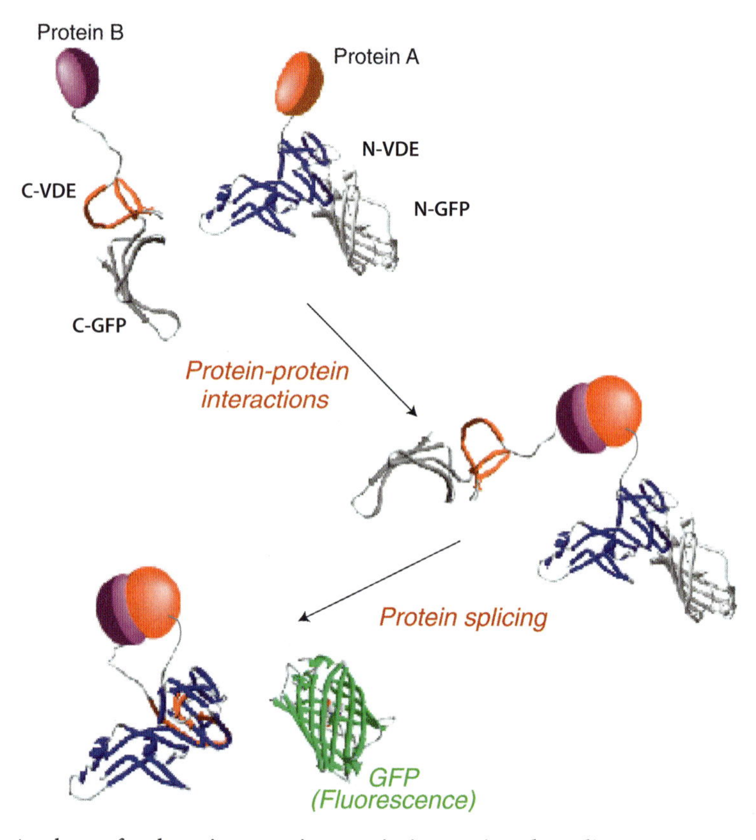

Fig. 3. A scheme for detecting protein–protein interactions by split GFP reconstitution. Ribbon diagrams of the N-terminal half of VDE (1–184 amino acids) and the C-terminal half of VDE (389–454 amino acids) are each connected to N-GFP and C-GFP fragments, respectively. Interacting proteins A and B are linked to opposite ends of the split VDE. Interaction between protein A and protein B accelerates the folding of N- and C-VDE and protein splicing occurs. The N- and C-GFP are thereby linked together by a peptide bond to yield the GFP fluorescence

of firefly luciferase revealed that the 3-D structure is folded into two compact domains. The C-terminal portion of the enzyme is separated from the larger N-terminal globular domain by a wide cleft, which is the location of the active site of the enzyme (Fig. 4; Conti et al. 1996). N- and C-terminal fragments of DnaE are connected to N- and C-terminal fragments of firefly luciferase, respectively. The free ends of the DnaE intein fragments are then fused to a pair of proteins of interest and the resulting fusions are expressed in mammalian cells. Upon interaction between the two proteins, the two DnaE fragments are brought close enough to fold and initiate splicing, restoring the luciferase by formation of a peptide bond. Reconstitution of luciferase is mon-

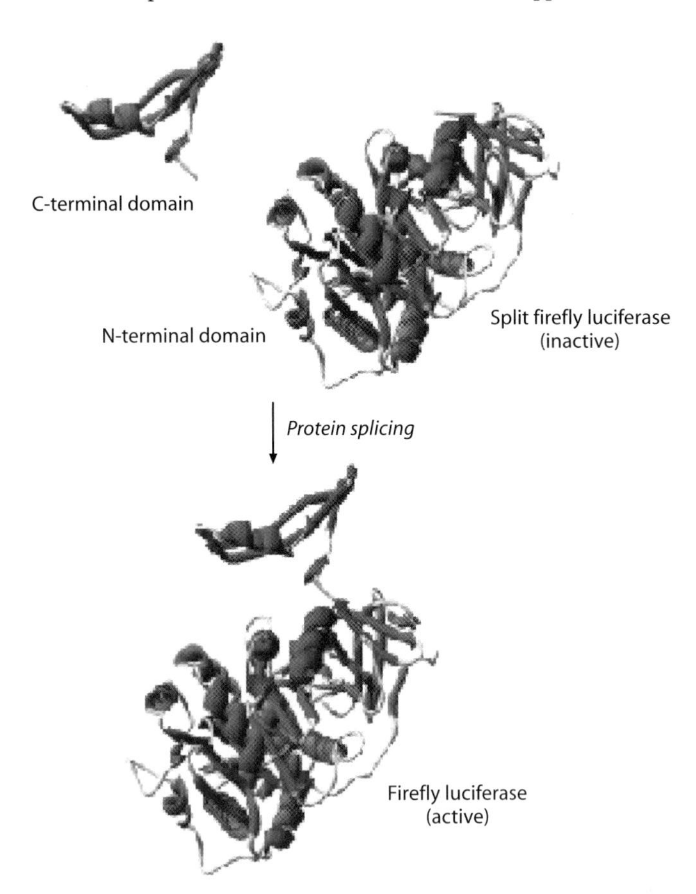

Fig. 4. 3-D structure of the split firefly luciferase. Split firefly luciferase, composed of N-terminal (1–437 amino acids) and C-terminal (438–544 amino acids) domains, does not possess bioluminescence activity. When the N- and C-terminal domains are linked together by protein splicing, its bioluminescence activity can be recovered in vivo

itored by its bioluminescence, of which intensity is again proportional to the number of interacting protein pairs. As a model system, an interaction between the insulin receptor substrate 1 and the SH2 domain of phosphatidylinositol-3-kinase has been demonstrated by using the split firefly luciferase reconstitution (Ozawa et al. 2001a).

Luciferase reconstitution by protein splicing was used to non-invasively monitor protein–protein interactions in living mice (Paulmurugan et al. 2002). The split reporters of firefly luciferase were fused to a pair of test proteins, MyoD and ID, which are known to interact strongly. Both proteins were transiently expressed in cultured cells, and, thereafter, the cells were implanted onto the backs of living mice. After injection of d-luciferin, the mice were placed in a light-shielding chamber, and photons emitted from the backs of the mice were collected by cooled CCD camera for a period of 1 min. The MyoD–ID interaction is induced by injecting TNF-α. The TNF-α-induced mouse showed significantly higher luminescence when compared with the mouse not receiving TNF-α.

These split GFP and split luciferase approaches have an advantage in that protein interactions can be monitored anywhere in the cells, whereas the traditional two-hybrid approaches are limited to interactions only in the nucleus. In addition, GFP or luciferase accumulates in a target cell until it degrades, and evidence of the interaction is thereby integrated in the cell. Imaging interacting protein pairs in living subjects may pave the way to functional proteomics in whole animals and provide a new tool for evaluating new pharmaceuticals targeted to modulate protein–protein interactions.

The basic concept of the above approach for detecting protein–protein interactions (Ozawa et al. 2000) was further applied to control protein splicing with a small molecule, rapamycin (Mootz and Muir 2002; Mootz et al. 2003). Rapamycin, a potent immunosuppressive agent, is known to bind a pair of proteins: the FK506-binding protein (FKBP) and the FKBP-rapamycin-associated protein (FRAP; Ho et al. 1996). The FKBP and the binding domain of FRAP (FRB) were used as a pair of protein interaction partners. The proteins were fused to a pair of N- and C-terminal fragments of an artificially split VMA intein (Fig. 5A). Upon addition of rapamycin, protein *trans*-splicing was triggered and the intein was excised. The two exteins were ligated to produce a new polypeptide with a potentially new function. This on/off switching of protein splicing with rapamycin was confirmed in vitro and in vivo.

In contrast to the *trans*-splicing based on the FKBP–FRB interaction, Buskirk et al. (2004) created another on/off switching system of protein splicing based on an *intra*molecular interaction, i.e. protein conformational change. A ligand-binding domain (LBD) of the human estrogen receptor (ER) binds a synthetic small molecule, 4-hydroxytamoxifen (4-HT), with high affinity. The ER was connected to the N- and C-terminal halves of RecA intein, yielding a 424-residue RecA(N)-ER-RecA(C) fusion (Fig. 5B). Using the fusion protein as a template, a cDNA library of mutated genes was obtained by error-prone polymerase chain reaction (PCR). The cDNAs were connected with the split GFP reporter, and screened based on the fluorescence in the presence and absence of 4-HT. A mutant RecA(N)-ER-RecA(C) fusion protein showed 4-HT-dependent activation of protein splicing. Using this intein, a clear-cut post-translational activation of KanR, LacZ, Ade2p and GFP upon addition of 4-HT was demonstrated.

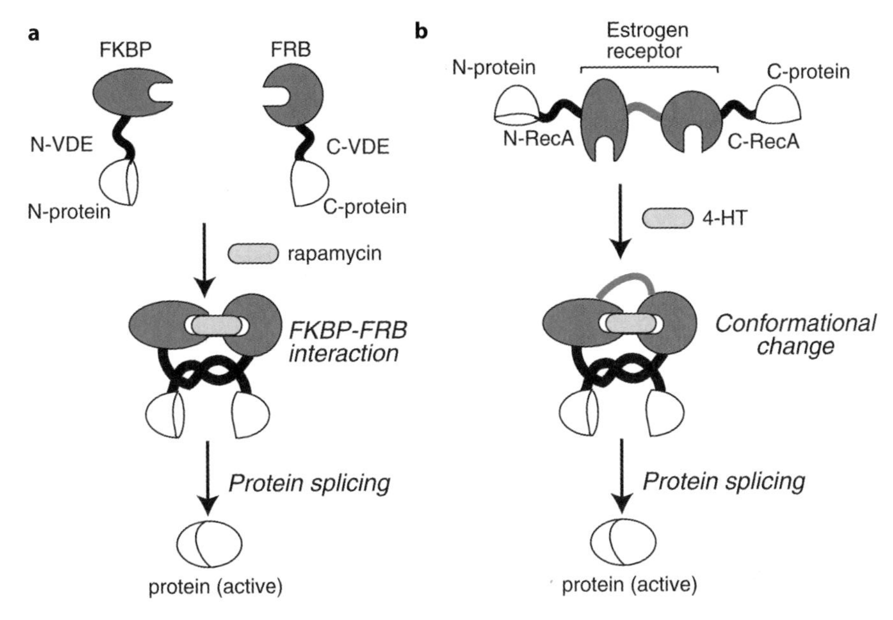

Fig. 5. Scheme for small molecule-induced conditional splicing. **a** FKBP and FRB are linked to N-VDE and C-VDE, respectively. Interaction between FKBP and FRB induced by rapamycin results in the folding of N- and C-VDE and protein splicing occurs. The N- and C-proteins (exteins) are linked together by a peptide bond. **b** N- and C-RecA are connected at the N- and C-terminal ends of the ligand-binding domain (LBD) in the estrogen receptor (ER), respectively. The fusion protein contains four mutations, V376A and R521G in LBD, and A34V and H41L in the RecA intein. Upon binding 4-hydroxytamoxifen (4-HT), the ER undergoes a conformational change, the N- and C-RecA fold correctly, and protein splicing occurs. The N- and C-proteins (exteins) are linked together by a peptide bond. As the exteins, KanR, LacZ, Ade2p and GFP were reconstituted by 4-HT-induced protein splicing

3.2 Protein Splicing in Intracellular Organelles

3.2.1 Identification of Organelle-Localized Proteins from cDNA Libraries

One of the most distinct features of eukaryotic cells, especially mammalian cells, is the compartmentalization of each protein. Protein localization is tightly bound to function, such that preferential localization of a protein is often essential for its function. Therefore, functional assays aimed at characterizing the cellular localization of proteins are very important for understanding complicated protein networks. The technique for identifying proteins localized in organelles largely relies on the isolation of compartments through cell fractionation and electrophoresis, combined with mass spectrometry. This biochemical method is useful for systematic identification, but it depends on the yield and purity of the intracellular organelle, and therefore the technique can be problematic for organelles that are difficult to isolate (Westermann and Neupert 2003).

Protein reconstitution technology opens a new avenue of genetic approaches for identifying mitochondrial proteins from large-scale cDNA libraries (Ozawa et al. 2003). The strategy is based on reconstitution of enhanced GFP (EGFP) by protein splicing with the DnaE intein (Fig. 6). A tandem fusion protein, containing a mitochondrial targeting signal (MTS) fused to the C-terminal fragments of the DnaE intein and EGFP, localizes to the mitochondrial matrix in mammalian cells. cDNA libraries generated from mRNAs are genetically fused to a sequence encoding the N-terminal fragments of EGFP and the DnaE intein. If test proteins expressed from the cDNA libraries contain a functional MTS, the fusion products translocate into the mitochondrial matrix, bringing the N- and C-terminal fragments of the DnaE intein close enough to fold correctly. This initiates protein splicing, thus linking the EGFP fragments with a peptide bond. This method has the advantage that only "mitochondria-positive" clones yield a fluorescence signal. These cells can then be isolated by fluorescence-activated cell sorting (FACS). Relevant genes can subsequently be identified by PCR and DNA sequencing. In this work, 27 proteins, including 10 novel proteins, were identified as mitochondrial (Ozawa et al. 2003).

This basic concept for identifying mitochondrial proteins was extended for designing a new indicator for identifying proteins translocating into the endoplasmic reticulum. A tandem fusion protein, containing an endoplasmic reticulum-targeting signal (ERTS) and the C-terminal fragments of DnaE and EGFP, localizes in the lumen of the endoplasmic reticulum. If test proteins expressed from the cDNA library contain an ERTS, the fusion products translocate into the endoplasmic reticulum, generating mature EGFP. Using the same procedure as the mitochondrial case, the fluorescent cells are collected by FACS, and relevant genes are identified. In this work, 110 non-redundant proteins targeting to the endoplasmic reticulum were identified (Ozawa et al. 2005). An important point of these genetic methods for identifying organelle-localized proteins is that they do not require purification of target organelles or separation of each protein. Therefore, the proteins localized in other organelles do not contaminate. Moreover, the methods provide accurate identification of gene products that are compartmentalized in the mitochondria or endoplasmic reticulum. This genetic approach will also allow the identification of proteins localized in the nucleus and peroxisomes by replacing the signal sequence attached to the C-terminal fragments of DnaE and EGFP with the targeting signals of each.

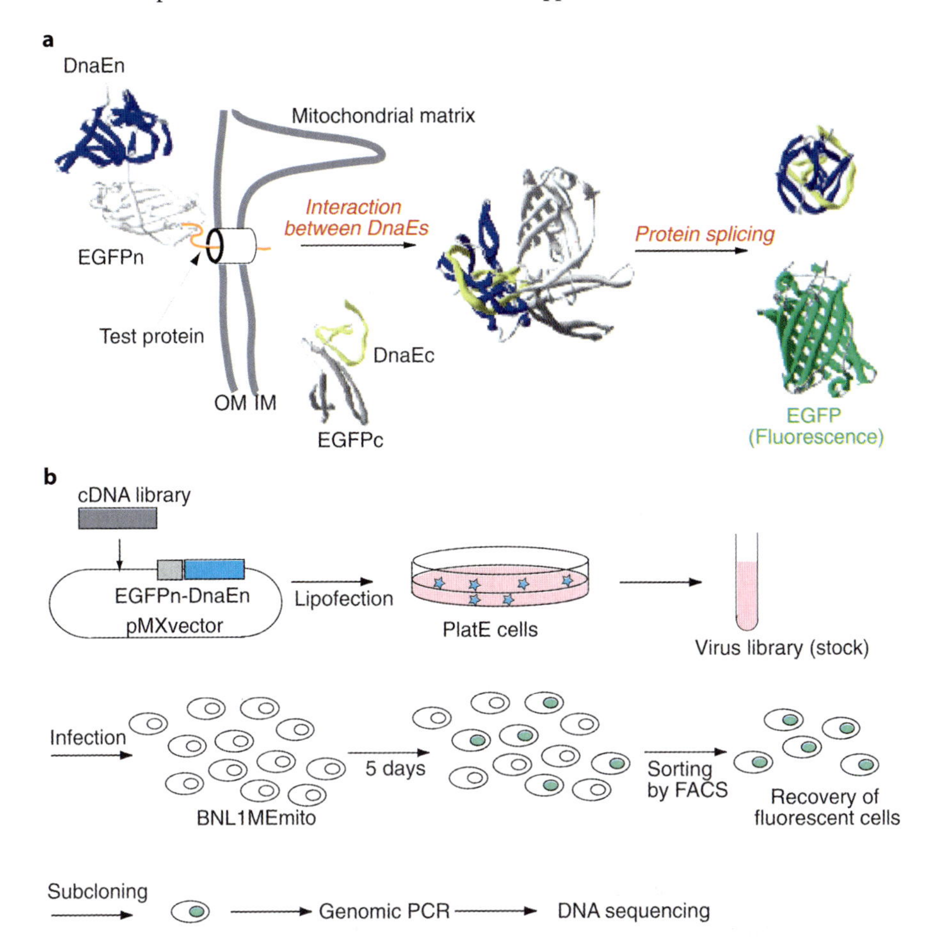

Fig. 6. Scheme showing how to identify mitochondrial proteins from cDNA libraries. **a** Principle for detecting translocation of a test protein into mitochondria using protein splicing of split-EGFP. EGFPc is connected with DnaEc and the mitochondrial targeting signal (MTS), which is predominantly localized in mitochondria. A test protein is connected with the EGFPn and DnaEn, which is expressed in the cytosol. When the test protein translocates into mitochondria, the DnaEn interacts with DnaEc, and protein splicing results. The EGFPn and EGFPc are linked together by a peptide bond, and the reconstituted EGFP recovers its fluorescence. **b** Strategy for identifying mitochondrial proteins. BNL1MEmito cells, which are expressing EGFPc-DnaEc, are infected with retrovirus libraries. Fluorescent cells are sorted by FACS, and subcloned into individual cells. cDNAs integrated in the genome are extracted by PCR, and identified by DNA sequencing

3.2.2 Detection of Protein Nuclear Transport

Protein movement inside living cells is an important dynamic event in eukaryotic cells. A typical example is protein nuclear transport, which plays a key role in regulating gene expression in response to extracellular signals. In order to non-invasively detect this molecular event in living subjects, a probe molecule using *Renilla* luciferase (Rluc) reconstitution by protein splicing has been developed (Fig. 7; Kim et al. 2004). Rluc has desirable features for a monomeric protein: small size (36 kDa), strong luminescence, and ATP is not nec-

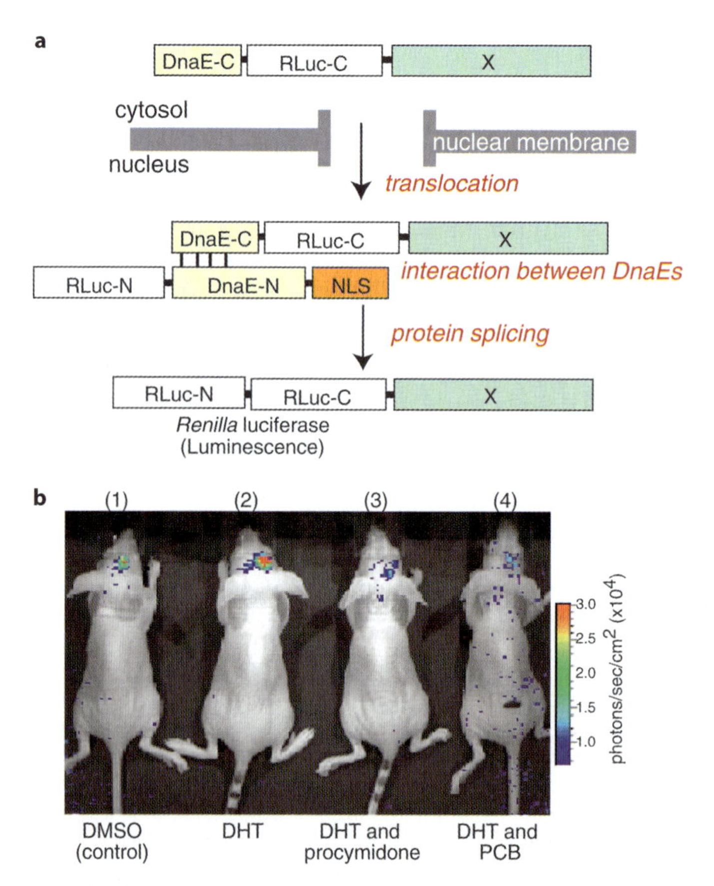

Fig. 7. Strategy for detecting protein translocations. **a** Principle of monitoring translocation of a particular protein (X) into the nucleus using protein splicing of split-*Renilla* luciferase (Rluc). RLuc-N (1–229 aa) is connected with DnaE-N and the nuclear localization signal (NLS), which is predominantly localized in the nucleus. DnaE-C is connected with RLuc-C (230–311 aa) and a protein X, which is localized in the cytosol. When the protein X translocates into the nucleus, the DnaE-C interacts with DnaE-N, and protein splicing results. RLuc-N and RLuc-C are linked by a peptide bond, and the reconstituted RLuc recovers its bioluminescence activity. **b** DHT-dependent translocation of AR in the mouse brain. The COS-7 cells expressing the probe were implanted in the forebrain of the mice. Of mouse groups 1–4, groups 1 and 2 were stimulated with 1% DMSO, whereas groups 3 and 4 were stimulated with procymidone and PCB, respectively. After DHT stimulation, the mice were imaged with a cooled CCD camera

essary for its activity (Lorenz et al. 1991). In addition, its substrate, coelenterate luciferin (coelenterazine), rapidly penetrates through cell membranes, making it suitable for in vivo imaging. The luciferase is split into N- and C-terminal fragments, and they are connected to N- and C-terminal fragments of the DnaE intein, respectively. The C-terminal fragment is permanently localized in the nucleus, while the N-terminal fragment is fused to a test protein in the cytosol. If the test protein translocates into the nucleus, the N-terminal Rluc can interact with the C-terminal Rluc in the nucleus, and full-length Rluc is reconstituted by protein splicing. In order to demonstrate the utility of this indicator, a well-known nuclear receptor, androgen receptor (AR), was examined. This receptor translocates from the cytosol into the nucleus upon binding to 5α-dihydrotestosterone (DHT). Quantitative analysis of AR translocation was shown with various exo- and endogenous chemical compounds in vitro. Moreover, the indicator enabled non-invasive in vivo imaging of AR translocation in the brains of living mice with a cooled CCD imaging system. This rapid and quantitative analysis in vitro and in vivo will provide a wide variety of applications for screening pharmacological or toxicological compounds and testing them in living animals (Kim et al. 2004).

3.2.3 *trans-Splicing in the Chloroplast in Plant Cells*

Split-reporter protein reconstitution by protein splicing with the DnaE intein (Ozawa et al. 2001a, 2003) has been further applied to growing safer transgenic plants. Genetic crossing in transgenic plants by pollen has been reported and is known to be a potential environmental risk (Bergelson et al. 1998). Genetically modified plants with a transgene integrated in the nucleus may be able to hybridize with a sexually compatible species to give rise to unexpected hybrids and their progeny.

Chin et al. (2003) have demonstrated the reconstitution of a herbicide-resistance protein from split genes in tobacco plants. An N-terminal portion of the herbicide-resistance gene *5-enol-pyruvylshikimate-3-phosphate synthase* (EPSPS) containing a chloroplast localization signal fused to the N-terminal DnaE intein fragment (EPSPSn-DnaEn) was integrated into the nuclear DNA of a *tabacum* plant. The remaining EPSPS gene fragment (EPSPSc) fused to the C-terminal DnaE intein fragment (EPSPSc-DnaEc) was integrated into the chloroplast genome by homologous recombination. The full-length EPSPS protein was generated in the chloroplast after translocation of the PESPSn-DnaEn fusion protein fragment to the plastid followed by trans-splicing. The resulting transgenic plants displayed improved resistance to the herbicide *N*-(phosphonomethyl)glycine (glyphosate), when compared with wild-type plants.

The point is that the chloroplasts of most crops are always maternally transmitted, and pollen-mediated gene transfer is limited to the gene in the nucleus. Therefore, pollen can spread only the nucleus-based fragment of the split gene of EPSPS far and wide, which is inactive. Thus, putting one part of a new gene into the chloroplast protects against transfer of the full gene to other plants. This system may be broadly applicable to all crop species for transgene containment, and a combination of approaches may prove most effective for environmentally safe transgenic crops.

3.3 Screening of Potential Antimycobacterial Agents

The split GFP reporter (Ozawa et al. 2000) was applied to the screening of antimycobacterial agents. Inhibitors of protein splicing could become highly specific antimycobacterial antibiotics because among many bacteria associated with a human host only mycobacteria include intein genes in their genomes (Paulus 2003). For example, the RecA DNA repair enzyme of *Mycobacterium tuberculosis* contains an intein. Protein splicing by the RecA intein restores the DNA-repairing enzyme, which is essential for the mycobacterial growth and survival (Davis et al. 1992).

Gangopadhyay et al. (2003) developed a high-throughput screening system for general protein-splicing inhibitors in vitro. The principle is based on split GFP reconstitution by protein splicing by the RecA intein. Insertion of the RecA intein at amino acid residue 129 of a GFP variant (GFPuv) causes the resulting fusion protein to be expressed entirely as inclusion bodies (Fig. 8). When the GFPuv-intein fusion protein is solubilized with urea and renatured, the fusion protein is able to undergo efficient protein splicing to yield GFPuv, leading to formation of the fluorescent chromophore. The formation of fluorescent GFPuv is thus sensitive to the presence of *anti*-splicing, and therefore potentially antimycobacterial, agents.

Conventional in vivo screening systems have the disadvantage that they involve monitoring the growth of bacteria and are therefore subject to interference by non-specific antibacterial agents. In contrast to the in vivo system, this in vitro system allows the examination of the inhibition of the splicing reaction specifically, and is therefore not susceptible to false signals that inhibit other reactions essential for growth.

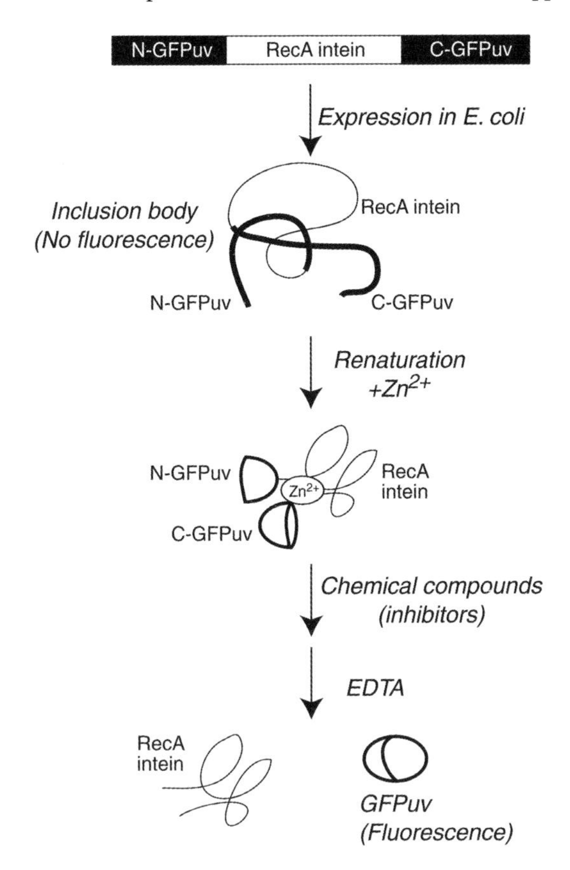

Fig. 8. In vitro screening system for protein splicing of RecA intein. A fusion protein composed of N-GFPuv, RecA intein, and C-GFPuv is expressed in *E. coli*, forming inclusion bodies. The inclusion bodies are solubilized with urea, refolded in the presence of zinc, and then possible inhibitors are added. Splicing is induced by addition of EDTA to neutralize zinc inhibition of protein splicing. The fluorescence intensity of the reconstituted GFP is measured. This method eliminates the isolation of compounds that simply interfere with refolding but are not intein-specific

4 Future Directions

Since the discovery of intein-based protein splicing, many inteins have been identified. Although details of the mechanism and kinetics of protein splicing have still to be worked out, it has been shown that inteins can become very powerful tools in the fields of chemistry and biology. A particularly compelling aspect of inteins is that the protein-splicing reaction occurs in vivo and in vitro. Applications maximizing this feature will continue to expand and provide exciting de novo insights into various fields. Further understanding of protein-splicing systems, discovery of new inteins, and development of modified inteins with novel properties will certainly advance intein technologies as a whole. Researchers are just beginning to uncover part of the treasures buried in inteins. Development of a variety of attractive and clever applications will increasingly speed up and emerge in the near future.

References

Bergelson J, Purrington CB, Wichmann G (1998) Promiscuity in transgenic plants. Nature 395:25

Brejc K, Sixma TK, Kitts PA, Kain SR, Tsien RY, Ormo M, Remington SJ (1997) Structural basis for dual excitation and photoisomerization of the *Aequorea victoria* green fluorescent protein. Proc Natl Acad Sci USA 94:2306–2311

Buskirk AR, Ong YC, Gartner ZJ, Liu DR (2004) Directed evolution of ligand dependence: small-molecule-activated protein splicing. Proc Natl Acad Sci USA 101:10505–10510

Chin HG, Kim GD, Marin I, Mersha F, Evans TC Jr, Chen L, Xu M-Q, Pradhan S (2003) Protein trans-splicing in transgenic plant chloroplast: reconstruction of herbicide resistance from split genes. Proc Natl Acad Sci USA 100:4510–4515

Chong S, Xu MQ (1997) Protein splicing of the *Saccharomyces cerevisiae* VMA intein without the endonuclease motifs. J Biol Chem 272:15587–15590

Conti E, Franks NP, Brick P (1996) Crystal structure of firefly luciferase throws light on a superfamily of adenylate-forming enzymes. Structure 4:287–298

David R, Richter MP, Beck-Sickinger AG (2004) Expressed protein ligation. Method and applications. Eur J Biochem 271:663–677

Davis EO, Jenner PJ, Brooks PC, Colston MJ, Sedgwick SG (1992) Protein splicing in the maturation of *M. tuberculosis* recA protein: a mechanism for tolerating a novel class of intervening sequence. Cell 71:201–210

Evans TJT, Xu M-Q (2002) Mechanistic and kinetic considerations of protein splicing. Chem Rev 102:4869–4884

Fields S, Song O (1989) A novel genetic system to detect protein-protein interactions. Nature 340:245–246

Gangopadhyay JP, Jiang SQ, Paulus H (2003) An in vitro screening system for protein splicing inhibitors based on green fluorescent protein as an indicator. Anal Chem 75:2456–2462

Grimm S (2004) The art and design of genetic screens: mammalian culture cells. Nat Rev Genet 5:179–189

Hirata R, Ohsumi Y, Nakano A, Kawasaki H, Suzuki K, Anraku Y (1990) Molecular structure of a gene, VMA1, encoding the catalytic subunit of H(+)-translocating adenosine triphosphatase from vacuolar membranes of *Saccharomyces cerevisiae*. J Biol Chem 265:6726–6733

Ho SN, Biggar SR, Spencer DM, Schreiber SL, Crabtree GR (1996) Dimeric ligands define a role for transcriptional activation domains in reinitiation. Nature 382:822–826

Kane PM, Yamashiro CT, Wolczyk DF, Neff N, Goebl M, Stevens TH (1990) Protein splicing converts the yeast TFP1 gene product to the 69-kD subunit of the vacuolar H(+)-adenosine triphosphatase. Science 250:651–657

Kawasaki M, Makino S, Matsuzawa H, Satow Y, Ohya Y, Anraku Y (1996) Folding-dependent in vitro protein splicing of the *Saccharomyces cerevisiae* VMA1 protozyme. Biochem Biophys Res Commun 222:827–832

Kim SB, Ozawa T, Watanabe S, Umezawa Y (2004) High-throughput sensing and noninvasive imaging of protein nuclear transport by using reconstitution of split *Renilla* luciferase. Proc Natl Acad Sci USA 101:11542–11547

Lorenz WW, McCann RO, Longiaru M, Cormier MJ (1991) Isolation and expression of a cDNA encoding *Renilla reniformis* luciferase. Proc Natl Acad Sci USA 88:4438–4442

Martin DD, Xu M-Q, Evans TC Jr (2001) Characterization of a naturally occurring *trans*-splicing intein from *Synechocystis* sp. PCC6803. Biochemistry 40:1393–1402

Mootz HD, Muir TW (2002) Protein splicing triggered by a small molecule. J Am Chem Soc 124:9044–9045

Mootz HD, Blum ES, Tyszkiewicz AB, Muir TW (2003) Conditional protein splicing: a new tool to control protein structure and function in vitro and in vivo. J Am Chem Soc 125:10561–10569

Nishiuchi Y, Inui T, Nishio H, Bodi J, Kimura T, Tsuji FI, Sakakibara S (1998) Chemical synthesis of the precursor molecule of the *Aequorea* green fluorescent protein, subsequent folding, and development of fluorescence. Proc Natl Acad Sci USA 95:13549–13554

Ormo M, Cubitt AB, Kallio K, Gross LA, Tsien RY, Remington SJ (1996) Crystal structure of the *Aequorea victoria* green fluorescent protein. Science 273:1392–1395

Ozawa T, Umezawa Y (2002) Peptide assemblies in living cells. Methods for detecting protein-protein interactions. Supramol Chem 14:271–280

Ozawa T, Nogami S, Sato M, Ohya Y, Umezawa Y (2000) A fluorescent indicator for detecting protein-protein interactions in vivo based on protein splicing. Anal Chem 72:5151–5157

Ozawa T, Kaihara A, Sato M, Tachihara K, Umezawa Y (2001a) Split luciferase as an optical probe for detecting protein-protein interactions in mammalian cells based on protein splicing. Anal Chem 73:2516–2521

Ozawa T, Nishitani K, Sako Y, Umezawa Y (2005) A high-throughput screening of genes that encode proteins transported into the endoplasmic reticulum in mammalian cells. Nucleic Acids Res 33:e34

Ozawa T, Takeuchi TM, Kaihara A, Sato M, Umezawa Y (2001b) Protein splicing-based reconstitution of split green fluorescent protein for monitoring protein-protein interactions in bacteria: improved sensitivity and reduced screening time. Anal Chem 73:5866–5874

Ozawa T, Sako Y, Sato M, Kitamura T, Umezawa Y (2003) A genetic approach to identifying mitochondrial proteins. Nat Biotechnol 21:287–293

Paulmurugan R, Umezawa Y, Gambhir SS (2002) Noninvasive imaging of protein-protein interactions in living subjects by using reporter protein complementation and reconstitution strategies. Proc Natl Acad Sci USA 99:15608–15613

Paulus H (2003) Inteins as targets for potential antimycobacterial drugs. Front Biosci 8: s1157–s1165

Pietrokovski S (1998) Modular organization of inteins and C-terminal autocatalytic domains. Protein Sci 7:64–71

Shioda T, Andriole S, Yahata T, Isselbacher KJ (2000) A green fluorescent protein-reporter mammalian two-hybrid system with extrachromosomal maintenance of a prey expression plasmid: application to interaction screening. Proc Natl Acad Sci USA 97:5220–5224

Westermann B, Neupert W (2003) 'Omics' of the mitochondrion. Nat Biotechnol 21:239–240

Wu H, Hu Z, Liu XQ (1998) Protein trans-splicing by a split intein encoded in a split DnaE gene of *Synechocystis* sp. PCC6803. Proc Natl Acad Sci USA 95:9226–9231

Yang F, Moss LG, Phillips GN Jr (1996) The molecular structure of green fluorescent protein. Nat Biotechnol 14:1246–1251

Zhang J, Campbell RE, Ting AY, Tsien RY (2002) Creating new fluorescent probes for cell biology. Nat Rev Mol Cell Biol 3:906–918

Intein Reporter and Selection Systems

David W. Wood, Georgios Skretas

1 Introduction

Three aspects of intein biology suggest a wide variety of uses in biotechnology. The first, described in this volume by Dassa and Pietrokovski, is that inteins are genetically mobile, and a characteristic exhibited by many inteins is the retention of splicing activity when artificially moved to foreign contexts. The second is that inteins can be used to reversibly inactivate non-native host proteins through internal and external gene fusions, allowing inteins to be used as molecular triggers in various reporter systems. This aspect of intein biology forms the foundation for this chapter and the systems described within. Finally, as explained by Mills and Paulus in this volume, the splicing reaction itself is fairly well understood and several intein structures have been solved (see Moure and Quiocho, this Vol.). This has allowed inteins to be modified for new biological applications, including those described in chapters by Chong and Xu; Tavassoli, Naumann and Benkovic; Ozawa and Umezawa; and Wood, Harcum and Belfort. The repertoire of available inteins has recently been augmented with examples of split inteins that can assemble and splice *in trans*, enabling the development of additional strategies for gene assembly and activation.

The ability of inteins to function in non-native contexts initially facilitated the development of several convenient model systems for intein study and characterization (Davis et al. 1991; Cooper et al. 1993; Xu et al. 1993; Chong et al. 1996). In time, however, it became clear that the flexibility and robustness of inteins suggested a wide variety of potential applications (Perler and Adam 2000). In pursuit of these applications, a number of additional systems have been developed to link intein activity to readily detectable or selectable

D.W. Wood (e-mail: dwood@princeton.edu), G. Skretas
Department of Chemical Engineering, Princeton University,
Princeton, New Jersey, USA

Nucleic Acids and Molecular Biology, Vol. 16
Marlene Belfort et al. (Eds.)
Homing Endonucleases and Inteins

phenotypes in expressing cells. In all cases, these systems are based on the genetic fusion of an intein to a well-characterized reporter protein. Splicing, or in some cases cleaving by the intein, activates the reporter protein, allowing the easy identification of conditions that affect the activity of the intein (Fig. 1).

Applications of these systems can be broken down into four broad categories. The first is simply to provide a tool to study intein activity under various conditions. Several early reporters provided initial insights into intein function in non-native host cells and proteins (Davis et al. 1992; Cooper et al. 1993), while more recent reporters have been used to examine the effects of specif-

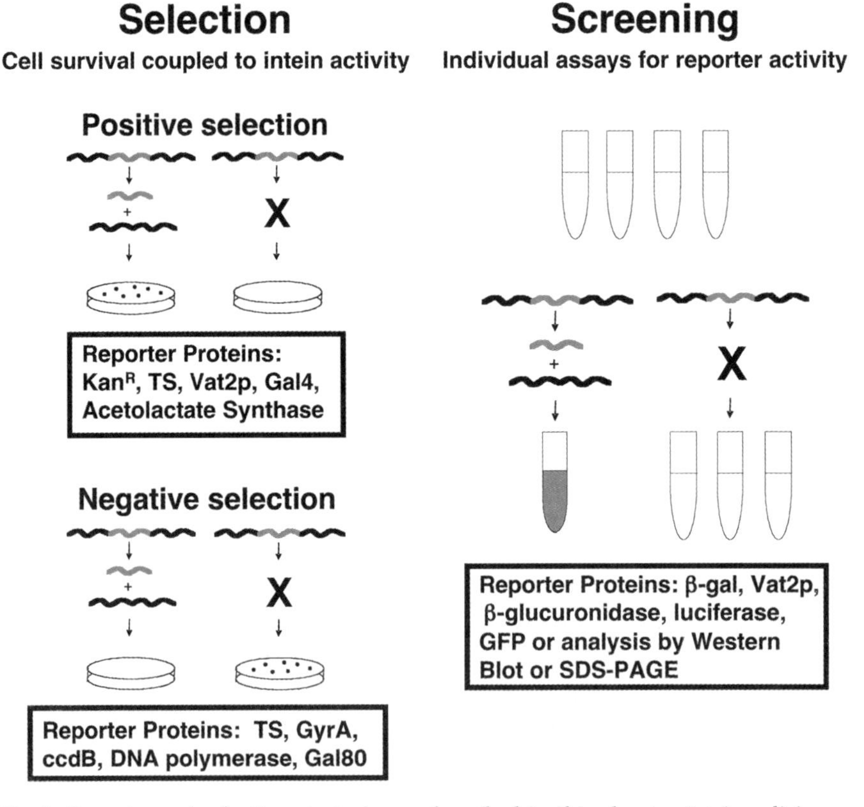

Fig. 1. Reporter and selection strategies as described in this chapter. Intein splicing or cleaving activates a reporter protein, leading to a change in growth phenotype (*selection*), or the presence of an assayable activity or change in molecular weight of the precursor (*screening*). Reporter proteins are designated for each strategy as discussed in the text.

ic intein and extein mutations on intein function (Xu et al. 1993; Chong et al. 1996; Xu and Perler 1996; Derbyshire et al. 1997; Kawasaki et al. 1997; Ghosh et al. 2001; Lew and Paulus 2002). A second application has been to use intein fusions to select open reading frames (ORFs) from large DNA libraries. In this case, the intein is linked to a selectable marker and the DNA library is inserted in such a way that non-ORF sequences will disrupt intein expression and function (Daugelat and Jacobs 1999). A third use for intein-reporter fusions is to facilitate the rapid screening of large chemical libraries for compounds that inhibit or otherwise affect intein splicing (Adam and Perler 2002; Lew and Paulus 2002; Gangopadhyay et al. 2003; Cann et al. 2004). Although there are many possible uses for such compounds, an important initial purpose for them would be to provide a new class of antibiotics against pathogenic intein-harboring bacteria (Belfort 1998; Paulus 2003; Liu and Yang 2004). Finally, an important application for many of these systems has been the rapid isolation of valuable intein mutants through directed evolution. Intein variants generated by this method have been successfully used for protein purification (Wood et al. 1999), and have shown potential as a general means to post-translationally activate arbitrary target proteins *in vivo* on demand (Buskirk et al. 2004; Zeidler et al. 2004; Skretas and Wood 2005).

This chapter is devoted to describing the development and use of reporter systems for intein splicing and cleaving. For each method, the underlying principle will be described, along with its most prevalent applications and significant results. The variety of these systems is increasing rapidly, and it is becoming clear that virtually any genetic selection scheme can be modified to include an intein-splicing component.

2 Systems for Direct Observation of Intein Function

The first inteins were discovered when discrepancies were noted between the lengths of certain genes and the size of their translated proteins (Hirata et al. 1990; Davis et al. 1991). It was quickly determined that the transcripts of these genes were not being processed, and subsequent phylogenetic analyses indicated the presence of large ORF insertions within their sequences. These results suggested that some form of splicing must be occurring at the protein level, and several conventional methods were used to examine and prove this hypothesis (Kane et al. 1990; Davis et al. 1992).

The initial studies of protein splicing relied heavily on Western blot analysis of the unspliced precursor and spliced products in the native context (Kane et al. 1990; Davis et al. 1991). It was eventually shown that splicing could occur with most of the native extein sequences deleted, and ultimately the intein's ability to splice was proven to be largely self-contained (Davis et al. 1992; Cooper et al. 1993). A breakthrough system for intein study was soon developed by inserting a hyperthermophilic *Pyrococcus* intein between two highly soluble domains, with one of them acting as an affinity tag for rapid purification of the precursor (Xu et al. 1993). Expression and purification of this fusion at low temperature provided an unspliced precursor, whose in vitro splicing could then be followed and analyzed via simple sodium dodecyl sulfate/polyacrylamide gel electrophoresis (SDS-PAGE). This technique was soon extended to the study of the *Saccharomyces cerevisiae* VMA1 intein (the Sce intein), resulting in several significant papers on intein manipulation and potential applications (Chong et al. 1996; Xu and Perler 1996). A critical result of this work was the identification of specific junction mutations that lead to isolated cleaving activity, leading to the first available system for intein-based affinity separations (Chong et al. 1997).

These initial methods, although cumbersome, can be viewed as primitive reporters for the splicing reaction. A critical capability of many of these early systems, which has not been matched in any subsequent enzyme-activation reporter system, is the ability to detect varying levels of precursor degradation through cleaving side reactions. These side reactions will generate a negative-splicing phenotype, but cannot be differentiated from the formation of a stable, unspliced precursor except through direct observation of the reaction products.

3 Selection and Reporter Systems for Intein Function

Direct observation of intein products through SDS-PAGE and by Western blot are still used in many cases, particularly for the verification of splicing with newly discovered inteins in their native hosts and for the observation of splicing from artificial precursors in *Escherichia coli* (Liu and Yang 2004; Mills et al. 2004; Southworth et al. 2004). They have even been used to screen small libraries of up to several hundred mutants for loss of splicing function (Kawasaki et al. 1997). However, these methods are impractical for evaluating libraries of thousands of mutant inteins, or in any application involving high-throughput screening. Therefore, a number of indirect methods have been developed where intein function is linked to an easily screened or selectable phenotype in expressing cells (Table 1). These techniques often constitute new intein applications by themselves, and have also facilitated the generation of new inteins with potentially useful characteristics.

Table 1. Intein reporter and selection systems

Reporter protein	Host	Inteins tested	Reporter/selection	Activity type	Reference	Comments
β-Galactosidase (α-fragment or full enzyme)	*E. coli*	*Mtu* RecA	Reporter	cis-splicing	Davis et al. (1992); Buskirk et al. (2004); Skretas and Wood (2005)	Should work with many types of β-galactosidase assays in many hosts
Yeast Vat2p	*S. cerevisiae*	*Sce* VMA	Positive selection or reporter	cis-splicing	Cooper et al. (1993); Buskirk et al. (2004)	Equivalent to VMA1 protein reporter
Thymidylate Synthase	*E. coli*	*Mtu* RecA	Positive or negative selection	cis-splicing or C-cleavage	Derbyshire et al. (1997); Wood et al. (1999); Skretas and Wood (2005)	Used to isolate fast-splicing and fast-cleaving mini-intein mutants
Aminoglycoside Phosphotrans-ferase (Kan^r protein)	*E. coli* or Mycobacteria	*Mtu* RecA	Positive selection	cis-splicing	Daugelat and Jacobs (1999); Buskirk et al. (2004)	ORFTRAP selection system and evolution of alosteric intein
GyrA	*E. coli*	Mycobacterial GyrA inteins	Negative selection	cis-splicing	Adam and Perler (2002)	Temperature-sensitive mutant or splicing inhibitor isolation
CcdB	*E. coli*	*Mtu* RecA	Negative selection	cis-splicing or C-cleavage	Lew and Paulus (2002)	Used to show robust behavior of *Mtu* intein
RB69 DNA polymerase	*E. coli*	*Tli* Pol-2[a]	Negative selection	cis-splicing	Cann et al. (2004)	Isolation of temperature-sensitive mutants and splicing inhibitors
Gal4/Gal80	*S. cerevisiae* or *Drosophila*	*Sce* VMA	Positive (Gal4) or negative (Gal80) selection	cis-splicing	Zeidler et al. (2004)	Should also work in mammalian cells

↓

Table 1. *(Continued)*

Reporter protein	Host	Inteins tested	Reporter/selection	Activity type	Reference	Comments
GFP	*In vitro*	*Mtu* RecA	Reporter	cis-splicing	Gangopadhyay et al. (2003)	Only reported cis-splicing *in vitro* system
GFP	*S. cerevisiae*	*Mtu* RecA (chimeric)	Reporter	cis-splicing	Buskirk et al. (2004)	Used to isolate ligand-triggered inteins
GFP	*E. coli*	*Sce* VMA (split intein)	Reporter	trans-splicing	Ozawa et al. (2001)	Reports protein-protein interactions
Acetolactate synthase isoform II	*E. coli*	*Ssp* dnaE[b]	Positive selection	trans-splicing	Ghosh et al. (2001)	Study of trans-splicing modulators *in vivo*
β-Glucuronidase	*Arabidopsis*	*Ssp* dnaE	Reporter	trans-splicing	Yang et al. (2003)	Demonstrates *in vivo* splicing in whole plants
Firefly luciferase	Mice *in situ*	*Ssp* dnaE	Reporter	trans-splicing	Paulmurugan et al. (2002)	Detects protein–protein interactions in living mice

[a] *Thermococcus litoralis Pol-2* DNA polymerase intein.
[b] *Synechocystis sp.* PCC6803 naturally split DnaE intein.

3.1 Insertion into the *lacZα* Gene

One of the first systems used to examine the ability of inteins to splice in non-native host contexts was based on the simple β-galactosidase complementation assay (Davis et al. 1992). The *Mycobacterium tuberculosis* RecA intein (Mtu RecA intein), along with 8 N-terminal and 20 C-terminal native extein residues, was cloned into the polylinker region of pUC19 to form a single ORF with exteins encoding the alpha fragment of β-galactosidase (LacZα). Intein splicing in this context would yield an active LacZα fragment, which could be readily detected by the widely used blue/white assay (Sambrook and Russell 2001). When this fusion-encoding plasmid was transformed into *E. coli* TG2 cells, the appearance of blue colonies on X-gal indicator plates confirmed that α-complementation was taking place. Furthermore, Western blots confirmed the presence of the excised intein.

Although this work appears to constitute the first successful intein-splicing reporter system, several key issues were left unresolved. One of these is the fact that the spliced extein product was never detected, although this product would presumably be required for the appearance of blue colonies. Furthermore, successful splicing would still leave 28 Mtu RecA extein residues within the LacZα fragment, suggesting that complementation can occur with significant insertions. This led the authors to wonder if the unspliced precursor might be able to provide α-complementation, in which case successful splicing would not be required. Control experiments to discount this possibility were not included in this early work. Nevertheless, this system was the first to suggest that inteins can inactivate foreign hosts through gene insertion and re-activate them by protein splicing, and subsequent studies using the full-length LacZ protein (Buskirk et al. 2004) and the α-fragment (Skretas and Wood 2005) support its reversible inactivation by intein insertion. Surprisingly, despite the simplicity and success of the LacZα reporter system, it has not yet become commonly used in the evaluation of new inteins or in the selection of intein mutants.

3.2 Insertion into the *Saccharomyces cerevisiae VAT2* Gene

Although the *S. cerevisiae* VMA1 intein (Sce intein) was the first to be reported (Hirata et al. 1990) and expressed in *E. coli* (Kane et al. 1990), its transfer into a new host protein was not reported until after the Mtu intein-LacZα system. However, the transfer of the Sce intein into the *VAT2* gene produced the first potential genetic selection for intein-splicing activity in yeast (Cooper et al. 1993). The *VAT2* gene encodes the 60 kDa subunit (Vat2p) of the vacuolar H^+-ATPase in yeast, and knockout strains for this gene can grow at pH 5.0, but not at pH 7.5 (Yamashiro et al. 1990). In principle, selection for growth at pH 7.5 using an intein-Vat2p fusion would provide a selection for active splicing. Furthermore, *ade2* strains have been developed to provide a color assay where *VAT2* knockouts form white colonies as opposed to red (Foury 1990).

This work was accompanied by several important discoveries. The first was that splicing could occur from fully non-native contexts (as long as the intein is followed by its obligatory C-terminal splice junction residue). Several junction mutations were also identified that abolished splicing or led to isolated cleaving reactions. Finally, one mutation was identified that slowed the splicing reaction to a point where unspliced precursor could be detected and shown to undergo splicing over time, thus allowing the first direct observation of intein splicing.

Ironically, despite the availability of the *VAT2* reporter system, all of the mutational studies associated with this work were carried out by Western blot analysis (Cooper et al. 1993). Indeed, as with the LacZα system, a negative control experiment to demonstrate that Vat2p activity was dependent on protein splicing was not initially carried out. However, subsequent investigators have used the original Sce intein *VMA1* host gene as a reporter for intein function in yeast, and it has been shown that successful splicing is required for activity in this context (Kawasaki et al. 1997; Nogami et al. 1997). This selection system yields phenotypes similar to the *VAT2* system, including the red/white color assay and pH sensitivity. It has also been shown that cells become sensitive to the presence of calcium ion in the non-splicing case (Kawasaki et al. 1997), thus providing an additional potential selection system.

3.3 Thymidylate Synthase System for Selection *in vivo*

A highly flexible and effective system for monitoring and selecting intein function is generated through intein fusions to the thymidylate synthase enzyme (TS; Derbyshire et al. 1997; Wood et al. 1999). This enzyme converts deoxyuridine monophosphate (dUMP) into deoxythymidine monophosphate (dTMP), which is ultimately converted into deoxythymidine triphosphate (dTTP) and incorporated into newly synthesized DNA (Carreras and Santi 1995). For each dTTP molecule produced, TS consumes one molecule of tetrahydrofolate cofactor. In the absence of exogenous thymine, a functional TS enzyme is absolutely required for DNA replication and cell growth, thus allowing a strong positive selection for TS function. On the other hand, in the presence of exogenous thymine and trimethoprim (a dihydrofolate reductase inhibitor), depletion of tetrahydrofolate by active TS arrests cell growth, thus providing a strong negative selection for TS activity (Belfort and Pedersen-Lane 1984). Furthermore, the positive selection for TS activity becomes more stringent at higher incubation temperatures (Derbyshire et al. 1997), while the negative selection can be tuned by the amount of trimethoprim added to the growth medium.

This selection system is made more useful by the fact that TS can be reversibly inactivated by both intein insertion (Derbyshire et al. 1997) and by external intein fusions to the N-terminus of the TS enzyme (Wood et al. 1999). In the internal fusion configuration, this system has been used as a reporter for splicing of artificially engineered Mtu RecA mini-inteins (Derbyshire et al. 1997), and has been used as a selection system to isolate engineered mini-inteins with native-like splicing activity (Wood et al. 1999). In the external fusion configuration, the intein is inserted between a maltose-binding protein and the TS enzyme to generate a highly soluble precursor protein (Wood et al. 1999). C-terminal cleavage by the intein releases and activates the TS enzyme, allowing the system to be used as a reporter for this reaction in mutated inteins and for the isolation of fast-cleaving mutant mini-inteins for use in bioseparations (Wood et al. 1999, 2000).

An important advantage of this system is the availability of relatively trivial positive and negative selections that can each be applied directly to mutant libraries. It has also been shown that this system can report extremely low levels of splicing by incubation at low temperature, but can also be used to select high levels of splicing and cleaving by incubation at higher temperatures (Derbyshire et al. 1997). This aspect, combined with the trimethoprim counterselection, provides the potential to allow the isolation of highly active and controllable inteins for both splicing and C-terminal cleaving applications.

3.4 Kanamycin Resistance: the ORFTRAP System

The ORFTRAP system links intein splicing to kanamycin resistance through the aminoglycoside phosphotransferase (*aph*) gene (Daugelat and Jacobs 1999). The goal of the ORFTRAP system is to provide a genetic selection to isolate large ORF fragments from libraries of digested chromosomal DNA. To construct the system, the Mtu RecA intein is inserted into the *aph* gene of plasmid pYUB53. DNA libraries are then cloned into an insert-permissive site in the intein such that ORF-encoding fragments will allow the entire precursor protein to be expressed, spliced and activated to confer a Kan^r phenotype in *E. coli* and *Mycobacteria* (Jacobs et al. 1991). However, if the insert is out of frame or contains a stop codon, the precursor will be truncated and a Kan^s phenotype will result. Screening transformants on kanamycin-containing medium thus allows the direct selection of clones containing ORF-encoding fragments.

The engineering of this system was somewhat difficult, as it required modification of the *aph* gene as well as the intein, and also required the identification of a highly insert-tolerant location in the intein. Furthermore, these modifications had to be designed without the availability of any solved intein structures. Construction began with insertion of the Mtu intein, plus a functionally required Cys residue at its C-terminus, into the *aph* gene at a unique SspI site. Although this insertion results in the addition of a Cys residue between the spliced exteins, this product is still able to confer a Kan^r phenotype in *E. coli* as a result of intein splicing. Linker-scanning mutagenesis was then combined with the kanamycin reporter to identify several permissive sites for insertions within the intein. The ORFTRAP vector was completed with the introduction of a two base pair frameshift immediately upstream of the most permissive site, thus preventing the original vector from generating a Kan^r phenotype.

A significant advantage of this system is that both the N- and C-exteins must be expressed as part of a single translation product in order to generate a positive phenotype. This aspect prevents the isolation of false positives arising from cryptic translational start sites within non-ORF fragments, as is common in simpler ORF-selection systems (Lutz et al. 2002). Despite some limitations, the authors have reported success in delivering screened ORFs from libraries of digested DNA and have been awarded a patent for this work (Jacobs and Daugelat 1999).

3.5 DNA GyraseA: Negative Selection for Splicing *in vivo*

The absence of inteins from multicellular organisms and their presence in vital genes of pathogenic bacteria suggest the potential for protein-splicing inhibitors to be used as therapeutics against several specific bacterial infections (Lew and Paulus 2002; Paulus 2003; Liu and Yang 2004). One of the most popular native gene hosts for inteins are DNA gyrases, and a number of mycobacterial species such as *Mycobacterium xenopi* and *M. leprae* have been found to contain intein insertions in their gyrase genes. In contrast to other inteins, such as the Mtu intein, which retain high levels of splicing activity when transferred to non-native protein hosts, many gyrase inteins have been found to require native extein sequences for efficient splicing (Telenti et al. 1997; Southworth et al. 1999). This aspect of the mycobacterial gyrase inteins requires special consideration in the construction of systems to report their function.

To address this difficulty, a reporter system was developed specifically for gyrase inteins through the use of the highly homologous *E. coli* gyrase A protein (GyrA). Active DNA gyrase is a heterotetrameric enzyme that catalyzes the negative supercoiling of DNA. Most wild-type gyrases are sensitive to quinolone drugs such as ofloxacin, which bind to the GyrA subunit and force it to form a covalent bond with the DNA substrate, leading to cell death (Hawkey 2003). Although GyrA can be rendered quinolone-resistant with a single amino acid substitution (GyrA^r), co-expression of quinolone-sensitive GyrA (GyrA^s) is dominant and will still lead to cell death in the presence of ofloxacin. The GyrA intein selection system exploits this dominant-lethal effect by linking intein-splicing activity to the formation of functional GyrA^s *in vivo* (Adam and Perler 2002). The purpose of the system is to identify conditions where intein splicing is suppressed, either by mutation or by the presence of compounds that inhibit intein activity.

To test the system, a number of functional mycobacterial GyrA inteins were cloned into the homologous insertion site of the *E. coli* GyrA^s, and these fusions were transformed into *E. coli* in strains that constitutively expressed GyrA^r (Adam and Perler 2002). It was shown that expression of the GyrA^s fusions in the presence of ofloxacin was lethal, while expression of the corresponding fusion with a non-functional mutant of the *M. xenopi* GyrA intein resulted in colony formation. Functional intein splicing was verified by Western blot in all cases where cell death had been observed, except for the *M. leprae* intein where a ligated extein product could not be detected. These results suggest that this *in vivo* selection has the ability to report levels of splicing that are too low to be detected by direct observation methods.

This selection system was further challenged by screening a mutant library of the *M. xenopi* GyrA intein for temperature-sensitive splicing activity. Intein variants that were suppressed at 37 °C but active at 30 °C were successfully identified by selection and phenotype testing on ofloxacin-containing agar plates. The selection system could also be converted into a high-throughput screening system by co-transformation with a green fluorescent protein expression plasmid. In this case, cell viability is easily detected through a fluorescence readout, allowing rapid screening of chemical libraries for potential splicing inhibitors. It is further likely that this reporter might be applied to any well-behaved intein for the detection of splicing repression.

3.6 Bacterial CcdB: Negative Selection for Splicing or Cleaving

An additional selection against *in vivo* intein splicing has been developed using the bacterial CcdB toxin (Lew and Paulus 2002). The CcdB protein inhibits DNA gyrase enzymes, and in its native context serves to prevent the loss of F-factor from bacterial hosts (van Melderen 2002). Recently, however, this protein has been incorporated into a number of commercial cloning vectors to allow selection for recombinant clones (Gabant et al. 1997). In this context, fusion of the LacZα protein upstream from CcdB produces a toxic product, but this toxicity can be eliminated by gene insertions that result in a large extension of the *lacZα* leader region. Furthermore, since the LacZα-CcdB fusion is under control of the *lac* promoter, its toxicity can be conveniently modulated by varying levels of isopropyl β-d-thiogalactoside (IPTG).

This system was verified by insertion of the Mtu RecA intein into the *lacZα* polylinker followed by transformation into a *lacI*^{q} *E. coli* strain. Insertion of a non-splicing Mtu RecA intein mutant allowed cell growth in the presence of IPTG, while the actively splicing wild-type intein resulted in cell death. This demonstrated that the LacZα-intein-CcdB fusion is non-toxic for the cells, but that toxicity is restored upon intein splicing. This reporter system was then used to demonstrate that the Mtu RecA intein remains functional irrespective of the nature of the ultimate residue of the N-extein, consistent with previous observations that this intein retains high levels of splicing efficiency when inserted into a variety of protein hosts. The presence of any of the 20 natural amino acids preceding the N-terminal splice junction resulted in arrested cell growth, although direct observation of the splicing products with SDS-PAGE revealed some variability as to splicing activity and cleaving side reactions (Lew and Paulus 2002).

Despite the fact that the system was tested only with the Mtu RecA intein, it is expected that it will be applicable to any protein-splicing element that remains functional in this context. The system is particularly convenient in that the inclusion of several N- and C-extein residues, which should maximize intein-splicing activity, will likely not interfere with the toxicity of the spliced product. Natural or artificial mini-inteins lacking a central endonuclease domain can also be used with this system since their size (usually in the 150 aa range) is considerably larger than the minimal size required for CcdB inactivation. Finally, since the CcdB protein in this system is not inactivated by an insertional intein fusion but by an end-to-end one, it is anticipated that this system may also serve as a reporter for C-terminal cleaving. In this case the C-terminal cleavage reaction would release the active CcdB protein in the absence of splicing.

3.7 T4 DNA Polymerase: Negative-Splicing Selection *in vivo*

Perler and coworkers have recently developed another *in vivo* selection against protein splicing by linking loss of intein activity to the survival of phage-infected bacterial cells (Cann et al. 2004). A genetically modified phage T4 strain (T4 gp43^{-}) was developed wherein its DNA polymerase carries several amber mutations. In non-suppressor strains the phage is non-viable and cannot bring about cell lysis, but it can be rescued through complementation by the wild-type DNA polymerase from the closely related phage RB69 (Wang et al. 1995). In this system, an intein-containing RB69 DNA polymerase is supplied to *E. coli* cells by simple plasmid transformation, and the transformants are subsequently infected with the defective T4 gp43$^-$ phage. Splicing by the intein activates the RB69 polymerase, allowing successful infection and cell lysis. However, under conditions where the intein cannot splice, the RB69 polymerase is not activated, allowing the cells to resist lysis and form colonies on phage-containing agar.

The system was tested by insertion of the *Thermococcus litoralis* Tli Pol-2 intein (Perler et al. 1992) into its homologous insertion site in the RB69 polymerase gene. Expression of this fusion in *E. coli*, followed by T4 gp43$^-$ infection, resulted in lysis of the bacterial cells and no colony formation on agar plates. In contrast, replacement of this intein with a splicing-deficient mutant permitted bacterial growth and colony formation. Remarkably, despite the hyperthermophilic origin of the Tli Pol-2 intein, it was able to supply enough active polymerase to suppress colony formation even at temperatures as low as 25 °C. The power of the system was demonstrated by the isolation of several temperature-sensitive mutants of the Tli Pol-2 intein, and it is anticipated that this system will hold similar utility in identifying inhibitors for the splicing reaction. As noted by the investigators, however, secondary screening steps might be required to account for evolved phage resistance in large-scale experiments.

3.8 Yeast *Gal4* Gene: Temperature-Sensitive Splicing *in vivo*

A reporter system that allows both screening and selection for intein activity was developed by linking intein splicing to transcriptional regulation of a selectable marker in *Saccharomyces cerevisiae* (Zeidler et al. 2004). The Sce intein was inserted into the DNA-binding domain of the yeast transcription factor Gal4 (Lohr et al. 1995) such that the resulting fusion produced active Gal4 only as a result of intein splicing. By linking intein-mediated Gal4 regulation to the *GAL2* galactose permease gene, a positive selection for splicing activity is generated on selective galactose medium. The system was verified by insertion of the wild-type Sce intein, which promoted growth on selective galactose medium while an inactive mutant did not.

This system was then used to evolve temperature-sensitive mutants of the Sce intein in the Gal4 context. A mutant library of the intein was prepared by error-prone PCR and cloned into the system, and several transformants with temperature-dependent growth phenotypes were isolated by selection at 18 °C followed by phenotypic screening at 30 °C. One of these mutants was found to contain a true temperature-sensitive intein, as it produced a stable precursor at the non-permissive temperature that exhibited temperature-dependent splicing activity. The validity of the temperature-sensitive intein was further verified by insertion into the related Gal80 protein, which represses Gal4 and thus provides a negative selection for protein splicing on galactose medium. In this case, wild-type and inactive inteins were used to verify the system, while insertion of the selected mutant intein produced the expected temperature-sensitive phenotypes.

A very exciting application of this system and its evolved intein is to control gene expression in higher organisms. To demonstrate this capability, the Gal4/Gal80 system was used to regulate a firefly luciferase reporter protein in *Drosophila melanogaster* Schneider cells (Zeidler et al. 2004). The temperature-sensitive intein was inserted into the Gal80 protein to provide repression of Gal4, and thus luciferase expression, only in the presence of intein splicing. The system was verified with wild-type and non-splicing mutant inteins, and shown to behave consistently with the expected intein-mediated activation of the Gal80 protein. As hoped for, the previously evolved temperature-sensitive Sce intein in the same context allowed luciferase expression to be controlled by varying the incubation temperature between 29 and 18 °C. These experiments were finally capped with an astonishing demonstration in transgenic *Drosophila* larvae, where the temperature dependence of the intein was shown to allow modulation of Notch protein expression in wing imaginal disks *in vivo*.

Although temperature-sensitive intein mutants have been previously reported, none of them have been shown to function consistently outside their original selection contexts (Adam and Perler 2002; Cann et al. 2004). The Gal4/Gal80 system is the first to show that controllable inteins developed in one context can be used to control other proteins in other contexts at the post-translational level. Furthermore, since this intein modulates the Gal4/Gal80 promoter system, in principle it can facilitate the conditional expression of any gene that can be placed under Gal4 transcriptional regulation (Phelps and Brand 1998). Several thousand cell lines incorporating Gal4 expression cassettes have already been generated in *Drosophila* (McGuire et al. 2004), and the Gal4 system has also been shown to work in mammalian systems (Lewandoski 2001).

3.9 Green Fluorescent Protein: Reporter for Splicing *in vitro*

A recently reported screening system for conditional intein splicing in vitro has been developed through the insertion of the Mtu RecA intein into a modified green fluorescent protein (GFP; Gangopadhyay et al. 2003). In this case, overexpression of the unspliced precursor protein in *E. coli* results in the formation of insoluble inclusion bodies. The inclusion bodies are dissolved in 8 M urea and the precursor is purified by metal-chelate chromatography, followed by refolding under optimized conditions. A key aspect of the system is that the refolding procedure can be carried out in the presence of 2 mM zinc, which is known to inhibit the intein-splicing reaction (Mills and Paulus 2001). This allows the formation of a stable unspliced precursor, which can then be induced to splice through the addition of EDTA. The splicing reaction can be followed by SDS-PAGE analysis or by the development of fluorescence over time.

A potential application of this system is high-throughput screening of compounds that might inhibit the intein-splicing reaction. In this pursuit, it holds the significant advantage that splicing can be analyzed in vitro, with a fully active unspliced precursor protein, and that it incorporates a reporter gene that is highly amenable to high-throughput applications. The in vitro nature of the system reduces the likelihood of false positives arising from secondary metabolic effects of test compounds *in vivo*, while at the same time it reduces the likelihood of false negatives arising from limitations in cellular uptake of library compounds. Furthermore, the use of the green fluorescent reporter protein should simplify the adoption of this system for a variety of well-established screening technologies.

3.10 *trans*-Splicing Intein Systems

An important application of intein splicing is the reconstitution of target proteins from separate components through *trans*-splicing and peptide ligation. These systems have allowed partial isotope labeling for NMR experiments (Yamazaki et al. 1998), reconstitution and activation of cytotoxic proteins (Evans et al. 1998), controlled protein splicing (Mootz et al. 2003), generation of cyclic peptide libraries (Scott et al. 1999), transgenic protein activation (Chen et al. 2001; Yang et al. 2003), and the detection of protein–protein interactions *in vivo* (Ozawa and Umezawa, this Vol.).

The study of *trans*-splicing and its applications has led to the development of a number of additional reporter systems, each designed for application-specific capabilities. The basic principle of each of these systems is the same however: a reporter gene is split into two parts, and each part is joined to one half of a *trans*-splicing intein gene for subsequent expression, assembly and trans-splicing. Many of these systems incorporate highly soluble, though nonfunctional, protein domains for direct observation of the splicing reaction via SDS-PAGE and Western blot (Martin et al. 2001; Mills and Paulus 2001; Nichols and Evans 2004). Others report the splicing reaction through activation of a convenient reporter protein. One of these is based on the reconstitution of a split GFP in *E. coli*, used to detect protein–protein interactions that guide the assembly and splicing of the split intein (Ozawa et al. 2001). Another example provides a genetic selection through the *trans*-splicing activation of *E. coli* acetolactate synthase isoform II, which rescues cells from growth inhibition by valine (Ghosh et al. 2001). This system has been used to study the effects of zinc on the trans-splicing reaction, and has allowed the identification of key residues and regions required for intein assembly and activation.

An important application of trans-splicing reporter systems has been the detection and reporting of the splicing reaction in eukaryotes, including plants and live mammals. In plant cells, assembly and splicing has been reported by the activation of β-glucuronidase, detected by a rather cumbersome histochemical staining procedure (Yang et al. 2003). However, this assay has shown conclusively that intein *trans*-splicing can occur in all parts of living plants, providing support for additional applications involving the assembly of more complex proteins and enzymes (Chen et al. 2001; Morassutti et al. 2002; Chin et al. 2003). A second system for the study of protein–protein interactions in living mice has been developed around the reconstitution and splicing of a split firefly luciferase protein (Paulmurugan et al. 2002). Protein–protein interactions guide the assembly of the intein and resulting luciferase activation, which can be detected using a cooled charge-coupled device camera. A discussion of these types of systems is included in the chapter by Ozawa and Umezawa in this volume.

4 Summary

Intein reporter and selection systems have already generated a variety of unique applications, which would not have been possible without the robust and flexible nature of the splicing reaction. It is also becoming clear that virtually any genetic reporter system can be modified to include a protein-splicing element, and new systems are rapidly being developed. At the same time, existing systems are being used to generate new inteins and applications. A very recent example is one where the kanamycin selection system has been combined with the GFP screen to develop an artificial, ligand-activated chimeric intein (Buskirk et al. 2004). This intein, which has been evolved to be dependent on the estrogen analog 4-hydroxytamoxifen, can in principle be used to controllably activate virtually any protein post-translationally on demand. This is one of a growing number of intein applications with important uses in many scientific and industrial fields, each of which has been developed through the use of systems to report and select intein function.

References

Adam E, Perler FB (2002) Development of a positive genetic selection system for inhibition of protein splicing using mycobacterial inteins in *Escherichia coli* DNA gyrase subunit A. J Mol Microbiol Biotechnol 4:479–487

Belfort M, Pedersen-Lane J (1984) Genetic system for analyzing *Escherichia coli* thymidylate synthase. J Bacteriol 160:371–378

Belfort M (1998) Inteins as antimicrobial targets: genetic screens for intein function. US patent 5,795,731, Health Research Incorporated (Albany, NY)

Buskirk AR, Ong YC, Gartner ZJ, Liu DR (2004) Directed evolution of ligand dependence: small-molecule-activated protein splicing. Proc Natl Acad Sci USA 101:10505–10510

Cann IK, Amaya KR, Southworth MW, Perler FB (2004) Bacteriophage-based genetic system for selection of nonsplicing inteins. Appl Environ Microbiol 70:3158–3162

Carreras CW, Santi DV (1995) The catalytic mechanism and structure of thymidylate synthase. Annu Rev Biochem 64:721–762

Chen L, Pradhan S, Evans TC Jr (2001) Herbicide resistance from a divided EPSPS protein: the split *Synechocystis* DnaE intein as an *in vivo* affinity domain. Gene 263:39–48

Chin HG, Kim GD, Marin I, Mersha F, Evans TC Jr, Chen L, Xu M-Q, Pradhan S (2003) Protein trans-splicing in transgenic plant chloroplast: reconstruction of herbicide resistance from split genes. Proc Natl Acad Sci USA 100:4510–4515

Chong S, Shao Y, Paulus H, Benner J, Perler FB, Xu M-Q (1996) Protein splicing involving the *Saccharomyces cerevisiae* VMA intein. The steps in the splicing pathway, side reactions leading to protein cleavage, and establishment of an in vitro splicing system. J Biol Chem 271:22159–22168

Chong S, Mersha FB, Comb DG, Scott ME, Landry D, Vence LM, Perler FB, Benner J, Kucera RB, Hirvonen CA, Pelletier JJ, Paulus H, Xu M-Q (1997) Single-column purification of free recombinant proteins using a self-cleavable affinity tag derived from a protein splicing element. Gene 192:271–281

Cooper AA, Chen YJ, Lindorfer MA, Stevens TH (1993) Protein splicing of the yeast TFP1 intervening protein sequence: a model for self-excision. EMBO J 12:2575–2583

Daugelat S, Jacobs WR Jr (1999) The *Mycobacterium tuberculosis* recA intein can be used in an ORFTRAP to select for open reading frames. Protein Sci 8:644–653

Davis EO, Sedgwick SG, Colston MJ (1991) Novel structure of the recA locus of *Mycobacterium tuberculosis* implies processing of the gene product. J Bacteriol 173:5653–5662

Davis EO, Jenner PJ, Brooks PC, Colston MJ, Sedgwick SG (1992) Protein splicing in the maturation of *M. tuberculosis* recA protein: a mechanism for tolerating a novel class of intervening sequence. Cell 71:201–210

Derbyshire V, Wood DW, Wu W, Dansereau JT, Dalgaard JZ, Belfort M (1997) Genetic definition of a protein-splicing domain: functional mini-inteins support structure predictions and a model for intein evolution. Proc Natl Acad Sci USA 94:11466–11471

Evans TC Jr, Benner J, Xu M-Q (1998) Semisynthesis of cytotoxic proteins using a modified protein splicing element. Protein Sci 7:2256–2264

Foury F (1990) The 31-kDa polypeptide is an essential subunit of the vacuolar ATPase in *Saccharomyces cerevisiae*. J Biol Chem 265:18554–18560

Gabant P, Dreze PL, van Reeth T, Szpirer J, Szpirer C (1997) Bifunctional lacZ alpha-ccdB genes for selective cloning of PCR products. Biotechniques 23:938–941

Gangopadhyay JP, Jiang SQ, Paulus H (2003) An in vitro screening system for protein splicing inhibitors based on green fluorescent protein as an indicator. Anal Chem 75:2456–2462

Ghosh I, Sun L, Xu MQ (2001) Zinc inhibition of protein trans-splicing and identification of regions essential for splicing and association of a split intein. J Biol Chem 276:24051–24058

Hawkey PM (2003) Mechanisms of quinolone action and microbial response. J Antimicrob Chemother 51 [Suppl 1]:29–35

Hirata R, Ohsumk Y, Nakano A, Kawasaki H, Suzuki K, Anraku Y (1990) Molecular structure of a gene, VMA1, encoding the catalytic subunit of H(+)-translocating adenosine triphosphatase from vacuolar membranes of *Saccharomyces cerevisiae*. J Biol Chem 265:6726–6733

Jacobs WR Jr, Daugelat S (1999) Vector constructs for the selection and identification of open reading frames. US Patent 5 981 182, Albert Einstein College of Medicine of Yeshiva University, Bronx, NY

Jacobs WR Jr, Kalpana GV, Cirillo JD, Pascopella L, Snapper SB, Udani RA, Jones W, Barletta RG, Bloom BR (1991) Genetic systems for mycobacteria. Methods Enzymol 204:537–555

Kane PM, Yamashiro CT, Wolczyk DF, Neff N, Goebl M, Stevens TH (1990) Protein splicing converts the yeast TFP1 gene product to the 69-kD subunit of the vacuolar H(+)-adenosine triphosphatase. Science 250:651–657

Kawasaki M, Nogami S, Satow Y, Ohya Y, Anraku Y (1997) Identification of three core regions essential for protein splicing of the yeast Vma1 protozyme. A random mutagenesis study of the entire Vma1-derived endonuclease sequence. J Biol Chem 272:15668–15674

Lew BM, Paulus H (2002) An *in vivo* screening system against protein splicing useful for the isolation of non-splicing mutants or inhibitors of the RecA intein of *Mycobacterium tuberculosis*. Gene 282:169–177

Lewandoski M (2001) Conditional control of gene expression in the mouse. Nat Rev Genet 2:743–755

Liu XQ, Yang J (2004) Prp8 intein in fungal pathogens: target for potential antifungal drugs. FEBS Lett 572:46–50

Lohr D, Venkov P, Zlatanova J (1995) Transcriptional regulation in the yeast GAL gene family: a complex genetic network. FASEB J 9:777–787

Lutz S, Fast W, Benkovic SJ (2002) A universal, vector-based system for nucleic acid reading-frame selection. Protein Eng 15:1025–1030

Martin DD, Xu M-Q, Evans TC Jr (2001) Characterization of a naturally occurring trans-splicing intein from *Synechocystis* sp. PCC6803. Biochemistry 40:1393–1402

McGuire SE, Roman G, Davis RL (2004) Gene expression systems in *Drosophila*: a synthesis of time and space. Trends Genet 20:384–391

Mills KV, Paulus H (2001) Reversible inhibition of protein splicing by zinc ion. J Biol Chem 276:10832–10838

Mills KV, Manning JS, Garcia AM, Wuerdeman LA (2004) Protein splicing of a *Pyrococcus abyssi* intein with a C-terminal glutamine. J Biol Chem 279:20685–20691

Mootz HD, Blum ES, Tyszkiewicz AB, Muir TW (2003) Conditional protein splicing: a new tool to control protein structure and function in vitro and *in vivo*. J Am Chem Soc 125:10561–10569

Morassutti C, de Amicis F, Skerlavaj B, Zanetti M, Marchetti S (2002) Production of a recombinant antimicrobial peptide in transgenic plants using a modified VMA intein expression system. FEBS Lett 519:141–146

Nichols NM, Evans TC Jr (2004) Mutational analysis of protein splicing, cleavage, and self-association reactions mediated by the naturally split Ssp DnaE intein. Biochemistry 43:10265–10276

Nogami S, Satow Y, Ohya Y, Anraku Y (1997) Probing novel elements for protein splicing in the yeast Vma1 protozyme: a study of replacement mutagenesis and intragenic suppression. Genetics 147:73–85

Ozawa T, Takeuchi TM, Kaihara A, Sato M, Umezawa Y (2001) Protein splicing-based reconstitution of split green fluorescent protein for monitoring protein-protein interactions in bacteria: improved sensitivity and reduced screening time. Anal Chem 73:5866–5874

Paulmurugan R, Umezawa Y, Gambhir SS (2002) Noninvasive imaging of protein-protein interactions in living subjects by using reporter protein complementation and reconstitution strategies. Proc Natl Acad Sci USA 99:15608–15613

Paulus H (2003) Inteins as targets for potential antimycobacterial drugs. Front Biosci 8: s1157–s1165

Perler FB, Adam E (2000) Protein splicing and its applications. Curr Opin Biotechnol 11:377–383

Perler FB, Comb DG, Jack WE, Moran LS, Qiang B, Kucera RB, Benner J, Slatko BE, Nwankwo DO, Hempstead SK, et al (1992) Intervening sequences in an *Archaea* DNA polymerase gene. Proc Natl Acad Sci USA 89:5577–5581

Phelps CB, Brand AH (1998) Ectopic gene expression in *Drosophila* using GAL4 system. Methods 14:367–379

Sambrook J, Russell DW (2001) Molecular cloning: a laboratory manual. Cold Spring Harbor Laboratory Press, Cold Spring Harbor, New York, USA

Scott CP, Abel-Santos E, Wall M, Wahnon DC, Benkovic SJ (1999) Production of cyclic peptides and proteins *in vivo*. Proc Natl Acad Sci USA 96:13638–13643

Skretas G, Wood DW (2005) Regulation of protein activity with small-molecule-controlled inteins. Protein Sci 14:523-532.

Southworth MW, Amaya K, Evans TC, Xu M-Q, Perler FB (1999) Purification of proteins fused to either the amino or carboxy terminus of the *Mycobacterium xenopi* gyrase A intein. Biotechniques 27:110–114, 116, 118–120

Southworth MW, Yin J, Perler FB (2004) Rescue of protein splicing activity from a *Magnetospirillum magnetotacticum* intein-like element. Biochem Soc Trans 32:250–254

Telenti A, Southworth M, Alcaide F, Daugelat S, Jacobs WR Jr, Perler FB (1997) The *Mycobacterium xenopi* GyrA protein splicing element: characterization of a minimal intein. J Bacteriol 179:6378–6382

Van Melderen L (2002) Molecular interactions of the CcdB poison with its bacterial target, the DNA gyrase. Int J Med Microbiol 291:537–544

Wang CC, Yeh LS, Karam JD (1995) Modular organization of T4 DNA polymerase. Evidence from phylogenetics. J Biol Chem 270:26558–26564

Wood DW, Wu W, Belfort G, Derbyshire V, Belfort M (1999) A genetic system yields self-cleaving inteins for bioseparations. Nat Biotechnol 17:889–892

Wood DW, Derbyshire V, Wu W, Chartrain M, Belfort M, Belfort G (2000) Optimized single-step affinity purification with a self-cleaving intein applied to human acidic fibroblast growth factor. Biotechnol Prog 16:1055–1063

Xu M-Q, Perler FB (1996) The mechanism of protein splicing and its modulation by mutation. EMBO J 15:5146–5153

Xu M-Q, Southworth MW, Mersha FB, Hornstra LJ, Perler FB (1993) In vitro protein splicing of purified precursor and the identification of a branched intermediate. Cell 75:1371–1377

Yamashiro CT, Kane PM, Wolczyk DF, Preston RA, Stevens TH (1990) Role of vacuolar acidification in protein sorting and zymogen activation: a genetic analysis of the yeast vacuolar proton-translocating ATPase. Mol Cell Biol 10:3737–3749

Yamazaki T, Otomo T, Oda N, Kyogoku Y, Uegaki K, Ito N, Ishino Y, Nakamura H (1998) Segmental isotope labeling for protein NMR using peptide splicing. J Am Chem Soc 120:5591–5592
Yang J, Fox GC Jr, Henry-Smith TV (2003) Intein-mediated assembly of a functional beta-glucuronidase in transgenic plants. Proc Natl Acad Sci USA 100:3513–3518
Zeidler MP, Tan C, Bellaiche Y, Cherry S, Hader S, Gayko U, Perrimon N (2004) Temperature-sensitive control of protein activity by conditionally splicing inteins. Nat Biotechnol 22:871–876

Industrial Applications of Intein Technology

DAVID W. WOOD, SARAH W. HARCUM, GEORGES BELFORT

Abstract

Intein-based bioseparations have become a widely used technique for the purification of single proteins at the research scale. Aspects of these purification methods suggest that they might eventually be used in more advanced applications, including large-scale protein production and high-throughput proteomic studies. Both of these applications require substantial analysis for further development, particularly concerning the economics and practicality of potential future systems. For scale-up to commercial protein production, an analysis has been made using widely accepted assumptions for process design and a computer model of a hypothetical process. It is clear that the intein process has the potential to be economically competitive, but will be more attractive with the use of pH- and temperature-controlled inteins with low-cost buffer systems. In the case of proteomic applications, prototypical devices have been constructed to demonstrate feasibility. Both of these studies indicate a strong potential for inteins to eventually join mainstream technologies in these areas.

D.W. Wood (e-mail: dwood@princeton.edu)
Department of Chemical Engineering, Princeton University, Princeton, New Jersey, USA

S. W. Harcum
Department of Bioengineering, Clemson University, Clemson, South Carolina, USA

G. Belfort
Department of Chemical and Biological Engineering,
Rensselaer Polytechnic Institute, Troy, New York, USA

Nucleic Acids and Molecular Biology, Vol. 16
Marlene Belfort et al. (Eds.)
Homing Endonucleases and Inteins

1 Introduction

As intein technologies become better developed, the potential arises for their application to a wide range of industrial problems. It is now clear that inteins will likely play a significant role in a number of industries, including pharmaceuticals, agriculture, biomaterials, specialty chemicals, and pure research. Indeed, intein-based bioseparation systems already constitute a major advance in the production of purified recombinant proteins for research use. These new purification methods are capable of reliable and efficient performance, and dozens of papers have now been published describing their use with a variety of protein targets at laboratory scale. The development of inteins that can be induced to cleave through changes in pH and temperature has also delivered new options for the design of large-scale purification processes. The power and flexibility of these systems suggest that they might become a core technology in bioseparations at both the research and manufacturing scales.

This chapter examines two significant applications emerging from research and development in the area of intein-facilitated bioseparations. The first is large-scale affinity separations based on IMPACT and other systems described by Chong and Xu in this volume. The scale-up of these methods, although promising, must be considered with regard to the associated practical and economic aspects that will ultimately dictate their success. A second significant potential application of inteins is in proteomics research. In this case, the reliability of intein-based purifications suggests that they might be applied to entire libraries of protein targets in a high-throughput format. Miniaturization is facilitated by the relative simplicity of intein-mediated bioseparations, allowing the development of a high-throughput microfluidic system based on a disposable platform. This device provides a research tool with significant advantages over existing methods for the production of purified protein libraries.

2 Scale-Up of Intein Processes

Although the advantages of intein-based bioseparations are well established at the laboratory scale, only a few very recent reports support their use at large scale (Sharma et al. 2003; Ma and Cooney 2004). Furthermore, the economic viability of inteins in the commercial production of any real therapeutic or commodity protein has yet to be proven directly. However, the practical issues associated with this type of process can be partially assessed through a theoretical analysis with a few basic assumptions. This is typically accomplished by comparing a hypothetical intein process to a well-characterized conventional process for a given product. This analysis can be further simplified through the use of several readily available engineering software tools.

2.1 Conventional Affinity Tag Processes

The development of conventional affinity tags in the mid-1980s provided a breakthrough method for the purification of arbitrary proteins at the research scale (Germino and Bastia 1984; LaVallie and McCoy 1995). With this method, purification of a given target protein is facilitated by the addition of an affinity tag sequence to the target protein gene at the genetic level. The expressed tag binds tightly and specifically to a corresponding immobilized ligand, thus facilitating the purification of its fusion partner through pre-optimized affinity purification protocols. Tags include the maltose- and chitin-binding domains, glutathione S-transferase, polyhistidine, and others, and research in this area continues to deliver new tags with new capabilities. The reliability of this method has made it particularly attractive in the investigation of newly discovered gene products, where it eliminates the need for new purification protocols to be developed for each new target. Furthermore, at the several milligram scale, the convenience of these gene fusions far outweighs the cost of subsequent tag removal.

In principle, conventional affinity tags might be used in the large-scale manufacture of highly purified proteins. Unfortunately, however, the cost associated with removal of the affinity tag is clearly prohibitive. Based on product literature (pMal fusion system, New England Biolabs, Beverly, MA, USA), the proteolytic cleaving of a given affinity tag can cost anywhere from $10,000 (using genenase) to $320,000 (using enterokinase) per gram of product delivered. At manufacturing scales of tens to hundreds of kilograms per year, these costs rapidly exceed the gross annual sales of even the most lucrative blockbuster drugs today. Finally, the addition of protease necessitates an additional purification step, and can complicate drug approval due to the highly bioactive nature of these enzymes.

2.2 Intein-Mediated Protein Purification

Modified inteins can effectively render affinity tags self-cleaving in response to temperature, pH, or thiol addition, thus eliminating the requirement for protease treatment in the recovery of native products from tagged fusion proteins. This has led to the development of several intein-mediated purification processes at the laboratory-scale for the recovery of recombinant proteins on a single column (Chong et al. 1997, 1998; Southworth et al. 1999; Wood et al. 1999; Xu et al. 2000; Zhang et al. 2001). The simplicity by which the cleavage reaction can be induced suggests that this technique may be appropriate for large-scale bioseparations. Furthermore, it has been demonstrated in at least one case that this method is feasible for proteins expressed under high cell-density conditions (Sharma et al. 2003), and a self-cleaving tag has also been successfully incorporated into a pilot-scale vortex-flow affinity capture scheme (Ma and Cooney 2004). Thus, the potential for economic feasibility of a large-scale intein-mediated purification system for recombinant proteins does exist.

2.2.1 Modeling Large-Scale Intein Bioseparations

In order to evaluate the intein-mediated system for scale-up, the laboratory-scale process must be considered from an economic prospective. Bioprocess simulators can aid this process by facilitating the economic assessment of a laboratory process at an industrial scale (Rouf et al. 2001; Shanklin et al. 2001; Petrides et al. 2002). To compare the economics of a conventional bioseparations process to one including an intein-mediated step based on the IMPACT system (New England BioLabs), a series of simulations were performed using the commercial software package SuperPro Designer (Intelligen, Inc., Scotch Plains, NJ, USA). The IMPACT system includes a modified *Saccharomyces cerevisiae* VMA1 intein, coupled to a chitin-binding affinity tag. Following purification of the fusion precursor protein, cleavage by the intein is induced through the addition of high concentrations of thiol compounds such as dithiothreitol (DTT). For the purposes of this study, recombinant α1-antitrypsin produced in *E. coli* was used as the hypothetical product protein around which each process was designed. The key difference between the two processes is the replacement of an immobilized-antibody affinity-purification step in the conventional process with an IMPACT-purification step in the intein process (Fig. 1). Both processes were simulated from seed cultures to freeze-dried bulk protein at a scale of 9000 kg/year. This level of production would be anticipated for an industrial enzyme or high-dose chronic therapeutic, such as α1-antitrypsin (de Serres 2002). By comparison, the current production level for human insulin is approximately 2000 kg/year (Ladisch 2001).

Several process assumptions were included for accurate analysis. For example, it was assumed that the quantities of recombinant protein expressed by the cells in both cases were equal and accounted for 10% of the total dry cell weight of the cell. The design parameters for the equipment in both processes, such as percent rejection in the filtration units of common components, were also assumed to be similar. For the ion-exchange, gel-filtration and affinity resins, conservative separation efficiency, reusability and cost estimates were used. The binding capacity and replacement frequency of the chitin resin were set according to values obtained from laboratory studies (5 mg/mL and 5 cycles, respectively). Cleavage efficiency of the fusion protein was set to the maximum value observed experimentally.

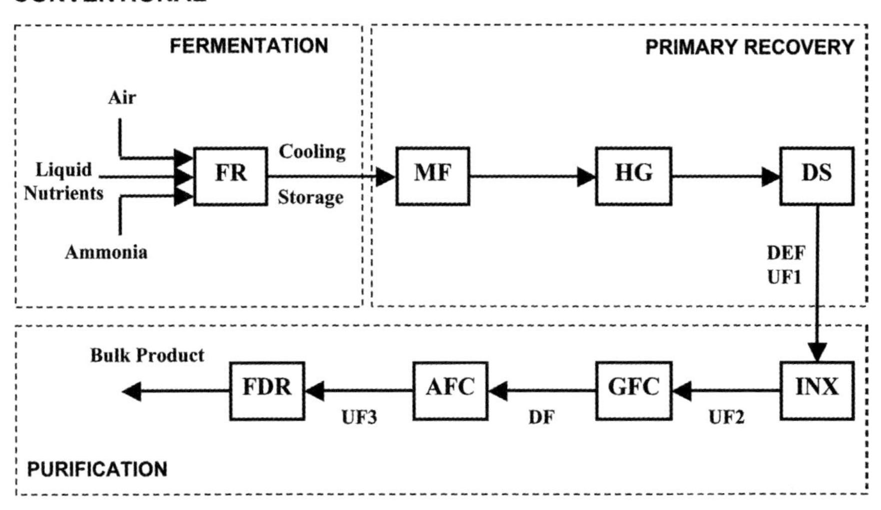

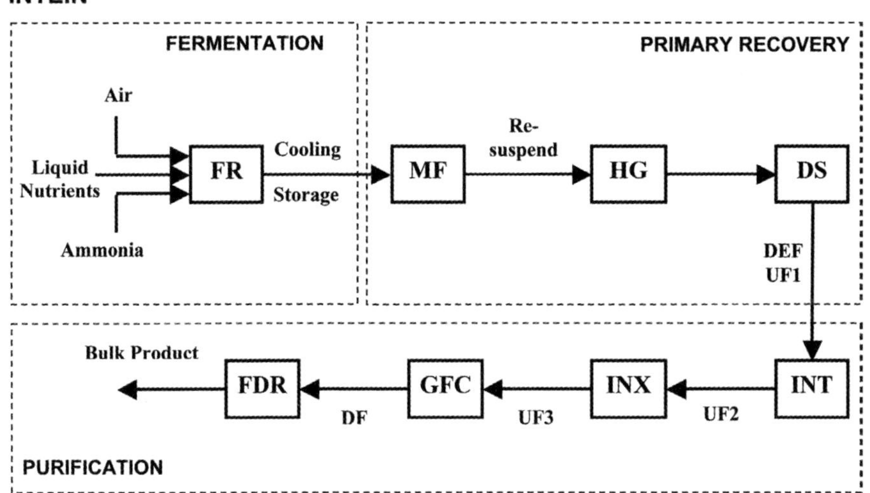

Fig. 1. Block diagram for the production of a recombinant protein. Conventional and intein processes are shown. *FR* Fermentation; *MF* microfiltration; *HG* homogenization; *DS* disk stack centrifugation; *DEF* dead-end filtration; *UF* ultrafiltration; *INX* ion-exchange chromatography; *GFC* gel-filtration chromatography; *DF* diafiltration; *AFC* affinity chromatography; *FDR* freeze drying; *INT* intein purification step

The basic economic rules for this type of comparison are well established and are summarized here. Raw material and consumable costs were estimated from current supplier list prices of the largest quantities available, while capital costs were determined based on factors commonly employed in bioprocessing plants (Harrison et al. 2003). The production cost ($/g) was calculated based on the annual revenue ($/year) required to obtain a uniform series disbursement for the total capital investment, with a 20% internal rate of return (IRR) over a 15-year period. Calculations included inflation and depreciation values based on a 2003 year of analysis. The plant construction period was 2.5 years, with a startup time of 6 months and a 75% operating capacity during the first year. Finally, the simulation for each process was performed in "design mode", thus allowing the SuperPro software to automatically size each process unit as required.

2.2.2 Economics of the IMPACT Process Scale-Up

Based on the stated assumptions, the overall product cost for the conventional process was calculated to be $38/g. However, incorporation of the scaled-up IMPACT step raised the cost to $87/g. Despite the fact that this is orders of magnitude cheaper than any conventional affinity-tag process, it is still economically unrealistic compared to most industrial large-scale protein-purification processes. For comparison, the total manufacturing costs estimated for Monsanto's bovine growth hormone are less than $5/g (Swartz 2001).

A detailed analysis of each process indicated that the operating costs were the major factor for the cost difference, with raw materials accounting for the most substantial increase in the cost of the intein process (Fig. 2A). Further analysis confirmed that the purification section of the intein process was the primary reason this process was not economically feasible, and the costs associated with DTT addition and the Tris-HCl buffer system were identified as major contributors to the price increase (Fig. 2B). The most expensive raw material on a mass basis was DTT, which accounted for 29% of all raw material costs for the intein process. The second most expensive raw material on a mass basis was Tris-HCl.

2.3 Economic Optimization of Intein-Based Bioseparations

Although the initial analysis points to DTT and Tris-HCl as the major contributors to the high cost of the intein process, these costs can be significantly inflated by limitations in other areas of the purification. For example, if the chitin-affinity resin has a low binding capacity, then the required scale for that

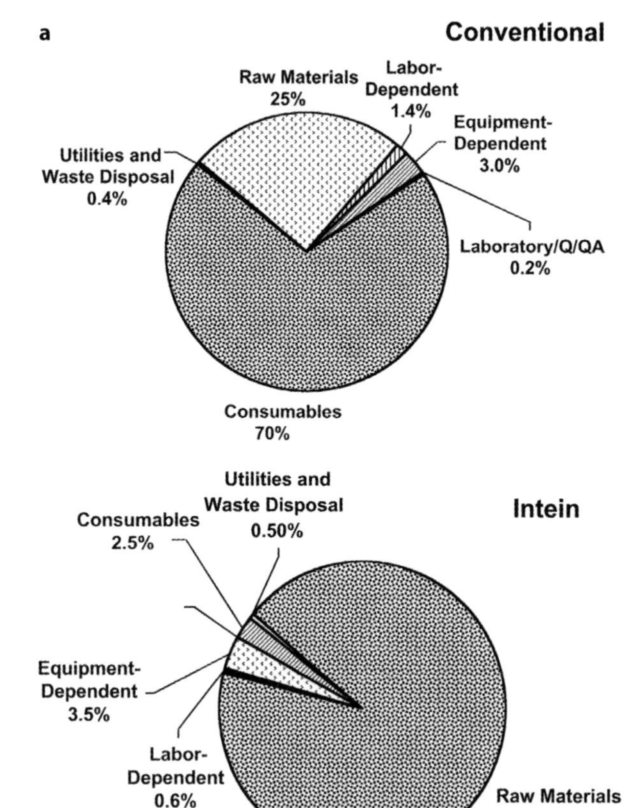

Fig. 2. Cost breakdowns for the conventional and intein processes. **a** Annual operating cost breakdown. Total annual operating expenditures were \$332 and \$704 million for the conventional and intein processes, respectively.

stage of the purification will increase. This leads to a "domino effect", where buffer consumption, equipment size and overall plant design can be greatly influenced. A full evaluation of the intein process must therefore include an impact analysis for hypothetical changes in several of its aspects. From this analysis, specific suggestions can be generated for further development and optimization, and realistic expectations for the long-term viability of this technology can be determined. For the intein process, the areas that should be examined have been grouped into three major areas: buffer composition, resin, and alternate intein-cleaving modes.

2.3.1 *Buffers*

The recommended buffer solutions for use with the IMPACT system at laboratory scale typically include HEPES or Tris-HCl as buffering agents, EDTA to chelate unwanted metals, and high concentrations of NaCl. The costs of the Tris-HCl and DTT (to induce cleavage) contribute most significant-

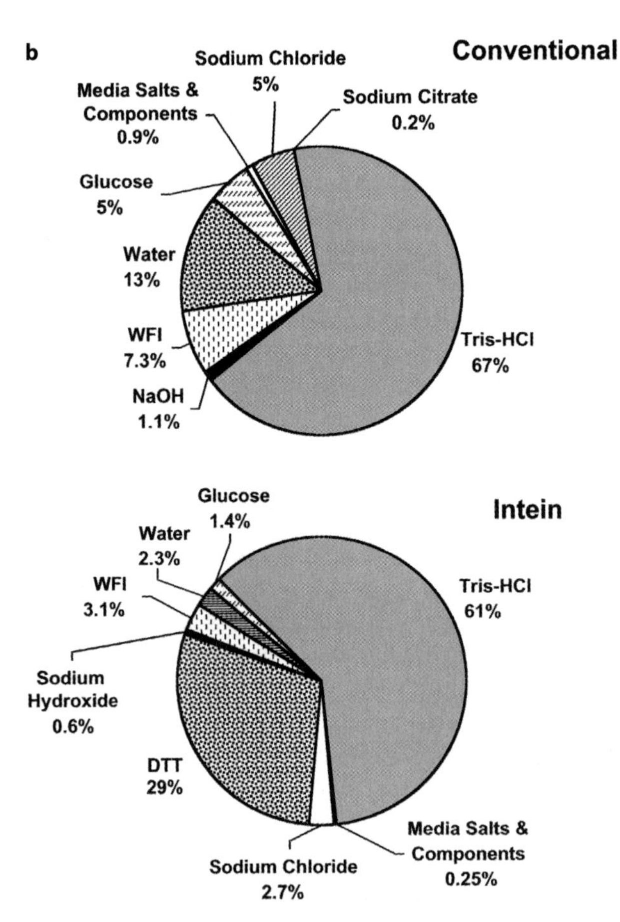

Fig. 2.
b Annual raw material cost breakdown. Total annual raw material costs were $83 and $653 million, respectively. *WFI* Water for injection

ly to the cost of the purified protein. The DTT cost alone is projected to be over $18/g of purified protein for a 100% efficient process. Therefore, the sensitivity to the price of Tris-HCl and DTT was determined for both the intein and conventional processes (Fig. 3A). The simulations indicate that the intein process is more sensitive to the price of all the buffer components than the conventional process, primarily due to the large volumes of buffer required for the intein washing and cleaving step. In addition, the intein process is more sensitive to the cost of Tris-HCl than DTT, again due to the large quantities of Tris-HCl required. A simple extrapolation of the data shown in Fig. 3A suggests that the intein process would become substantially more economical than the conventional process if the overall buffer costs can be cut by more than 70%. This would require inexpensive buffering components, such as phosphate, and possible reductions in EDTA and NaCl. Furthermore, the elimination of DTT will likely be necessary, suggesting the use of alternate cleaving agents or pH-inducible cleaving inteins.

2.3.2 Resins

The two most common laboratory resins used with intein-mediated purification systems are amylose and chitin. The reported binding capacity for these resins is on the order of 2–3 mg of fusion protein per milliliter column volume. This is significantly lower than most commercial affinity resins, which have a binding capacity between 15–20 mg protein per milliliter column volume. However, the cost of the chitin resin is relatively inexpensive compared to most conventional affinity resins, with a small-scale retail cost of approximately $1900/l. The amylose resin is more expensive with a small-scale retail cost of approximately $3750/l. By comparison, traditional affinity chromatography resins can cost up to $6000/l or more, whereas simpler ion exchange resins usually cost $300/l (Harrison et al. 2003). These comparisons suggest several approaches for improving the economics of the intein process. In particular, increases in the binding capacity of the chitin and amylose resins were examined, as well as decreases in the size of the affinity tag and intein.

It is expected that the development of a high-capacity chitin resin might substantially increase its cost, and therefore the effects on product cost of resin cost and capacity were simulated over a range of values (Fig. 3B). These simulations indicate that an increase in the binding capacity would result in a significant decrease in the product cost, primarily due to a decrease in the required volume of the IMPACT step. Further, product cost was relatively insensitive to resin cost, even for a tenfold increase. Resin reusability was examined and found to not be a significant factor on the product cost for the more inexpensive chitin resins (data not shown). These results suggest that further research into resins with higher binding capacities would be of great economic benefit to the intein process, even if the resin cost increased significantly.

A significant difference between the simulations for the two processes was that the intein process required significantly more biomass to achieve the same production capacity. While the conventional process required only the expression of target protein, the intein process required the expression of the affinity-tagged intein-fusion protein. The cells were assumed to have a fixed capacity to express recombinant protein, although exact values for a given product protein would have to be determined experimentally. Some proteins have been observed to have higher target protein expression levels when fused to an intein-CDB tag (Chong et al. 1998), while inclusion of an N-terminal maltose-binding domain tag is also generally thought to improve expression and solubility (Kapust and Waugh 1999). To determine the effect of target protein size on product cost, the process was simulated with various ratios of product protein size to intein affinity-tag size. In each case, the quantity of recombinant fusion protein was maintained at 10% of the dry cell weight. The simulation results clearly indicate that the product cost is very sensitive to the target protein size (Fig. 3C), and that the use of smaller intein-affinity tags can substantially decrease the overall product cost.

2.3.3 Alternate Intein-Cleaving Modes

The development of inteins that are induced to cleave by a pH and/or temperature shift has the potential to eliminate the need for DTT in intein-based

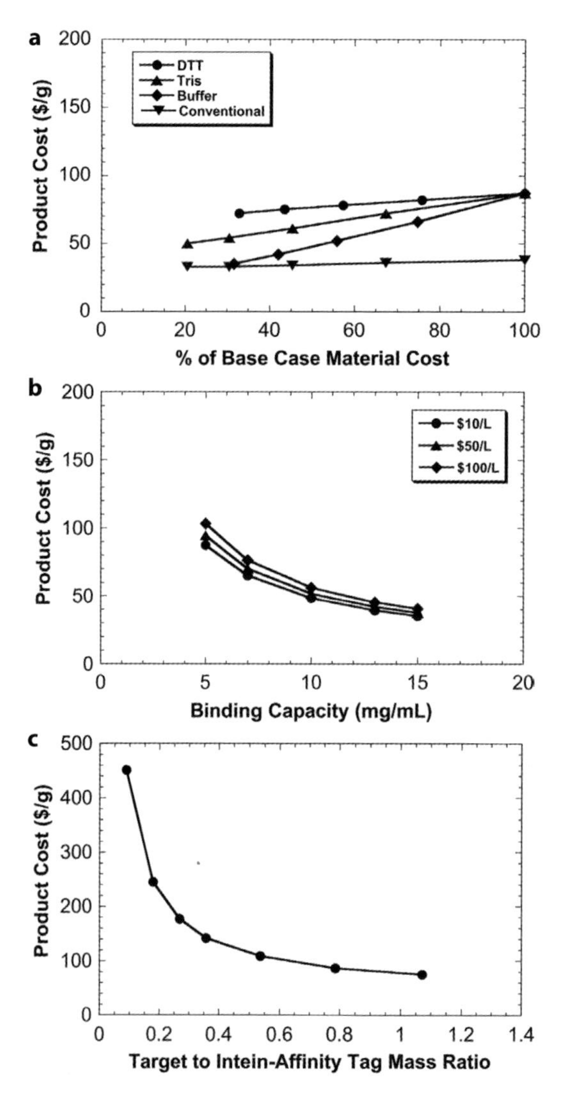

Fig. 3. Cost analysis for the conventional and intein processes. **a** Effect of raw material costs on product cost. The buffer cost was reduced by lowering both Tris-HCl and DTT costs, simultaneously. For the conventional process, only the effect on cost due to Tris-HCl was examined. **b** Effect of chitin resin binding capacity on product cost for the intein process. **c** Effect of target protein size on product cost for the intein process. The relative size of the target protein compared to the intein-affinity tag was examined to account for various intein-affinity tag sizes as well as various recombinant protein sizes

bioseparations (Southworth et al. 1999; Wood et al. 1999). To examine the process economics associated with these inteins, simulations were conducted based on pH-induced cleaving in a phosphate buffer system, with a resin cost of \$100/l and binding capacity of 15 mg/ml. Not surprisingly, the removal of DTT from the process resulted in significant improvements in cost. The overall product cost decreased to \$13.2/g compared to the base-case intein process cost of \$87/g. An increase in the cost of the chitin resin to \$1000/l only increased the product cost by \$2.8/g, underlining again the importance of resin capacity and buffer price over resin cost. As part of this series of simulations, the incubation time for the cleavage reaction was varied from 1 to 14 h. Surprisingly, however, this did not significantly affect the overall product cost (data not shown), perhaps because the intein-cleaving reaction only represents a small fraction of the overall process time from fermentation to delivery of purified bulk product.

2.4 Economics of an Optimal Large-Scale Intein Process

Based on the results of these sensitivity analyses, several recommendations can be made to economically optimize the scale-up of intein processes. For the basic IMPACT process, these include use of a cheaper phosphate buffer system, increasing the capacity of the affinity resin to 15 mg/ml, decreasing the chitin resin cost to \$100/l, and increasing the resin reusability to 100 cycles. The overall product cost for this improved intein process was \$21/g compared to the conventional process at \$38/g and the base-case intein process at \$87/g. The raw material costs would still account for 84% of the annual operating costs. However, the annual operating costs were reduced from \$704 to \$159 million. DTT and phosphate buffer costs were still the two largest raw material costs at 50 and 18%, respectively, compared to 29 and 61% for the Tris-HCl base case.

In addition to the changes proposed above, the intein process can be modified to include pH-inducible cleaving. In this case, the DTT is eliminated completely from the process, resulting in an additional 50% savings in raw materials cost, and a final product cost of \$13.2/g. The purification stage of the resulting process is the cheapest of all the intein processes, and is substantially cheaper than the conventional affinity separation process. However, the other stages of the process are still somewhat more expensive than the conventional process due to the additional biomass capacity required to produce the tagged protein (Fig. 4).

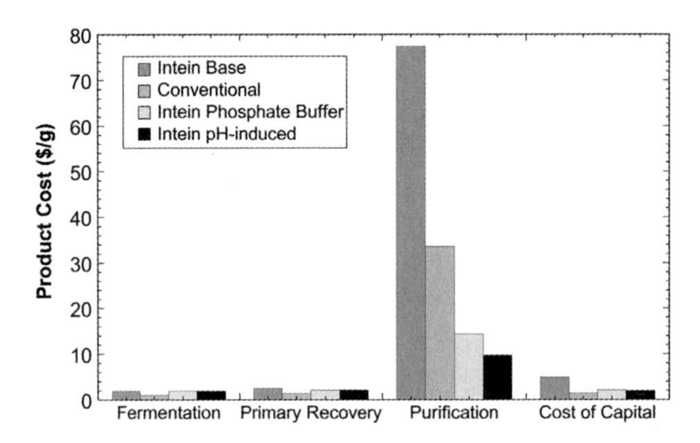

Fig. 4. Capital distribution for the production of a recombinant protein by conventional or intein-mediated processes. The intein pH-induced case is also based on the use of phosphate buffer

2.5 Additional Considerations for Intein Process Scale-Up

A major strength of the intein-based purification process is that it retains the flexibility of conventional affinity-tag technology. This aspect is likely to motivate the development and use of smaller affinity tags with strong affinities for inexpensive resins. Smaller inteins have also been shown to improve process efficiency due to decreases in the overall affinity-tag size, and this effect will be accentuated in cases where more expensive eukaryotic expression systems are required. Although structural studies indicate that conventional inteins will never be smaller than approximately 140 amino acids, a precise understanding of the structural and functional aspects of the cleaving reaction may allow much smaller pseudo-inteins to be developed. Ironically, however, the development of high-capacity resins may be more important to the commercial adoption of large-scale intein processes than any future improvements in the inteins themselves.

The combination of strong affinity-binding of the tagged target protein with an irreversible on-column cleaving reaction distinguishes the intein process from any conventional chromatography or adsorption processes. These unique features suggest several opportunities for a variety of optimized large-scale process configurations. In particular, expanded bed adsorption and vortex flow adsorption (Ma and Cooney 2004) have been proposed. These methods minimize the pressure drop found in traditional chromatographic columns, improve contact between the affinity tag and the resin, and are well suited to the use of inteins. In addition, the precise controllability of pH-triggered inteins has allowed the development of predictive models for process time, product peak shape, and product concentration in conventional chromatographic column process configurations (Wood et al. 2000).

A final aspect of inteins that will be critical to their adoption is the availability of highly controllable inteins that can be used in a wide range of expression hosts. Ideally, these inteins will be strongly repressed during protein production and purification, but will be easily induced to cleave through a pH shift or the addition of a cheap and biologically benign small molecule. Although it was not clearly examined here, premature cleaving has a dramatic negative effect on process efficiency. Prematurely cleaved tags occupy sites on the affinity resin, thus decreasing the direct yield of uncleaved precursor and the apparent capacity of the resin itself. This is a significant problem that has not been solved, particularly in the development of pH-inducible inteins. More importantly, no small-molecule-inducible cleaving inteins have yet been developed.

With the development of new inteins and cheaper affinity chemistries, the economic attractiveness of inteins for large-scale applications will be undeniable. However, some obstacles will likely remain. The pharmaceutical industry is notoriously conservative in the adoption of new technologies. Validation of intein processes will require proof that the product protein will have no uncleaved precursors, and that the cleaved tags can be easily and completely removed from the product stream. This represents a new set of challenges relative to the conventional purification of untagged proteins. In the commodity enzyme industry, purity is often not a major concern, and extremely cheap large-scale methods are generally employed to produce acceptable products. In this case, highly optimized intein processes may become economical due to their capability to significantly increase purity at a relatively low cost.

3 Scale-Down of Intein Processes

As mentioned above, inteins are expected to have wide use in proteomics research for recovery, isolation and analysis using miniaturized channels on a microfluidics platform and for orienting and tagging proteins in microarrays and on chips. Scaling down intein technology will have application in meeting the requirements of fast, low-volume, and parallel processing for proteomics analysis. Miniaturization is a key component with any high-throughput technology and requires substantial engineering input. Drug discovery, sensors development, and assay integration are examples of potential applications of this technology to biomedicine, biodefense, and biotechnology. After summarizing why inteins could be useful in microfluidics, we describe the use of inteins in recovering and detecting proteins in miniaturized fluidics channels. Then, in a second application of inteins, we illustrate the use of intein-mediated biotinylation of proteins for protein microarray analysis and screening.

3.1 Microfluidics

The main goal of microfluidics devices for scale-down bioprocessing is to transfer, recover and analyze very small quantities of fluids containing a desirable protein on a miniaturized platform, i.e., chip or disk, such that the final product is of high yield and purity. In some cases it may be necessary to crystallize the final product. As Fish and Lilly (1984) have shown with large-scale bioprocessing for the production of pharmaceuticals, the fewer the number of isolation and recovery steps or unit processes the better, i.e. less product is lost overall. Hence, adsorptive processes and specifically bio-affinity capture techniques are particularly attractive since they exhibit high selectivity and can be easily miniaturized (small beads with high ligand densities).

For small-scale massively parallel applications on a fluidics platform, it is imperative that the purification process be as simple as possible with a minimum number of recovery steps and minimum addition of cofactors or other chemicals. Tripartite fusion proteins, with inteins as controllable linkers, binding domains for selective affinity adsorption and the protein of interest, are attractive for microfluidics bioprocessing for the following reasons:

1. Cleavage or splicing is non-protein-specific, i.e., the same controllable intein linker can be used with different product proteins. This broadens the applicability of the process to any binding domain and product of interest.
2. Cleavage or splicing occurs without additives or cofactors (besides hydrogen ions for temperature/pH-controllable inteins) and does not require exogenous proteases. Also, external triggers such as temperature can be universally used to simultaneously cleave desired proteins from their fusions.
3. Binding of the tripartite fusion to the solid substrate (bead or synthetic microporous membrane) is specific for a chosen binding domain, yet generalizable for all product proteins, i.e., the same ligand-adsorbent can be used for different proteins.

Miao et al. (2005) have reported on a single-step fusion-based affinity purification of proteins from *E. coli* lysates with pH-controllable linkers and a chitin-binding domain (CBD) in a fluidic device. The linkers were previously derived from self-splicing inteins (Wood et al. 1999). Two different linkers were generated to solve two distinct separation problems: one (called the CM or cleaving mutant mini-intein) for rapid single-step affinity purification of a wide range of proteins (Wood et al. 2000), and the other (called the SM or splicing mutant mini-intein) specifically for the purification of cytotoxic proteins (Wu et al. 2002). The two problems addressed by scale-down of the intein cleavage and splicing in a fluidics channel are protein purification by intein-mediated cleavage and purification of a cytotoxic protein by insertional inactivation and protein splicing (Fig. 5A,B). A rotating compact disk (CD) format was chosen due to its simplicity in effecting fluid movement through centrifugal force without the complications associated with electro-osmosis and other pumping methods. The design of the fluidic device is shown in Fig. 6. The fluidics platform was based on standard photolithography and wet chemical etching techniques of a silicon substrate to produce a fluidics channel with two reservoirs at each end and containing a barrier that retained the beads but allowed the fluid to pass. The silicon channel was covered and sealed with a film of silicone rubber [poly(dimethyl siloxane), PDMS] and angled holes drilled through the PDMS above the reservoirs. Several channels and covers were then adhered to a blank CD, placed in a rotating device, filled with affinity beads and *E.coli* lysate in the inner reservoir, and spun to drive (pump) the fluid through the affinity bed to the outer reservoir. An elution buffer was then passed through the bed and the recovered proteins analyzed. Scale-down factors of ~200-fold were achieved by separating in a ~25-μl bed volume. This work is widely applicable to small-scale massively parallel proteomic separations.

After expressing the tripartite fusion protein (step 1 in Fig. 5A), the clarified cell lysate is loaded onto a chitin column and allowed to selectively bind to the column (step 2). Then the pH is shifted so that the intein (CM) selectively cleaves at its C-terminus (step 3), allowing the purified product protein to be released into the eluate. This fluidic separation has been applied to the α-subunit of RNA polymerase and to the DNA-binding domain of the intron endonuclease I-TevI.

In the second example of single-step affinity purification and recovery on a microfluidics platform from an *E. coli* lysate, a toxic protein was recovered directly from a fluidics column by selective controlled splicing of the linker domain from a tripartite fusion, allowing post-assembly of the desired toxic protein (Wu et al. 2002). In this case, the toxic protein was inactivated by insertion of SM, the controllably splicing mini-intein. Affinity purification was then achieved via the CBD which was itself inserted into the SM (Wu et al. 2002; Fig. 5B). Purification steps 2 and 3 are essentially the same as those used for the intein cleavage process. Again, microfluidic eluate was compared with material purified in a column, and found to be of equivalent purity and specificity. This fluidic separation was applied to the toxic intron endonuclease I-TevI. The yield of protein from the microfluidics platform was approximately equivalent to that from the laboratory-scale preparation (Wu et al. 2002). Also, the endonuclease activity of I-TevI from the fluidics purification exhibited approximately the same activity as that using a laboratory column. These experiments demonstrated not only that this cytotoxic protein can be purified at small scale, but also that the purified protein is highly active.

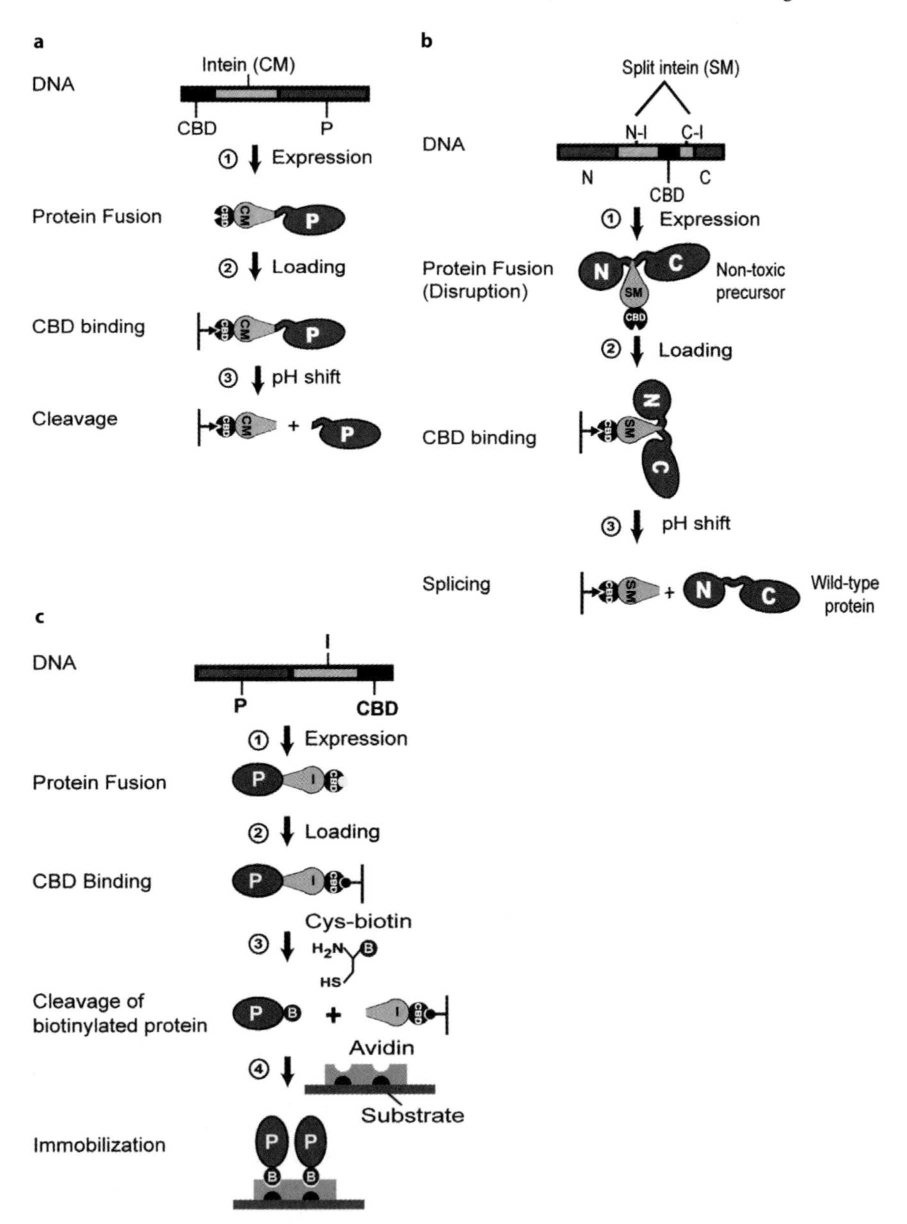

Fig. 5. Intein-mediated purification systems and oriented immobilization of proteins through intein-mediated biotinylation on a fluidics platform. **a** Strategy for purification of a target protein (*P*) using controllable CM intein (*I*) cleavage with a chitin-binding domain (*CBD*) in a tripartite fusion protein. *Step 1* Expression of the tripartite fusion precursor protein in vivo. *Step 2* After cell lysis and clarification in chitin column buffer (pH 8.5), the cell lysate is loaded onto a chitin column. *Step 3* After washing the column, the pH is shifted from 8.5 to 6.0 to trigger C-terminal cleavage and the cleaved products are eluted after 24 h at 4°C. **b** Strategy for intein-mediated purification of a cytotoxic protein (*N-C*) using

3.2 Protein Micro-arrays

During this post-genomic era, attention is being focused on the proteome in order to understand the role of proteins in vivo and to develop drug targets, biomarkers and treatment tools for various diseases. Screening methods that rely on bead-based and biochip technologies are offered as possible solutions. The efficacy (sensitivity, reproducibility and reliability) of the latter processes depends on the efficient capture of proteins on solid substrates. Strategies for orienting proteins or peptides to substrates are often very tedious, requiring many steps, and binding to the surface (ligand) is often not very tight or selective (such as with His-tag to Ni-NTA). Inteins can also be used to modify proteins so that they can attach to solid substrates in a desirable orientation. Yao and his group have reported on the use of intein-mediated site-specific biotinylation of proteins (maltose-binding protein, enhanced green fluorescent protein, and glutathione S-transferase) in vitro and in vivo and subsequent immobilization onto avidin-functionalized glass slides (Lesaicherre et al. 2002; Lue et al. 2004). Their process is summarized in Fig. 5C. After expression and loading of the tripartite fusion protein onto a chitin column (steps 1 and 2), cleavage at the N-terminus of the intein through addition of cysteine-biotin (step 3) releases the biotin-conjugated protein for immobilization onto an avidin-coated substrate (step 4). Based on SDS-PAGE, the efficiency of the biotinylation was 90–95%. The strategy generated "a protein array on which the proteins were oriented optimally and were able to retain their native activity suitable for subsequent biological screening" (Lesaicherre et al. 2002).

controllable SM intein splicing with CBD in a tripartite fusion protein. *Step 1* The intein-disrupted non-toxic precursor protein is expressed in vivo after induction with IPTG. *Step 2* After cell lysis and clarification in chitin column buffer (pH 8.5), the cell lysate is loaded onto a chitin column. *Step 3* After washing the column, the pH is shifted from 8.5 to 7.5 to trigger splicing. The products are eluted after 24 h at 4 °C. They include full-length I-TevI and the N- and C-terminal cleavage products (after Miao et al. 2004). **c** Strategy for purification, reaction (biotinylation) and oriented immobilization of a target protein (*P*) using controllable N-terminal intein (*I*) cleavage with a chitin-binding domain (*CBD*) in a tripartite fusion protein. *Step 1* Expression of the tripartite fusion precursor protein in vivo. *Step 2* After cell lysis and clarification in chitin column buffer, the cell lysate is loaded onto a chitin column. *Step 3* After washing the column, cysteine-biotin (*Cys-biotin*) is added to the column buffer to trigger N-terminal cleavage and to covalently attach biotin (*B*) through the cysteine reaction to the protein. *Step 4* After eluting the cleaved products from the column, they are added to an avidin-coated glass slide to generate a protein microarray. (After Lesaicherre et al. 2002)

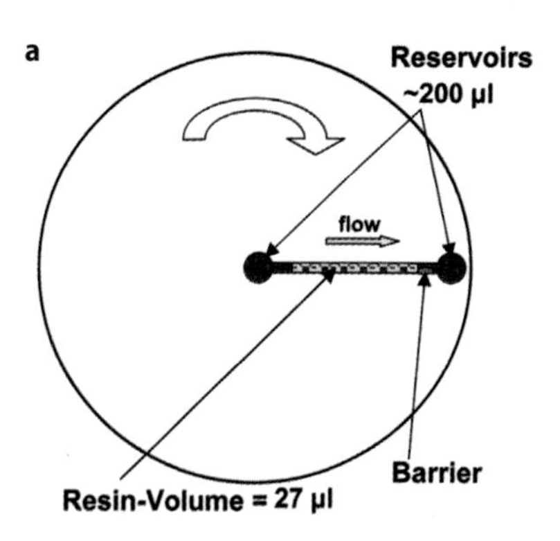

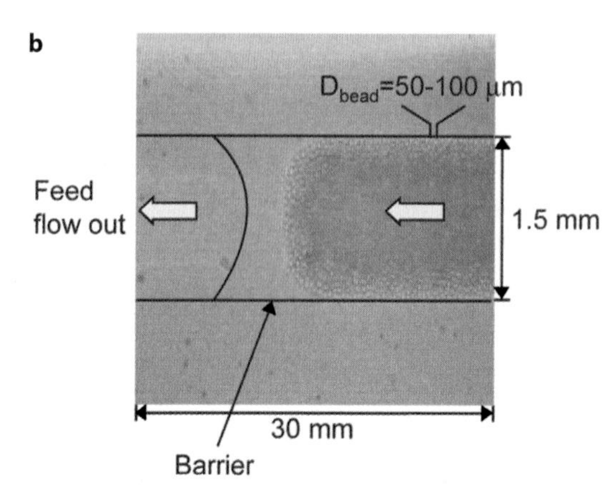

Fig. 6. Fluidics channel on a rotating compact disk (CD) for intein-mediated protein purification. **a** Plan view of the scaled-down fluidic channel on a rotating CD with poly(dimethyl siloxane) (PDMS) covers and filled with affinity beads. The direction of fluid flow is from the loading (feed or inner) reservoir to the eluting (product or outer) reservoir (*arrow*) and is driven by centrifugal force. The reservoirs at both ends of the channel are drilled at an angle in the PDMS cover to reduce liquid overflow. The rotational speed is 250 rpm. **b** Optical micrograph of the fluidics channel with 50–100 µm diameter chitin-coated affinity beads. The depth of the column is 0.6 mm. Other dimensions are given in the figure

4 Summary

As intein bioseparations technologies become better developed, their applicability to a number of industrially relevant problems will increase. Based on a simple examination of a hypothetical bioprocess built on an intein purification step, it is clear that inteins have the potential to be both convenient and economical at large scale. This is particularly true for inteins that can be induced to cleave by simple changes in pH and temperature. At a smaller scale, the simplicity with which inteins allow the purification of various target proteins makes it a powerful system in proteomics. The development of smaller, more controllable inteins, combined with advances in binding chemistries and process configurations, will undoubtedly increase the opportunities for intein use in the future.

References

Chong S, Mersha FB, Comb DG, Scott ME, Landry D, Vence LM, Perler FB, Benner J, Kucera RB, Hirvonen CA, Pelletier JJ, Paulus H, Xu M-Q (1997) Single-column purification of free recombinant proteins using a self-cleavable affinity tag derived from a protein splicing element. Gene 192:271–281

Chong S, Montello GE, Zhang A, Cantor EJ, Liao W, Xu M-Q, Benner J (1998) Utilizing the C-terminal cleavage activity of a protein splicing element to purify recombinant proteins in a single chromatographic step. Nucleic Acids Res 26:5109–5115

De Serres FJ (2002) Worldwide racial and ethnic distribution of α1-antitrypsin deficiency: summary of an analysis of published genetic epidemiologic surveys. Chest 122:1818–1829

Fish NM, Lilly MD (1984) The interactions between fermentation and protein recovery. Bio/Technology 2:623–627

Germino J, Bastia D (1984) Rapid purification of a cloned gene product by genetic fusion and site-specific proteolysis. Proc Natl Acad Sci USA 81:4692–4696

Harrison RJ, Todd P, Rudge SR, Petrides DP (2003) Bioseparations science and engineering. Oxford Univ Press, New York

Kapust RB, Waugh DS (1999) *Escherichia coli* maltose-binding protein is uncommonly effective at promoting the solubility of polypeptides to which it is fused. Protein Sci 8:1668–1674

Ladisch MR (2001) Bioseparations engineering. Wiley, Hoboken, NJ

LaVallie ER, McCoy JM (1995) Gene fusion expression systems in *Escherichia coli*. Curr Opin Biotechnol 6:501–506

Lesaicherre ML, Lue RY, Chen GY, Zhu Q, Yao SQ (2002) Intein-mediated biotinylation of proteins and its application in a protein microarray. J Am Chem Soc 124:8768–8769

Lue RY, Chen GY, Hu Y, Zhu Q, Yao SQ (2004) Versatile protein biotinylation strategies for potential high-throughput proteomics. J Am Chem Soc 126:1055–1062

Ma J, Cooney CL (2004) Application of vortex flow adsorption technology to intein-mediated recovery of recombinant human alpha1-antitrypsin. Biotechnol Prog 20:269–276

Miao J, Wu W, Spielmann T, Belfort M, Derbyshire V, Belfort G (2005) Single-step affinity purification of toxic and non-toxic proteins on a fluidics platform. Lab Chip 5:248-253.

Petrides DP, Koulouris A, Lagonikos PT (2002) The role of process simulation in pharmaceutical process development and product commercialization. Pharmaceut Eng 22:56–65

Rouf SA, Douglas PL, Moo-Young M, Scharer JM (2001) Computer simulation for large scale bioprocess design. Biochem Eng J 8:229–234

Shanklin T, Roper K, Yegneswaran PK, Marten MR (2001) Selection of bioprocess simulation software for industrial applications. Biotechnol Bioeng 72:483–489

Sharma S, Zhang A, Wang H, Harcum SW, Chong S (2003) Study of protein splicing and intein-mediated peptide bond cleavage under high-cell-density conditions. Biotechnol Prog 19:1085–1090

Southworth MW, Amaya K, Evans TC, Xu MQ, Perler FB (1999) Purification of proteins fused to either the amino or carboxy terminus of the *Mycobacterium xenopi* gyrase A intein. Biotechniques 27:110–114, 116, 118–120

Swartz JR (2001) Advances in *Escherichia coli* production of therapeutic proteins. Curr Opin Biotechnol 12:195–201

Wood DW, Wu W, Belfort G, Derbyshire V, Belfort M (1999) A genetic system yields self-cleaving inteins for bioseparations. Nat Biotechnol 17:889–892

Wood DW, Derbyshire V, Wu W, Chartrain M, Belfort M, Belfort G (2000) Optimized single-step affinity purification with a self-cleaving intein applied to human acidic fibroblast growth factor. Biotechnol Prog 16:1055–1063
Wu W, Wood DW, Belfort G, Derbyshire V, Belfort M (2002) Intein-mediated purification of cytotoxic endonuclease I-TevI by insertional inactivation and pH-controllable splicing. Nucleic Acids Res 30:4864–4871
Xu MQ, Paulus H, Chong S (2000) Fusions to self-splicing inteins for protein purification. Methods Enzymol 326:376–418
Zhang A, Gonzalez SM, Cantor EJ, Chong S (2001) Construction of a mini-intein fusion system to allow both direct monitoring of soluble protein expression and rapid purification of target proteins. Gene 275:241–252

Subject Index

K

L

R

Printing: Krips bv, Meppel, The Netherlands
Binding: Stürtz, Würzburg, Germany